MW01201250

# High-Power GaAs FET Amplifiers

For a complete listing of the *Artech House Microwave Library*,
turn to the back of this book

# High-Power GaAs FET Amplifiers

John L. B. Walker

Editor

Artech House

Boston • London

**Library of Congress Cataloging-in-Publication Data**

Walker, John L. B., editor
High-power GaAs FET Amplifiers
Includes bibliographical references and index.
ISBN 0-89006-479-2
1. Power amplifiers. 2. Metal semiconductor field effect transistors.
3. Gallium arsenide semiconductors. 4. Microwave integrated circuits.
I. Title

TK7871.58.P6W35 1993 92-37801
621.3815'35—dc20 CIP

**British Library Cataloguing in Publication Data**

Walker, John L. B., editor
High-power GaAs FET Amplifiers
I. Title
621.3815

ISBN 0-89006-479-2


**685 Canton Street**
**Norwood, MA 02062**



**International Standard Book Number: 0-89006-**
**Library of Congress Catalog Card Number:**

**10 9 8 7 6 5 4 3 2**

*To Paula, John, Jenny, and James*

*In memorium to Octavius Pitzalis,*
*so tragically killed in an automobile accident,*
*11 January 1993*

# Contents

Preface xi

Chapter 1 Introduction and Basic Theory 1
1.1 Introduction 1
1.2 Analysis of Ideal GaAs FET Class A and B Amplifiers 3
1.2.1 Device Physics Versus Circuit Design Viewpoint 3
1.2.2 Class A Amplifiers 4
1.2.3 Single-Ended Class B Amplifiers With Resistive Load 11
1.2.4 Single-Ended Class B Amplifiers With Tuned Load 15
1.2.5 Push-Pull Class B Amplifiers 18
1.2.6 Comparison of the Various Types of Power Amplifiers 21
1.2.7 Validity of the Preceding Analysis 21
1.3 The Concept of Power per Millimeter of Gatewidth 23
1.4 Small-Signal (Linear) and Large-Signal (Nonlinear) Models for an FET 24
1.5 Overview of Design Techniques 30
1.6 Bandwidth Limitations of Reactively Matched Amplifiers 35
1.6.1 Output Matching Network 35
1.6.2 Input Matching Network 38
References 40

Chapter 2 High-Power GaAs FETs 43
2.1 Introduction 43
2.1.1 The Development of the High-Power GaAs FET 43
2.1.2 Current Technology for the High-Power GaAs FET 45
2.2 High-Power FET Design: FET Channel Cross-Section Design 48
2.2.1 The Flow of the Design Process 48
2.2.2 Designing the Epitaxial Wafer Structure 50

2.2.3 Determination of the Gate Length 56
2.2.4 The Scaling Law 58
2.2.5 Breakdown Voltage and Recess Structure 66
2.2.6 Parasitic Resistance 72
2.2.7 Equivalent Circuits 74
2.3 High-Power FET Design: FET Pattern Design 77
2.3.1 The Flow of the Design Process 77
2.3.2 Output Power and Total Gatewidth 77
2.3.3 Determination of the Unit Gatewidth 80
2.3.4 Chip Size 83
2.3.5 Determination of the Number of Pads 84
2.3.6 Pattern Layout 87
2.3.7 Chip Backside Structure 90
2.4 Thermal Properties 94
2.5 Manufacturing 97
2.5.1 Epitaxial Wafer Growth 97
2.5.2 Flow of the Wafer Manufacturing Process 103
2.5.3 Isolation 103
2.5.4 Ohmic Electrode Formation 105
2.5.5 Gate Electrode Formation 108
2.5.6 Protective Layers 111
2.5.7 Overlay Wiring 112
2.5.8 Backside Processing and Via-Hole Connections 112
2.6 Evaluation 113
2.6.1 Evaluation of dc Characteristics 113
2.6.2 Output Power Measurement 114
2.6.3 Measurement of Distortion Features 118
2.6.4 Impedance Measurement and Load-Pull Measurement 120
2.7 Current FETs 123
2.7.1 Standard FET Chips 123
2.7.2 Internally Matched FETs 128
2.7.3 MMIC Power Amplifiers 134
2.8 Trends in Technology 136
2.8.1 Material Technology: The InP MISFET 137
2.8.2 The Heterojunction FET: HEMT and Heterostructure MISFET 139
2.8.3 The Heterojunction Bipolar Transistor 141
2.9 Conclusion 142
References 143

Chapter 3 Computer-Aided Design of GaAs FET Power Amplifiers 147
3.1 Introduction 147

3.2 GaAs FET Nonlinear Models 148
3.2.1 The MESFET Large-Signal RF Equivalent Circuit 149
3.2.2 The MESFET Static dc Model 162
3.2.3 General Guidelines for Large-Signal Model Extraction 172
3.3 A Large-Signal Amplifier Simulation 173
3.3.1 The FLK202XV Large-Signal Model 173
3.3.2 Simulation of the 3.7- to 4.2-GHz, 1-W Class AB Amplifier 174
References 188

Chapter 4 High-Power GaAs FET Amplifier Design 189
4.1 Introduction 189
4.2 Budgeting Transmitting Chain RF Performance 189
4.3 Performance Characterization and Modeling 192
4.3.1 Pulsed RF Testing 192
4.3.2 Bias Points and Class of Operation 194
4.3.3 Small-Signal Modeling 195
4.4 Design Techniques 198
4.4.1 Load-Pull 198
4.4.2 Nonlinear CAD 199
4.4.3 Modified Cripps Method 199
4.5 Scaling 204
4.6 Matching Network Design 208
4.6.1 Output and Interstage Network Load Line Analysis 209
4.6.2 Harmonic Termination Effects 210
4.6.3 Stability Considerations 212
4.7 Thermal Considerations 212
4.8 Gate Current and Insertion Phase 215
4.9 Dual-Gate FET Power Amplifier 216
References 225

Chapter 5 Thermal Effects and Reliability 227
5.1 Introduction 227
5.2 Thermal Fundamentals 228
5.3 Thermal Calculations for Practical FETs 229
5.4 Pulsed Operation 240
5.5 Measurement of Thermal Resistance and Channel Temperature 243
5.6 Reliability 248
5.6.1 Failure Mechanisms 248
5.6.2 Reliability Statistics 250
5.6.3 Reliability Testing 257
5.7 Conclusion 260
References 260

Chapter 6 Combining Techniques 263
6.1 Introduction 263
6.2 Distributed Amplifier Power Combining 264
6.2.1 Small-Signal Analysis 264
6.2.2 Effect of Resistive Terminations and Loss Within the FET on Small-Signal Analysis 271
6.2.3 Large-Signal Analysis 275
6.3 Passive Power Combining/Dividing Networks 281
6.3.1 Two-Way In-Phase Power Combiner/Divider Networks 282
6.3.2 Two-Way Quadrature-Phase Power Combiner/Divider Networks 285
6.3.3 *N*-Way Power Combiner/Divider Networks 289
6.4 Power Combining Methods 292
6.4.1 Corporate Power Combining 292
6.4.2 Serial Power Combining 302
6.4.3 *N*-Way Power Combining 305
Appendix 6A 307
References 311

Chapter 7 Systems Applications of GaAs FET Power Amplifiers 315
7.1 Introduction 315
7.2 Satellite Applications 316
7.2.1 Reliability 316
7.2.2 Active Phased Arrays 317
7.2.3 Power Combined Amplifiers 319
7.2.4 Mobile Tactical 321
7.2.5 Earth Terminals 328
7.3 Terrestrial Telecommunications 331
7.3.1 Line-of-Sight Links 332
7.3.2 Linearized Amplifiers 332
7.4 Radar and EW Applications for High-Power GaAs FET Amplifiers 334
References 350

About the Authors 353

Index 357

# *Preface*

This book has its origins in the European Microwave Conference held in Wembley, United Kingdom, in September 1989. In that year the editor organized and chaired a one-day workshop on "High-Power Solid-State Amplifiers" as part of the conference. In practice, the only type of solid-state device considered by the speakers was the GaAs field effect transistor (FET). This workshop was well attended and proved to be a very stimulating event. Altogether, there were 11 formal presentations by speakers from Japan, the United States, France, and the United Kingdom. Shortly afterward, the editor was approached by Artech House to consider writing or editing a book on the subject. The latter option was the only viable one! Similar workshops are also a regular feature of the IEEE International Microwave Symposium and these also prove to be lively and stimulating events.

While several books are available that have a section devoted to GaAs FET power amplifiers, the depth of coverage is, in general, rather shallow and many important aspects are not covered at all. To the best of the editor's knowledge, there is no book available that comprehensively covers all aspects of high-power GaAs FET amplifiers; hence, we thought it worthwhile to produce this book, which attempts to bring together all the relevant material.

As with any book of this sort in which the chapters are written by different authors, there is an inevitable difference in style and symbolism as well as some overlap between the chapters when first received by the editor. The objective during the editing process has been to make this book as coherent as possible and to minimize the impression that it is a collection of isolated chapters. Accordingly, each chapter has been edited to remove as much duplication as possible, and an attempt has been made to achieve a consistent symbolism throughout the book—a task that would have been formidable prior to the advent of word processors.

Chapter 1 is a general overview of GaAs FET power amplifiers, and Chapter 2 describes in detail the design and fabrication of power GaAs FETs. Chapter 3 considers how modern computer-aided design tools can be applied to the design of a GaAs FET power amplifier, while Chapter 4 describes the more traditional design methods and concludes with a practical example. Chapter 5 is devoted to the thermal design and analysis of a GaAs FET and the closely allied topic of reliability. Chapter 6 describes the methods that can be employed to combine amplifiers to obtain increased power output and, finally, Chapter 7 considers the application of GaAs FET power amplifiers in communications, electronic warfare, and radar systems.

This book should be of use not only to students, academics, and industrialists engaged in high-power amplifier design, but also to people involved in device design, fabrication, and modeling as well as system designers.

## ACKNOWLEDGMENTS

I would like to acknowledge the tremendous amount of time and effort put in by all the coauthors. All are employed by industrial organizations and are under significant pressure to achieve the commercial objectives of their organizations, particularly in this time of economic depression. Accordingly, writing a chapter for a book inevitably has to be low on their list of priorities and thus it represents a very large personal commitment. Without that commitment this book would not have been possible.

I would also like to thank Thorn-EMI Electronics for permission to undertake the task of editing and contributing to this book and for allowing me the use of facilities within the company.

Finally, I would like to acknowledge the patience and understanding of my wife and family during the many evenings spent working on this book.

John L. B. Walker
August 1992

# Chapter 1
# Introduction and Basic Theory

**J. L. B. Walker**
*Thorn-EMI Electronics*

## 1.1 INTRODUCTION

At the start of the 1970s, the designer of solid-state power amplifiers had a very limited choice of active devices at his or her disposal, namely, silicon bipolar transistors, GaAs Gunn diodes, and silicon and GaAs Impatt diodes. Silicon bipolar transistors were usable in the amplification mode only up to about 4 GHz, a situation that has not changed significantly in the intervening 20 years. Above 4 GHz designers were restricted to using Gunn or Impatt diodes for amplification with all the accompanying disadvantages, namely, limited bandwidth, a high noise figure, low efficiency, and large size as a result of having to incorporate circulators to separate incident and reflected signals.

All of this changed, however, by the middle of the decade when the first commercially available GaAs field effect transistors (GaAs FETs) appeared offering usable gain for amplification up to the X-band. These were small-signal devices initially, but device designers were quick to realize their potential for power amplification and power GaAs FETs soon became available commercially.

Today, GaAs FETs and their derivatives such as *high-electron mobility transistors* (HEMTs) have completely replaced diodes in small-signal and low-noise applications. They have accomplished a similar feat in the power amplification field also except in low duty cycle, high peak pulse power applications where Impatts still have an unrivalled advantage over other solid-state devices. Discrete power GaAs FETs with power outputs ranging from 25W at 4 GHz to 1W at 20 GHz can now be purchased commercially from a number of vendors (mostly Japanese).

In the mid-1970s research commenced on the development of GaAs *monolithic microwave integrated circuits* (MMICs) and this has had a profound effect on all aspects of microwave solid-state component design including power amplifiers. In an MMIC the passive as well as the active circuitry is fabricated on a semi-insulating GaAs substrate. The driving force behind the development of MMICs was the observation that such systems as *direct broadcast by satellite* (DBS) and phased-array radars would not become an economic reality unless the cost of the basic microwave components could be reduced substantially. GaAs MMICs were perceived as the means of achieving this objective because they eliminated the labor-intensive steps of hybrid assembly, tuning, and testing. MMICs also offer other advantages such as increased reliability as a result of reducing the number of interconnections, decreased size and weight, and greater uniformity of performance between samples. However, the cost of developing GaAs MMICs is high compared to their hybrid counterparts, and tuning of the finished product is impossible with the result that far greater care has to be exercised at the design stage to model accurately all of the parasitic, cross-coupling, and, for power amplifiers, nonlinear and thermal effects because one cannot afford mistakes.

By the mid-1980s the technology for manufacturing GaAs MMICs was sufficiently mature that standard MMIC building blocks became commercially available, for example, the 2- to 6-GHz, 10-dB gain small-signal amplifier from Anadigics [1]. Also, several companies were able to offer a "foundry" service, which allows customers to design their own circuits using the foundry's design manual. This circuit is then fabricated by the foundry and the processed chips are returned to the customer for use.

As a result of these developments, the engineer requiring a solid-state power amplifier today has many options at his or her disposal, ranging from a conventional MIC construction using either packaged or unpackaged discrete power GaAs FETs to a full custom foundry MMIC design. In general, state-of-the-art electrical performance requires MIC construction, but the monolithic approach is applicable for large volume, lower cost, less technically demanding situations. Nevertheless, impressive results have been achieved from monolithic power amplifiers, including 11W over 3 to 6 GHz with 11-dB gain and 12% power-added efficiency [2], 3.5W over 5 to 6 GHz with 8-dB gain and 37% power-added efficiency [3], 2.5W over 9 to 10 GHz with 14.5-dB gain and 36% power-added efficiency [4], and 0.5W over 2 to 18 GHz with 5-dB gain and 14% power-added efficiency [5].

This book is devoted exclusively to the design of power GaAs FETs and power amplifiers employing these devices. Electrical, thermal, and reliability aspects are considered for both hybrid and monolithic realizations. Although the book is centered around the GaAs FET, most of the material applies equally well to the various types of HEMTs currently being developed, and much of the material also applies to heterojunction bipolar transistors (HBT) with appropriate modifications.

## 1.2 ANALYSIS OF IDEAL GaAs FET CLASS A AND B AMPLIFIERS

### 1.2.1 Device Physics Versus Circuit Design Viewpoint

The analysis of ideal GaAs FET Class A and B amplifiers can be considered either from the circuit designer's or the device physicist's viewpoint. The circuit designer wants a device in which the mutual transconductance $g_m$ is independent of the gate-source voltage $V_{gs}$, so that, among other things, distortion is minimized. However, such a device requires a nonuniform channel doping profile [6], specifically one in which the doping level rises as the interface between the active layer and the substrate is approached.

The device physicist, on the other hand, usually bases his or her analysis on a flat doping profile because this simplifies device calculations. In addition, nonuniform doping profiles are more difficult to specify and to produce so the material supplier usually aims to provide flat profile active layers, and most commercially available FETs are of this type. However, with a flat doping profile $g_m$ falls, and thus the gain decreases, while the gate is biased more negatively with respect to the source so that small-signal high-gain amplifier stages are biased close to $V_{gs} = 0$, that is, close to the point where maximum drain current flows. High-power stages, on the other hand, have to be biased close to $V_{gs} = -|V_P|/2$ where $V_P$ is the pinch-off voltage (the modulus sign is inserted to avoid any ambiguity since $V_P$ is variously specified as either a negative or a positive number) so that the drain current can be fully modulated for maximum power output, but the small-signal gain from a given FET is inherently lower at this bias point. Flat doping profiles result in $g_m$ having a $V_{gs}^{-1/2}$ dependency and thus such FETs inherently generate distortion components in the output spectrum, even in Class A amplifiers under relatively small-signal conditions.

An interesting compromise between the two viewpoints has been analyzed by Kushner [7]. He considered the case for which $g_m$ decreases linearly from its maximum value at $V_{gs} = 0$ to zero at pinch-off. Such a characteristic more closely approximates reality than does the constant $g_m$ case and it results in simple analytic expressions for the various parameters such as power output, efficiency, etc., but one would never deliberately set out to produce such an FET because it not only generates a second harmonic component under small-signal Class A conditions but it also requires a nonuniform doping profile for its realization. The linear $g_m$ case inevitably has different values for the theoretical maximum efficiency, power output, etc., compared with those for the constant $g_m$ case. The interested reader is referred to Kushner's article for further details.

In this chapter the circuit designer's viewpoint of a constant $g_m$ will be adopted because this not only considerably simplifies the calculations by eliminating the voltage dependency of $g_m$, but also because this has to be the ultimate goal in real applications.

### 1.2.2 Class A Amplifiers

The circuit of an ideal Class A amplifier is shown in Figure 1.1. The gate dc bias circuitry has been omitted for simplicity, and the parallel tuned circuit at the output is not required for Class A operation. The choke is assumed to present an infinite impedance to any RF signal while the series dc blocking capacitor is designed to be a short circuit to any RF signal. We further assume in this analysis that the FET has infinite input and output impedances and that it has the idealized IV characteristics shown in Figure 1.2.

For Class A operation the bias point for maximum output power is $I_d = I_F/2$, $V_{ds} = V_S$, $V_{gs} = (-|V_P| + V_\phi)/2$, as shown in Figure 1.2, in which case the dc power delivered by the supply and dissipated in the FET is

$$P_{dc} = V_S I_F/2 \tag{1.1}$$

Note that the dc power supplied is independent of the RF input or output power level, which is one of the major drawbacks of Class A amplifiers because they consume this amount of power even with no applied signal.

Most GaAs FET data sheets specify $I_{dss}$, which is, by definition, the drain current that flows when $V_{gs} = 0$ volts, but since the gate contact is a Schottky barrier the channel is not fully open in this condition due to the internal barrier voltage $V_\phi$ of the Schottky diode. The drain current can be increased by up to 20% above $I_{dss}$ by forward biasing the gate with respect to the source by about 0.5V to partially overcome the internal barrier voltage and thus increase the channel open-

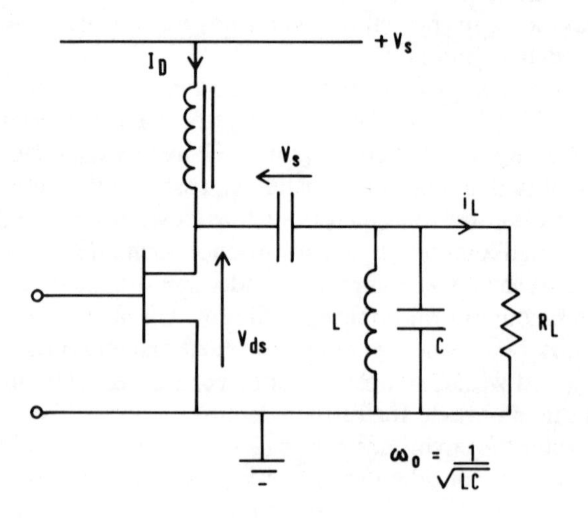


**Figure 1.1** Circuit diagram of a Class A or B amplifier.

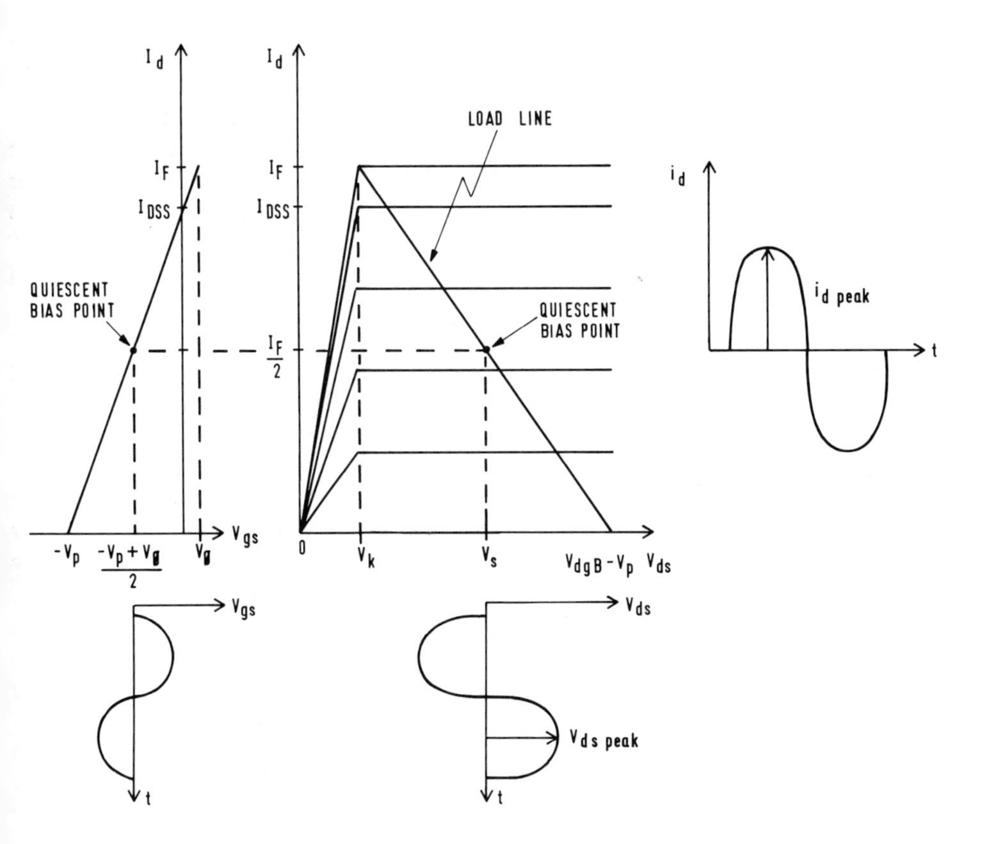


**Figure 1.2** Voltage-current waveforms for a Class A amplifier.

ing. This enables a larger RF current swing to occur as Figure 1.2 shows and hence higher output power is achieved. However, one must be very careful to control the RF input signal to ensure that excessive gate current does not flow during the positive half of the RF cycle because this would be detrimental to FET reliability. In practical applications thermal and reliability considerations may preclude the use of $I_F/2$ as a bias point and one may be forced to use $I_{dss}/2$ (or lower) as the bias point.

The voltage and current waveforms of Figure 1.2 show that the RF power delivered to the load is given by

$$P_{RF} = v_{d\text{speak}}\, i_{d\text{peak}}/2 \tag{1.2}$$

Maximum power output occurs when $v_{d\text{speak}}$ and $i_{d\text{peak}}$ assume their greatest values.

The minimum instantaneous value of $v_{dspeak}$ must not fall below the knee voltage $V_K$ and the maximum instantaneous voltage must not exceed the breakdown voltage; but breakdown is normally specified in terms of the maximum drain to gate voltage $V_{dgB}$ when the gate is biased to pinch-off, that is, $V_{gs} = -|V_P|$. Hence, the maximum drain to source voltage is $V_{dgB} - |V_P|$ and so the maximum value for $v_{dspeak}$ is

$$v_{dspeak} = (V_{dgB} - |V_P| - V_K)/2 \tag{1.3}$$

The maximum value for $i_{dpeak}$ is $I_F/2$ and hence from (1.2) the maximum RF output power is given by

$$P_{\mathrm{RFmax}} = I_F(V_{dgB} - |V_P| - V_K)/8 \tag{1.4}$$

where $P_{\mathrm{RFmax}}$ is the maximum *linear* RF output power that can be achieved from an FET, that is, when the output voltage and current are pure single-frequency sinusoids and where an $x$-dB increase in input power results in an $x$-dB increase in output power.

The power added efficiency is, by definition, given by

$$\eta = (1 - 1/G)P_{\mathrm{RF}}/P_{\mathrm{dc}} \tag{1.5}$$

where $G$ is the RF gain of the FET at the particular value of $P_{\mathrm{RF}}$ being achieved. Power-added efficiency is used as the measure of efficiency for transistors rather than the simple ratio of RF output power over dc input power to discriminate against low-gain devices in which a significant fraction of the output power comes from the RF input power. However, for multistage amplifiers with moderate gain, power-added efficiency is virtually identical to the simple definition of efficiency. Since for a Class A amplifier $P_{\mathrm{dc}}$ is independent of $P_{\mathrm{RF}}$, then it can be seen immediately from (1.5) that a 10-dB reduction in output power results in a tenfold reduction in power added efficiency, assuming $G$ remains constant, which is a problem in some applications (see Chap. 7). From (1.1) and (1.4), we can easily deduce that the maximum power-added efficiency is given by

$$\eta_{\mathrm{max}} = (1 - 1/G)(V_{dgB} - |V_P| - V_K)/4V_S \tag{1.6}$$

From Figure 1.2 and (1.3) it can be seen that the supply voltage $V_S$ for maximum output power is given by

$$V_S = V_K + v_{dspeak} = (V_{dgB} - |V_P| + V_K)/2 \tag{1.7}$$

Once again, thermal and reliability considerations may dictate that a lower value of supply voltage be used. Combining (1.6) and (1.7) results in the maximum power-added efficiency being given by

$$\eta_{\max} = \frac{1}{2}\left(1 - \frac{1}{G}\right)\frac{(1 - \alpha)}{(1 + \alpha)} \tag{1.8}$$

where

$$\alpha = \frac{V_K}{V_{dgB} - |V_P|} \tag{1.9}$$

In the limit of infinite gain and zero knee voltage, the classical result of 50% maximum power-added efficiency is obtained for a Class A amplifier. For a typical power GaAs FET with $V_K = 1.5$V, $V_{dgB} = 25$V, $V_\phi = 0.7$V, and $V_P = 3.5$V, then $\alpha = 0.7$ and the maximum power-added efficiency is reduced to 43%. A much more serious problem is the drastic reduction in power-added efficiency that occurs with low-gain power FETs, for example, an FET with only 6-dB gain reduces the maximum efficiency by 25% to only 32%. Practical effects such as circuit losses and nonoptimum load impedances will reduce the efficiency still further.

The achievement of the maximum linear RF output power given by (1.4) requires that the FET have a load impedance $R_L$ in Figure 1.1 given by

$$R_L = v_{dspeak}/i_{dpeak} = (V_{dgB} - |V_P| - V_K)/I_F \tag{1.10}$$

This value of load impedance is invariably different from the value required to achieve a good impedance match and hence power amplifiers may exhibit a high output VSWR depending on their type.

A plot of the instantaneous drain-source voltage versus drain current superimposed on the device IV characteristics is termed a *load line*, as illustrated in Figure 1.2. For a Class A amplifier, the slope of the load line happens to be identical to the optimum RF load impedance given by (1.10), but this is not necessarily the case for other classes of amplifiers. Also, as discussed in Section 4.4.3, the load line becomes an ellipse if the load is not a pure resistance. Practical FETs always have some reactive components between the drain and source terminals of the device, and hence the load line at the device terminals is never a straight line. However, if the reactive elements within the FET are considered to be part of the output matching network, then in the ideal case the load line seen by the voltage-controlled current generator is still a straight line, but the voltage across it and the current through it are not the same as the voltage and current at the device terminals.

The maximum linear output power as given by (1.4) occurs when the RF gate-source voltage amplitude is

$$v_{gs} = \frac{|V_P| + V_K}{2} \tag{1.11}$$

If the input RF voltage is increased beyond this value (assuming that this does not result in avalanche breakdown in the negative half-cycle nor excessive gate current in the forward half), then the RF output power will increase, but the drain current will now be a distorted sine wave with symmetrical clipping of the negative and positive half-cycles as shown in Figure 1.3. The clipped drain-current waveform can be expressed in a Fourier series form as

$$I_{ds}(t) = a_0 + \sum_{n=1}^{\infty} a_n \cos n\omega t + \sum_{n=1}^{\infty} b_n \sin n\omega t \tag{1.12}$$

where

$$a_1 = \hat{I}\,[1 - \frac{2}{\pi}\cos^{-1} x + \frac{1}{\pi}\sin(2\cos^{-1} x)]$$

$$a_n = \begin{cases} 0, \text{ for } n \text{ even} \\ \hat{I}\left\{\dfrac{4x}{n\pi}\sin(n\cos^{-1} x) - \dfrac{2}{\pi}\left[\dfrac{\sin(n+1)\cos^{-1}x}{n+1} + \dfrac{\sin(n-1)\cos^{-1}x}{n-1}\right]\right\}, \text{ for } n \text{ odd} \end{cases} \tag{1.13}$$

$$b_n = 0,$$

$$x = I_F/2\hat{I}$$

Now $\hat{I}$ is proportional to $v_{gs}$, that is, to the square root of the RF input power $P_{in}$, thus $\hat{I} = k\sqrt{P_{in}}$ where $k$ is a constant. If one defines $P_{in,lin}$ as the input power that results in $v_{gs}$ having the amplitude given by (1.11), then $\hat{I} = I_F/2$ when $P_{in} = P_{in,lin}$. Thus,

$$x = I_F/2\hat{I} = \sqrt{P_{in,lin}/P_{in}} \tag{1.14}$$

Also $\hat{I}^2 R_L/2 = GP_{in}$ where $G$ is the linear gain, and the output power at the fundamental is given by $P_{out} = a_1^2 R_L/2$. Making use of these relations and substituting (1.14) into (1.13) results in the output power being given by

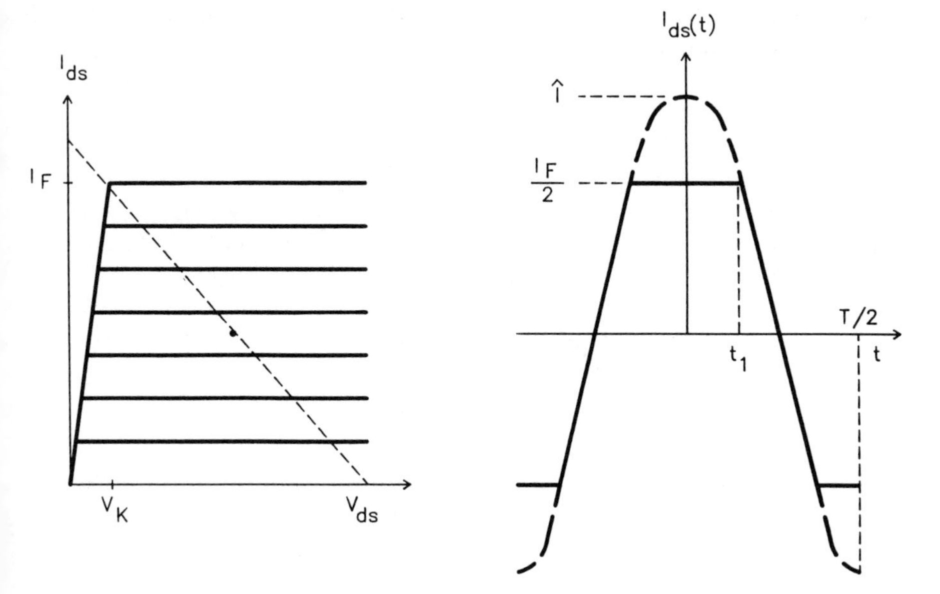


**Figure 1.3** Drain current waveform of an overdriven Class A amplifier.

$$P_{out} = GP_{in} \leq P_{RFmax}, \qquad P_{in} \leq P_{in,lin}$$

$$P_{out} = GP_{in}\left\{1 - \frac{2}{\pi}\cos^{-1}\sqrt{P_{in,lin}/P_{in}} + \frac{1}{\pi}\sin\left[2\cos^{-1}\sqrt{P_{in,lin}/P_{in}}\right]\right\}^2, \qquad P_{in} \geq P_{in,lin} \tag{1.15}$$

This equation enables one to predict the output power of an FET as a function of the input power in both the linear and nonlinear regions. Equation (1.15) is compared with experimental data [8] obtained at 10 GHz on a Fujitsu FLC081 power FET operated at $V_{ds}$ = 10V in Figure 1.4, where we see that (1.15) is in excellent agreement with the experimental data. If $P_{1dB}$ is defined as the output power at which the gain has decreased by 1 dB from its small-signal linear value then the following observations can also be made from Figure 1.4:

1. $P_{1dB}$ is 1 dB higher than $P_{RFmax}$—a coincidence!
2. $P_{1dB}$ occurs when $P_{in} = P_{in,lin}$ + 2 dB.
3. The saturated output power $P_{sat}$ is 1.1 dB higher than $P_{1dB}$.

Saturation occurs when the drain current is a squarewave, in which case Fourier analysis shows that

$$P_{sat} = (16/\pi^2)P_{RFmax} \tag{1.16}$$

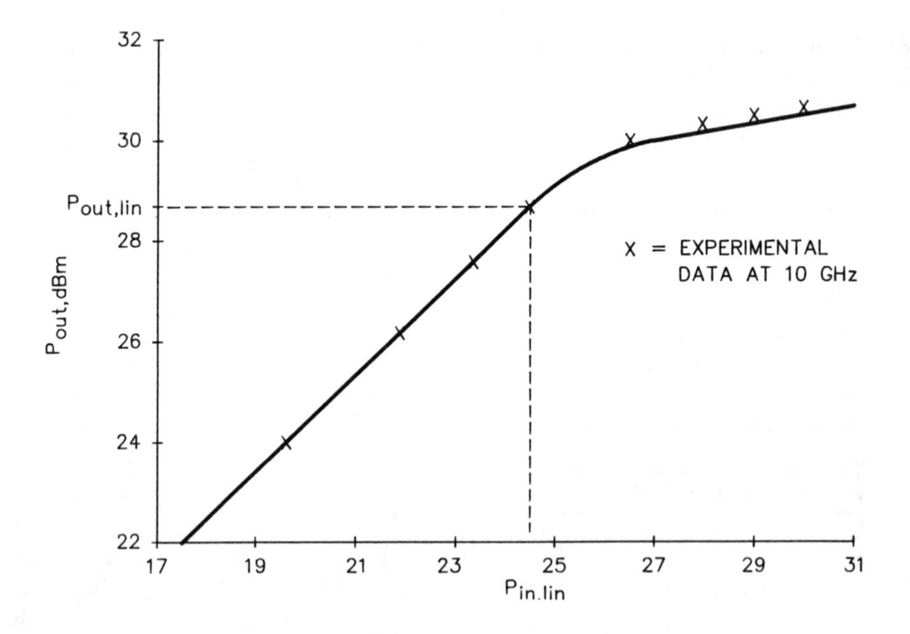


**Figure 1.4** Theoretical and experimental output power characteristic of a Class A GaAs FET.

Note that gate current must flow during the positive half-cycle whenever $P_{in} > P_{in,lin}$ if the input is a sinusoid and this may be detrimental to long-term reliability. Also, the supply voltage should be reduced below the value given by (1.7) if the device is operated outside the linear region so as not to exceed the maximum permissible gate-drain voltage. This in turn reduces the absolute values of $P_{RFmax}$, $P_{1dB}$, and $P_{sat}$, but (1.15) still describes the output power characteristic. However, if the input is a squarewave rather than a sinusoid then (1.16) is the output power that can be achieved without having to decrease the supply voltage and, furthermore, negligible gate current will flow during the positive half-cycle. Note that the output matching circuit must allow the harmonic current components to flow. The preceding analysis has assumed that the FET sees a purely resistive load at all frequencies as given by (1.10).

The power-added efficiency $\eta$ for $P_{in} > P_{in,lin}$ is given by

$$\eta = \frac{1}{2}\left(1 - \frac{1}{G}\right)\left(\frac{1-\alpha}{1+\alpha}\right)\frac{P_{in}}{P_{in,lin}}\left[1 - \frac{2}{\pi}\cos^{-1}\sqrt{P_{in,lin}/P_{in}} + \frac{1}{\pi}\sin\left(2\cos^{-1}\sqrt{P_{in,lin}/P_{in}}\right)\right]^2 \tag{1.17}$$

In the limit of infinite small-signal gain, the efficiency tends asymptotically to

81.6%, which is the classical result for a Class A amplifier driven with a squarewave input. However, if the gain of the FET is finite, then the power-added efficiency does not increase asymptotically as the input power level is increased but reaches a maximum before decreasing. For example, if the gain is 6 dB, then the efficiency peaks at about the 1-dB compression point. That the efficiency must reach a peak rather than continually increase if the gain is finite can be deduced by noting that in class A the efficiency is proportional to the difference between the RF input and output power. When the input power exceeds $P_{\text{in,lin}}$, the input power increases faster than the output power with the result that an optimum input power exists that maximizes the efficiency [9].

Finally, Snider [9] has examined a Class A amplifier in which the load impedance is $4/\pi$ times the value given by (1.10). With this value of load impedance the current cannot be fully modulated if a pure sinusoidal voltage output is desired [i.e., the output power is less than the value given by (1.4) and the RF gate-source voltage is less than the value given by (1.11)]. However, if the RF gate-source voltage is increased to the value given by (1.11), then the output voltage waveform becomes a squarewave even though the current is still a sinusoid, provided that the output circuit also presents an open circuit to all the harmonics. In this case the maximum RF output power and efficiency are 27% higher than the values given by (1.4) and (1.8), respectively. The advantage of this mode of Class A operation compared to the overdriven case is that the higher output power and efficiency are achieved without any extra voltage stress being placed on the device compared to normal linear Class A operation.

### 1.2.3 Single-Ended Class B Amplifiers With Resistive Load

The only difference between a single-ended Class B amplifier with resistive load and the preceding Class A amplifier is that the quiescent bias point is changed to $I_d = 0$, $V_{ds} = V_S$, $V_{gs} = -|V_P|$. The device IV characteristics, together with the voltage and current waveforms, are shown in Figure 1.5.

The RF drain current waveform is a half-sinewave given by

$$i_d = \begin{cases} i_{d\text{peak}} \sin \omega_0 t & 0 < \omega_0 t < \pi \\ 0 & \pi < \omega_0 t < 2\pi \end{cases} \tag{1.18}$$

with $i_{d\text{peak}} \leq I_F$. This waveform can be Fourier analyzed into

$$i_d = i_{d\text{peak}} \left( \frac{1}{\pi} + \frac{1}{2} \sin \omega_0 t - \frac{2}{\pi} \sum_{n=2,4,\ldots}^{\infty} \frac{1}{n^2 - 1} \cos n\omega_0 t \right) \tag{1.19}$$

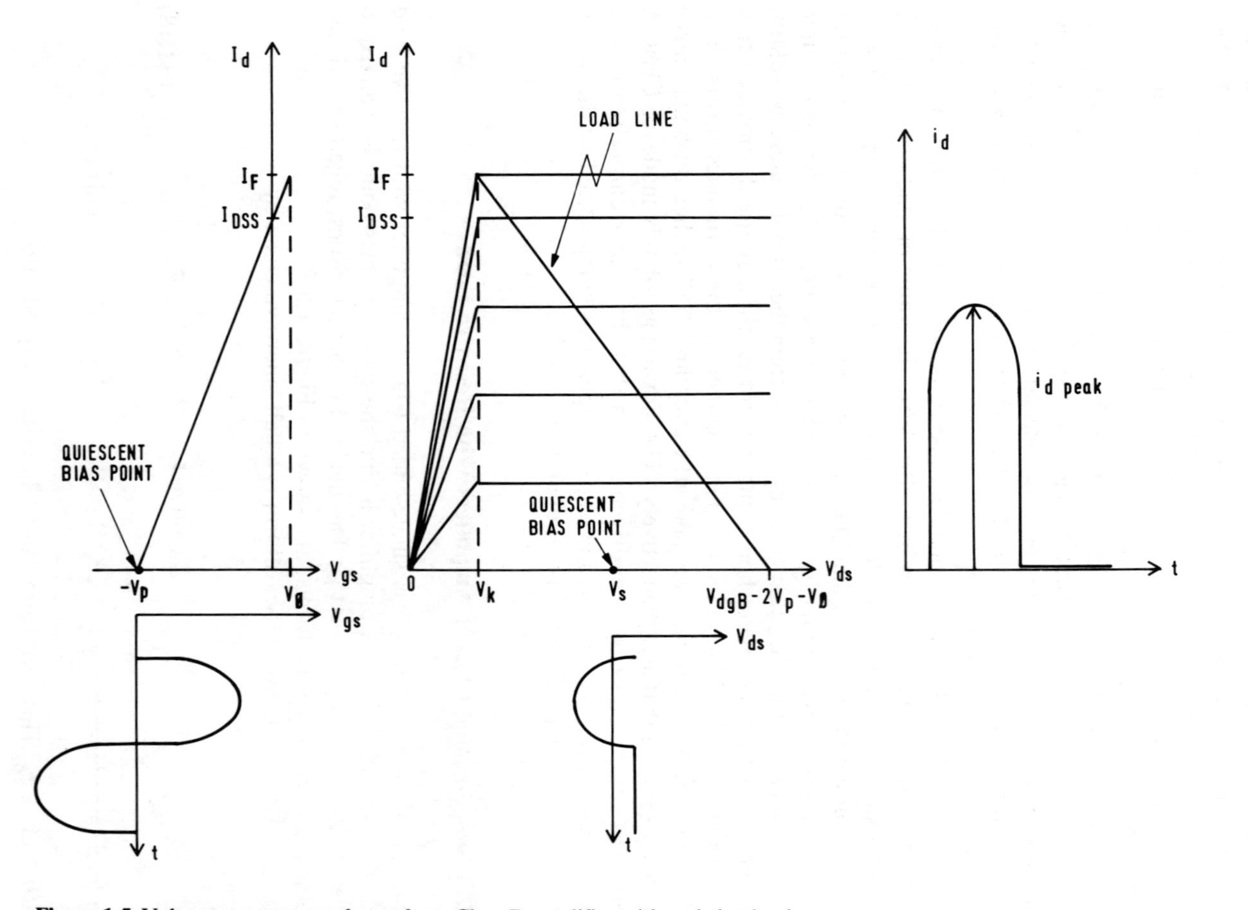


**Figure 1.5** Voltage-current waveforms for a Class B amplifier with resistive load.

that is, the current waveform contains a dc component that is a function of the RF power output, a component at $\omega_0$, and only even order harmonics. Note that a Class B amplifier always generates harmonics even when operated in its "linear" region! Linearity in this case means only that the output power at the fundamental is linearly related to the input power.

The current $i_L$ flowing through the load is identical to that given by (1.19) except that it is inverted and the dc term is eliminated: $i_L = -(1 - 1/\pi)i_d$. The voltage across the load is obviously given by $R_L i_L(t)$ and the voltage across the FET is thus $V_S + R_L i_L(t)$ resulting in the load line shown in Figure 1.5. Note that the quiescent bias point does not lie on the load line and that the load line is not unique. The load line drawn applies when the FET is delivering its maximum output power. If the drive level is reduced, then the load line is parallel to the one shown but displaced toward the origin.

For maximum power output we require the voltage swing across the FET to be maximized. However, when the drain-source voltage reaches its maximum value the gate-source voltage is at its peak negative value of $-2|V_P| - V_\phi$ (assuming that the gate-source breakdown voltage exceeds this value); hence, the maximum permitted value of $V_{ds}$ is $V_{dgB} - 2|V_P| - V_\phi$. The minimum acceptable value for $V_{ds}$ is $V_K$ as before. Using these values and the maximum and minimum values of $i_L$ results in $P_L$ and $V_S$ being determined by the following two simultaneous equations:

$$\left.\begin{aligned} V_S - R_L I_F(1 - 1/\pi) &= V_K \\ V_S + R_L I_F/\pi &= V_{dgB} - 2|V_P| - V_\phi \end{aligned}\right\} \tag{1.20}$$

Solving these equations yields an optimum load impedance of

$$R_L = (V_{dgB} - 2|V_P| - V_K - V_\phi)/I_F \tag{1.21}$$

which is almost the same as for Class A, with the supply voltage being given by

$$V_S = (1 - 1/\pi)\,[V_{dgB} - 2|V_P| - V_\phi + V_K/(\pi - 1)] \tag{1.22}$$

The load impedance happens to be the slope of the load line in this case also.

The maximum fundamental linear output power can be obtained directly from (1.19) and (1.21):

$$P_{\text{RFmax}} = (V_{dgB} - 2|V_P| - V_K - V_\phi)I_F/8 \tag{1.23}$$

which is slightly lower than in Class A. This difference can be minimized by selecting an FET with a low pinch-off voltage. Note that all the power in the harmonics is also dissipated in the load and this is one of the disadvantages of the resistively terminated single-ended Class B amplifier. For example, the second harmonic is only 7.4 dB below the fundamental. This problem can be overcome by using either a tuned load or a push-pull configuration as described in Sections 1.2.4 and 1.2.5, respectively.

The dc power delivered by the supply is given by

$$P_{dc} = V_s i_{d\text{peak}}/\pi \tag{1.24}$$

Unlike Class A, $P_{dc}$ is dependent on the RF output power level with no power consumed in the absence of an RF signal. This is an extremely important advantage of Class B operation.

Using (1.22), (1.23), and (1.24) and the definition of power-added efficiency given by (1.5), the maximum power-added efficiency is given by

$$\eta_{\max} = \frac{\pi^2}{8(\pi - 1)}(1 - 1/G)\frac{1 - \beta}{1 + [\beta/(\pi - 1)]} \tag{1.25}$$

where

$$\beta = \frac{V_K}{V_{dgB} - 2|V_P| - V_\phi} \tag{1.26}$$

In the limit of infinite gain and zero knee voltage the maximum efficiency is 57.6%, which is 7.6% more than in linear Class A operation. Using the same values for $V_K$, $V_{dgB}$, $V_\phi$, and $V_P$ as in the Class A example results in $\beta = 0.087$ and the maximum power-added efficiency is reduced to 50.5%. However, a 10-dB reduction in output power results in the efficiency being reduced by a factor of $\sqrt{10}$ compared with a tenfold reduction in Class A, which is an important advantage. As in the Class A amplifier, the output power can be increased by operating the amplifier in its nonlinear region, for example the saturated output power is 2.1 dB higher than $P_{\text{RFmax}}$ and it can be shown [9] that (1.15), although derived for a Class A amplifier, is still an exact description of the output power characteristic of a resistively terminated class B amplifier. Also, like a Class A amplifier, the efficiency does not increase monotonically with input power but, in fact, reaches a peak of 1.13 times $\eta_{\max}$ when the input power is 5.3 dB higher than $P_{\text{in,lin}}$ but falls to $32/\pi^3$ times $\eta_{\max}$ at saturation.

There is, however, one serious drawback to Class B operation compared to Class A; Class B amplifiers require twice the RF input voltage for a given RF output power. If the FET has the same input impedance under the two different bias conditions, the implication is that the ideal GaAs FET has 6 dB less gain in Class B than in Class A and hence one requires an FET to have at least a 10-dB gain in Class A in order to achieve an acceptable performance in Class B. For this reason virtually no work has been reported on Class B GaAs FET amplifiers above X-band. In practice the situation is far more complicated than this. For a start, $g_m$ falls dramatically for most FETs as $V_{gs}$ approaches pinch-off unless, as mentioned in Section 1.2.1, a nonuniform doping profile is used with the result that virtually no GaAs FET amplifiers use pure Class B bias conditions. In practice Class AB bias is used with a quiescent current of about 10 to 20% of $I_{dss}$. The slight degradation in efficiency resulting from a nonzero quiescent current is more than compensated for by the increased gain. An additional complication in calculating the gain reduction in Class B is caused by the fact that the gate-source capacitance $C_{gs}$ decreases as $V_{gs}$ is made more negative and thus a larger fraction of the input voltage is dropped across $C_{gs}$, which reduces the amount of gain reduction. Nevertheless, the general statement can be made that a given GaAs FET will exhibit less gain in Class B than in Class A.

### 1.2.4 Single-Ended Class B Amplifiers With Tuned Load

Figure 1.1 is also the circuit for this type of amplifier if the parallel tuned circuit resonant at $\omega_0$ is now included where $\omega_0$ is the frequency of the RF input signal. It is assumed that this tuned circuit presents an infinite impedance at $\omega_0$ but a short circuit at all harmonics. The device IV characteristics, together with the voltage and current waveforms, are shown in Figure 1.6. For pure Class B operation, the quiescent bias point is $I_d = 0$, $V_{ds} = V_s$, $V_{gs} = -|V_P|$.

The RF drain current waveform is still given by (1.19), but the shunt parallel tuned circuit provides a short-circuit path for all the harmonics so that only the fundamental current component flows into the load. This results in the voltage across the FET being a pure sinusoid at $\omega_0$. Because there are no odd-order harmonic components, it is unnecessary to provide a short circuit for these components; therefore, a practical realization of the parallel tuned circuit is a shunt short-circuit transmission line, which is a quarter wavelength long at $\omega_0$. This presents the required open circuit to the fundamental and a short circuit to all even-order harmonics, but it does mean that optimum Class B performance from this single-ended circuit is only achieved over a narrow bandwidth. A harmonic reaction amplifier [10] is an alternative method of attempting to realize the optimum load impedance at the fundamental and the harmonics, but it is also narrowband.

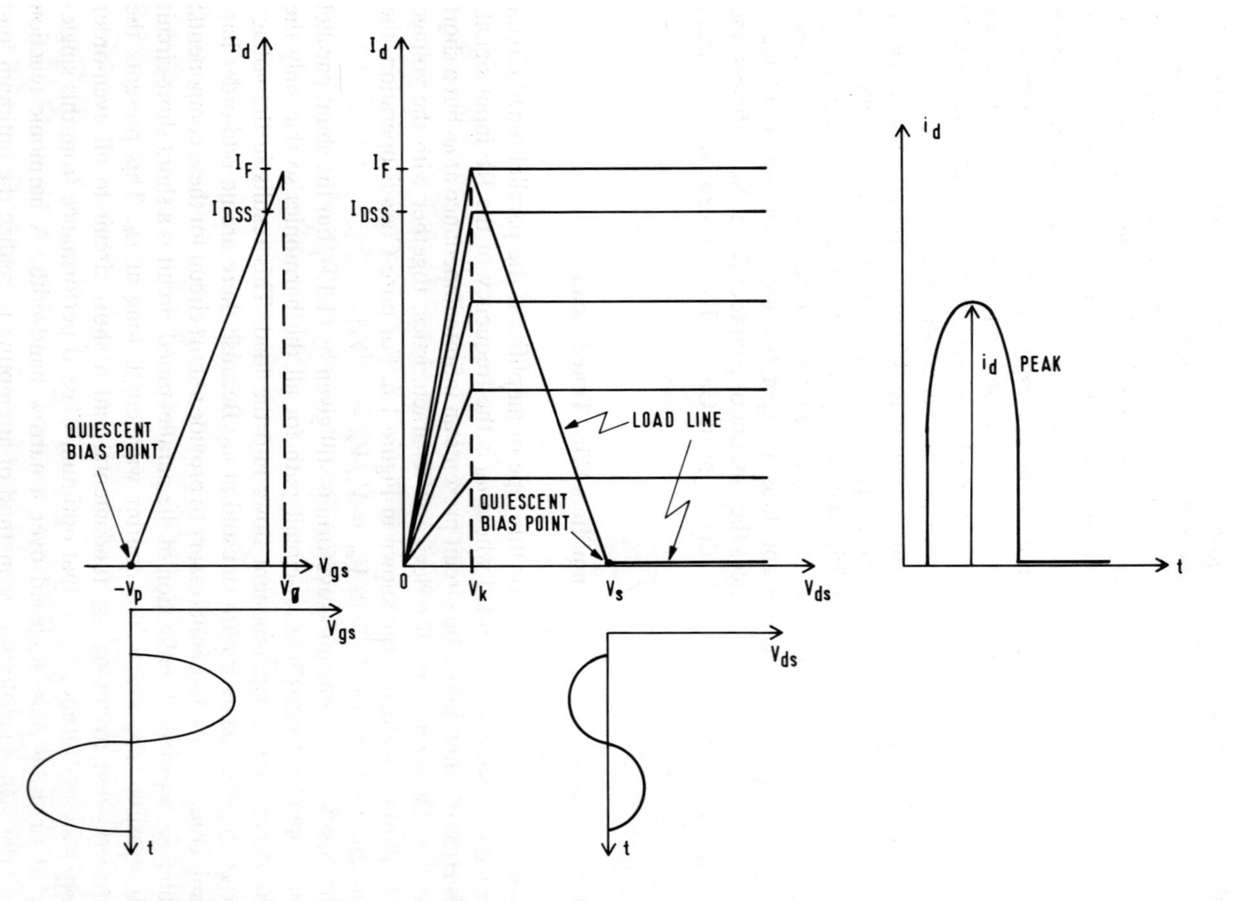


**Figure 1.6** Voltage-current waveforms for a Class B amplifier.

As in the case of a resistive load, the maximum permissible drain-source voltage is $V_{dgB} - 2|V_P| - V_\phi$ and the minimum acceptable value is $V_K$; hence, the amplitude of the sinusoidal voltage across the load is given by

$$V_{\text{peak}} = (V_{dgB} - 2|V_P| - V_K - V_\phi)/2 \tag{1.27}$$

and consequently $V_s$ is given by

$$V_S = (V_{dgB} - 2|V_P| + V_K - V_\phi)/2 \tag{1.28}$$

The maximum amplitude of the fundamental drain current component, which is also the current flowing through the load, is $I_F/2$ and hence the maximum linear RF output power is still given by (1.23); that is, the presence of the tuned circuit has no effect on the maximum output power. Similarly, the optimum load impedance is unchanged from the value given by (1.21), but in this case the value of $R_L$ is not given by the slope of the load line shown in Figure 1.6. However, the tuned circuit does increase the efficiency because the supply voltage given by (1.28) is lower than that for the resistive load case given by (1.22). The dc power delivered by the supply is still given by (1.24) and thus the maximum power added efficiency is given by

$$\eta_{\max} = \frac{\pi}{4}(1 - 1/G)\frac{(1 - \beta)}{(1 + \beta)} \tag{1.29}$$

with $\beta$ given by (1.26). In the limit of infinite gain and zero knee voltage, one arrives at the classical result of 78.5% maximum power-added efficiency for a Class B amplifier. Using the same values for $V_K$, $V_\phi$, $V_{dgB}$, and $V_P$ as in the Class A example results in the maximum power-added efficiency being reduced to 66.0%. Higher power output and efficiency are possible by overdriving the amplifier. In fact, it can be shown that the power output and efficiency characteristics of an overdriven Class B amplifier with a tuned load are identical to those of a Class B amplifier with a resistive load. Thus an absolute maximum efficiency of 88.7% is possible in the limit of infinite gain and zero knee voltage. Snider [9] has shown that it is possible to achieve 100% efficiency if a slightly higher value of load resistance at the fundamental is used provided that the output matching circuit simultaneously provides a short circuit to all even harmonics and an open circuit to all odd harmonics.

Geller and Goettle [11] have reported excellent results for an amplifier of this type; specifically, they achieved 1W over 3.7 to 4.2 GHz with an 11-dB gain and 54% minimum efficiency over the entire bandwidth using a Class AB bias of 10% $I_{DSS}$, the best efficiency in the band being 65%. This performance was achieved using a standard commercially available Fujitsu power GaAs FET (FLK 202XV) specified as having a 6-dB gain at 14.5 GHz, but screened for a low pinch-off voltage of 1.5V, high breakdown voltage (17V), and low knee voltage (1V). The use at 4 GHz of an FET designed for 14-GHz applications ensures minimal efficiency degradation in Class B from the gain reduction phenomenon, and the selection of low pinch-off and knee voltages ensures a low value for $\beta$, which keeps the efficiency high. Also, the efficiency versus power output curve showed almost exactly the theoretical $\sqrt{10}$ per 10-dB reduction. Geller and Goettle's design is analyzed in detail in Section 3.3. Also, Bahl et al. [12] have reported a Class B MMIC that gave 1.7W over 5 to 6 GHz with an 8-dB gain and 70% minimum power-added efficiency. In this design the second harmonic is short circuited as required, but the design differs slightly from the ideal in that the third harmonic is short circuited as well rather than open circuited. Once again, a low pinch-off voltage (2V) was used.

The major disadvantage of the single-ended tuned load Class B amplifier is its intrinsically narrow bandwidth compared with the resistive load type described in Section 1.2.2. The tuned load clearly provides only a short circuit to the harmonics over a narrow bandwidth. If the tuned load is replaced by a bandpass filter, then broader bandwidth with respect to suppression of harmonic energy in the load can be achieved, but even this approach is restricted to less than an octave of bandwidth because the second harmonic of the lower band-edge frequency, which we want to suppress, coincides with the upper band-edge frequency, which we want to pass. Also, bandpass filters will not, in general, present a short circuit to all harmonics over the band of interest, which leads to efficiency variations over the passband. Harmonic suppression is not only desirable from the user's viewpoint, it is also intimately connected with achieving high efficiency by reflecting the harmonic energy with the correct phase so as to achieve a more favorable voltage waveform across the FET. The solution to achieving the optimum voltage waveform and harmonic suppression over broad bandwidths is to use a phasing rather than a filtering scheme, such as that used in the push-pull amplifiers described in the next section.

### 1.2.5 Push-Pull Class B Amplifiers

A push-pull amplifier is shown in Figure 1.7, together with the voltage and current waveforms at various points in the circuit. As before, it is assumed that the series dc blocking capacitors appear as a short circuit to any RF signal, while the chokes are assumed to be an open circuit. The input 0–180-deg power splitter drives the two transistors in antiphase so that only one transistor is conducting at any given time as both devices are biased at pinch-off. Fourier analysis of the two drain current waveforms gives

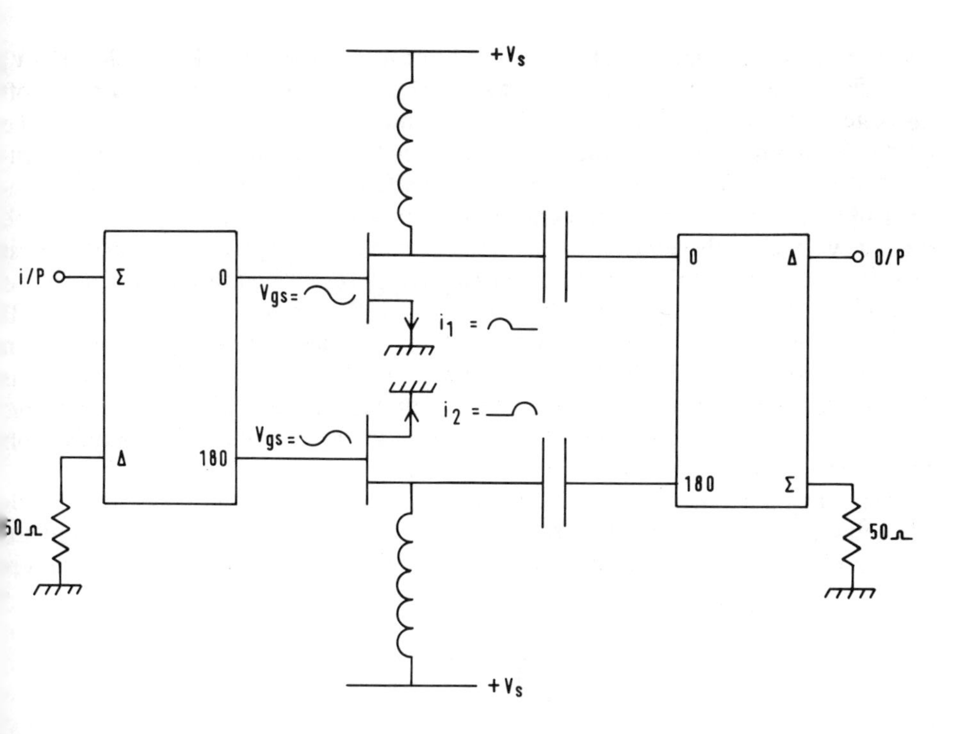


**igure 1.7** Circuit diagram of a Class B push-pull amplifier with resistive load.

$$
\begin{aligned}
i_{d1} &= i_{d\text{peak}}\left(\frac{1}{\pi} + \frac{1}{2}\sin\omega_0 t - \frac{2}{\pi}\sum_{n=2,4,\ldots}^{\infty}\frac{1}{n^2-1}\cos n\omega_0 t\right)\\
i_{d2} &= i_{d\text{peak}}\left(\frac{1}{\pi} - \frac{1}{2}\sin\omega_0 t - \frac{2}{\pi}\sum_{n=2,4,\ldots}^{\infty}\frac{1}{n^2-1}\cos n\omega_0 t\right)
\end{aligned}
\tag{1.30}
$$

hich shows that the fundamental components are in antiphase while the harmonics re in phase. The dc components pass through the chokes while the ac components ass through the capacitors and are combined in the output 0–180-deg coupler such hat the fundamental components add up in phase at the $\Delta$ port and the harmonics dd up in phase at the $\Sigma$ port and are dissipated in the dummy load, assuming that multioctave coupler is used. This type of circuit can offer very broadband perormance and it has twice the output power capability of its single-ended counterart, but the same gain since the input power has to be doubled to achieve the ame voltage swing at the FET gates. Since each FET sees a purely resistive load t all frequencies, the circuit shown in Figure 1.7 is the push-pull version of the esistively loaded single-ended amplifier described in Section 1.2.2 and so the optimum load impedance, supply voltage, and power-added efficiency are given by (1.21), (1.22), and (1.25), respectively. Note that an impedance transformer needs to be connected between each FET and the output coupler to transform $50\Omega$ to the optimum load impedance. While a conventional balanced amplifier using quadrature couplers provides a substantial improvement in VSWR [13], no such improvement occurs when a 180-deg coupler is used. Also, the power output and efficiency are directly reduced by any loss in the output coupler, thus single-ended amplifiers will always be more efficient than their push-pull counterparts. However, Lane et al. [14] have reported achieving 2W at 40% power-added efficiency over 9.2 to 10.2 GHz with a 5-dB gain from a push-pull Class B amplifier.

An alternative configuration for a push-pull amplifier is shown in Figure 1.8. The fundamental current components add up in phase in the center-tapped primary to produce an output current that flows through the load, while the harmonics are out of phase and produce no net current in the primary and hence no harmonic power is dissipated in the load. The harmonic currents flow to ground through the center tap, thus the harmonics are terminated in an effective short circuit and so this circuit is the push-pull counterpart of the tuned load single-ended amplifier

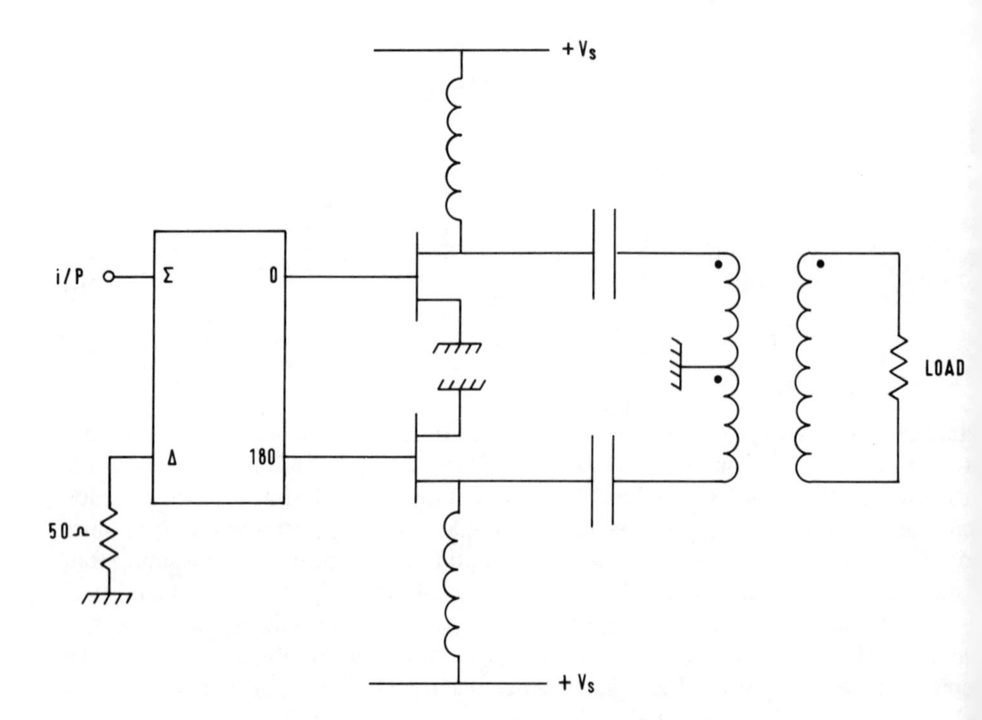


**Figure 1.8** Circuit diagram of a Class B push-pull amplifier with tuned load.

described in Section 1.2.3 (i.e., it has twice the output power but the same efficiency).

The ideal transformer shown in Figure 1.8 is realized at microwave frequencies by a *balun*, which is a device for converting a balanced load to an unbalanced one and vice versa. For this application one needs a multioctave, low-loss, and preferably planar balun such as the one described by Pavio and Kikel [15].

### 1.2.6 Comparison of the Various Types of Power Amplifiers

The key performance parameters of the various types of amplifiers are summarized in Table 1.1. Assuming that negligible harmonic power must be dissipated in the load, then theoretically the highest output power with the highest efficiency over the widest bandwidth is provided by a push-pull Class B amplifier using a balun for combining, but the efficiency and power output are greatly reduced by any loss in the balun. The highest gain, on the other hand, is provided by a Class A amplifier. The other types of Class B amplifiers provide intermediate levels of performance in one or more of the parameters, but in some applications their performance may be entirely adequate and their simpler construction is a positive advantage.

### 1.2.7 Validity of the Preceding Analysis

The analysis in Sections 1.2.2 to 1.2.5 uses a set of dc characteristics to predict the performance at microwave frequencies. The validity of this analysis needs to be examined. First, the dc characteristics can be measured in a number of different ways, but one should use short-pulse IV measurements [16]. There are two reasons for this. First, an FET has a number of surface states and deep level traps which cannot empty and fill at microwave frequencies or under short-pulse conditions but which can do so at the slow sweep speeds used on curve tracers. Second, it is important for the IV characteristics to be measured in such a way that the thermal situation is representative of that which will occur in a practical microwave amplifier. Thus for best accuracy one should use a short-pulse IV test centered on the intended quiescent bias point.

An additional aspect needing examination is the effect of the finite output impedance of the FET. The preceding analysis assumed that the FET had an infinite output impedance; in practice, the device has a finite output resistance in parallel with a capacitance. The capacitance can always be tuned out at a spot frequency—at least in theory—by resonating it with an external inductance, while the resistance merely means that a fraction of the output current is dissipated within the FET itself rather than in the load. This effect is frequency independent, provided one is operating above the frequency range where conductance dispersion occurs

**Table 1.1**
Performance Parameters of Class A and B Amplifiers

| | *Class A* | *Class B Single-Ended Resistive Load* | *Class B Single-Ended Tuned Load* | *Class B Push-Pull Resistive Load* | *Class B Push-Pull Tuned Load* |
|---|---|---|---|---|---|
| Maximum output power | $I_F(V_{dgB} - \|V_P\| - V_K)/8$ | $I_F(V_{dgB} - 2\|V_P\| - V_K - V_\phi)/8$ | $I_F(V_{dgB} - 2\|V_P\| - V_K - V_\phi)/8$ | $I_F(V_{dgB} - 2\|V_P\| - V_K - V_\phi)/4$ | $I_F(V_{dgB} - 2\|V_P\| - V_K - V_\phi)/4$ |
| Maximum power-added efficiency | $\frac{1}{2}(1 - 1/G)\left(\frac{1-\alpha}{1+\alpha}\right)$ | $\frac{\pi^2}{8(\pi-1)}\left(1-\frac{1}{G}\right)\left(\frac{1-\beta}{1+\beta/(\pi-1)}\right)$ | $\frac{\pi}{4}\left(1-\frac{1}{G}\right)\left(\frac{1-\beta}{1+\beta}\right)$ | $\frac{\pi^2}{8(\pi-1)}\left(1-\frac{1}{G}\right)\left(\frac{1-\beta}{1+\beta/(\pi-1)}\right)$ | $\frac{\pi}{4}\left(1-\frac{1}{G}\right)\left(\frac{1-\beta}{1+\beta}\right)$ |
| Efficiency decrease/10-dB decrease in output power | 10 | $\sqrt{10}$ | $\sqrt{10}$ | $\sqrt{10}$ | $\sqrt{10}$ |
| Optimum RF load resistance | $(V_{dgB} - \|V_P\| - V_K)/I_F$ | $(V_{dgB} - 2\|V_P\| - V_K - V_\phi)/I_F$ | $(V_{dgB} - 2\|V_P\| - V_K - V_\phi)/I_F$ | $(V_{dgB} - 2\|V_P\| - V_K - V_\phi)/I_F$ | $(V_{dgB} - 2\|V_P\| - V_K - V_\phi)/I_F$ |
| Gain $G$ (dB) | x | x − 6 | x − 6 | x − 6 | x − 6 |
| dc power input with zero RF input | $V_S I_F/2$ | 0 | 0 | 0 | 0 |
| Supply voltage $V_S$ | $(V_{dgB} - \|V_P\| + V_K)/2$ | $\left(\frac{\pi-1}{\pi}\right)(V_{dgB} - 2\|V_P\| - V_\phi) + V_K/\pi$ | $(V_{dgB} - 2\|V_P\| + V_K - V_\phi)/2$ | $\left(\frac{\pi-1}{\pi}\right)(V_{dgB} - 2\|V_P\| - V_\phi) + V_K/\pi$ | $(V_{dgB} - 2\|V_P\| + V_K - V_\phi)/2$ |

[16], but its effect should be allowed for when predicting the performance of power FETs.

As a practical demonstration of the validity of the preceding analysis, Geller and Goettle [11] used an FET with $V_{dgB} = 17\text{V}$, $V_K = 1\text{V}$, $V_P = 1.5\text{V}$, $V_\phi = 0.7\text{V}$, and $I_F = 800$ mA. The theoretical output power from such an FET in Class B is 1.24W, while 1.1W was achieved in practice at 4.1 GHz, which is 0.5 dB lower than theoretical. This 0.5-dB loss must account for circuit losses as well as any finite output resistance within the FET. The theoretical power-added efficiency is 61.9% while 65.1% was achieved in practice. The large signal gain at 4.1 GHz was 10.4 dB and it should be noted that a Class AB bias of about $I_{dss}/10$ was used, which degrades the accuracy of the theoretical calculations slightly. Nevertheless, we can see that the preceding analysis predicts results in close agreement with experimental values.

## 1.3 THE CONCEPT OF POWER PER MILLIMETER OF GATEWIDTH

The analysis of Section 1.2 shows that the maximum RF output power from a given FET is

$$P_{\text{RFmax}} = I_F(V_{dgB} - m|V_P| - V_K - nV_\phi)/8 \tag{1.31}$$

where $m = 1$ and $n = 0$ for Class A, and $m = 2$ and $n = 1$ for Class B bias. Now $I_F$ is given by [16]

$$I_F = qN_D v_{\text{sat}} h w_g \tag{1.32}$$

where $q$ = electron charge, $N_D$ = doping density, $v_{\text{sat}}$ = saturated electron drift velocity, $h$ = epilayer thickness, and $w_g$ = gatewidth. Hence, the power output per unit gatewidth is given by

$$P_{\text{RFmax}}/w_g = qN_D v_{\text{sat}} h(V_{dgB} - m|V_P| - V_K - nV_\phi)/8 \tag{1.33}$$

Obviously, device designers try to maximize the right-hand side of (1.33), but since $V_{dgB}$, $V_P$, and $V_K$ are themselves functions of $N_D$, $v_{\text{sat}}$, and $h$, among other things, the determination of the optimum power FET structure is a very complex process. This aspect is dealt with in much greater detail in Chapter 2.

Thus, the power output per unit gatewidth for a given FET cross-sectional structure is, at least in principle, a constant, a typical value being around 0.5 W/mm. This concept is used extensively in GaAs MMICs because gatewidth is a design parameter at the monolithic circuit designer's disposal. The concept obviously has negligible relevance to the hybrid circuit designer using discrete devices. The concept needs to be used with caution, however, since the power output per millimeter is not truly a constant—one cannot expect a 10-mm device to have 10 times the output power of a 1-mm device due, among other things, to resistive losses and phase changes along the gate length (although we try to minimize these effects by using multiple fingers). Also, this concept completely ignores the gain of the device, and the gatewidth often has to be restricted to maintain an acceptable gain. Nevertheless, used with caution, the concept forms a useful aid in the design of monolithic power amplifiers. However, claims made by device designers must be examined carefully because it is not unknown for unrealistically high values of $P_{RF}/w_g$ to be published based on short gatewidth devices.

## 1.4 SMALL-SIGNAL (LINEAR) AND LARGE-SIGNAL (NONLINEAR) MODELS FOR AN FET

The analysis of Section 1.2 is a "black box" approach assuming an ideal device with infinite input and output impedances. We now determine the effects of using real devices with complex finite input and output impedances.

Figure 1.9 shows a schematic cross section of an idealized FET. The application of gate and drain bias voltages causes the depletion region to increase in volume compared with the zero bias case and to assume the shape shown. If an ac signal is now superimposed on the gate dc bias voltage, then the depletion layer boundary will move in sympathy with the ac voltage, causing the drain current also to have an ac component (i.e., the current flowing between the drain and source is determined by a voltage-controlled current generator $g'_m$ as shown in Figure 1.9)

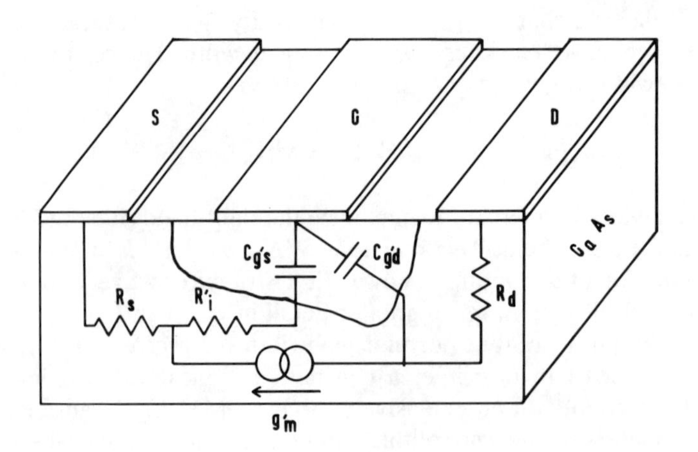


**Figure 1.9** Schematic cross section of an idealized GaAs FET.

The depletion region obviously has a capacitance associated with it resulting in the components $C'_{gs}$ and $C'_{gd}$, and outside the depletion region are the parasitic series resistances $R_S$ and $R_d$. It is also necessary to include an intrinsic channel resistance $R'_i$ in the equivalent circuit, as show in Figure 1.9. A detailed explanation of the origin of $R'_i$ can be found in Ladbroke [16]. The circuit shown in Figure 1.9 is only a partial equivalent circuit for a GaAs FET; the full equivalent circuit will be discussed later.

The values of $C'_{gs}$, $C'_{gd}$, $R'_i$, and $g'_m$ depend on the depletion depth $d$, which for the purposes of this discussion we assume depends only on that fraction of the total gate-source voltage that is dropped across the depletion region $V_{\text{dep}}$, which is given by

$$V_{\text{dep}} = V_{gs}\left(1 + \frac{v'_{gs}}{V_{gs}} \cos \omega t\right) \tag{1.34}$$

where $V_{gs}$ is the dc gate-source bias voltage and $v'_{gs} \cos \omega t$ is that fraction of the RF voltage applied to the gate which is dropped across the depletion region. The assumption that these elements depend only on $V_{\text{dep}}$ is an oversimplification because the elements also depend on the drain-source voltage, but this assumption is made here in order to illustrate the principles involved as simply as possible. Taking $C'_{gs}$ as an example, $C'_{gs}$ is a function of $V_{\text{dep}}$ and is thus a nonlinear and time-varying circuit element. However, since $V_{\text{dep}}$ is a periodic function, then so also is $C'_{gs}$ and thus it may be expanded into the Fourier series

$$C'_{gs} = \sum_{n=0}^{\infty} C_n \cos n\omega t \tag{1.35}$$

where $C_n$ is determined from

$$C_n = \frac{1}{2\pi} \int_0^{2\pi} C'_{gs} \cos n\omega t \, d(\omega t) \tag{1.36}$$

Since $C'_{gs}$ is a function of $v'_{gs}/V_{gs}$ then so also are the coefficients $C_n$.

Under small-signal conditions we assume that the ratio $v'_{gs}/V_{gs}$ is sufficiently small that all the terms in the Fourier series of (1.35) can be neglected except for $C_0$, which can be expanded into the power series

$$C_0 = C_{gs} + C_{gs1}\left(\frac{v'_{gs}}{V_{gs}}\right) + C_{gs2}\left(\frac{v'_{gs}}{V_{gs}}\right)^2 + \ldots \tag{1.37}$$

Again, under small-signal conditions we assume that the ratio $v'_{gs}/V_{gs}$ is sufficiently

small that only the first term needs to be retained. Thus under small-signal conditions the nonlinear time-varying capacitance $C'_{gs}$ is approximated by the time
invariant, voltage-independent capacitance $C_{gs}$, which is the time-averaged valu
of $C'_{gs}$ and depends only on the dc bias voltage. Similarly, $C'_{gd}$, $R'_i$, and $g'_m$ ar
approximated by their time-averaged values $C_{gd}$, $R_i$, and $g_m$ and under these con
ditions the equivalent circuit is now a linear circuit. As mentioned earlier, th
equivalent circuit shown in Figure 1.9 is only a partial one; the full equivalen
circuit is shown in Figure 1.10. An excellent derivation of this equivalent circui
can be found in Ladbroke [16]. A number of properties of an FET can be deduce
immediately from the equivalent circuit by visual inspection. First, the input an
output impedances are obviously finite and complex compared to the infinite valu
of an ideal device. Second, the equivalent circuit elements $R_s$ and $C_{gd}$ give rise t
internal feedback within the FET and make the transistor nonunilateral, that is
$S_{12}$ is nonzero, which can make the transistor unstable for some combinations o
source and load impedances. Third, only a portion of the gate-source voltage i
dropped across $C_{gs}$ due to the potential divider action of $C_{gs}$ in series with $L_g$, $R_g$
$R_i$, and $R_s$. In fact, the voltage across $C_{gs}$ is halved, and consequently so also is th
output current when the frequency is doubled (i.e., the gain decreases by 6 dB
octave in the microwave frequency range).

Under large-signal conditions the assumption that all terms in the Fourie
series other than $C_0$ can be neglected and that only the first term in the powe
series of (1.37) needs to be retained is no longer valid. The values o
$C'_{gs}$, $C'_{gd}$, $g'_m$, and $R'_i$ in Figure 1.9 must all be treated as nonlinear, time-varyin
circuit elements. Taking $C'_{gs}$ as an example again, the current flowing through thi
capacitor is given by

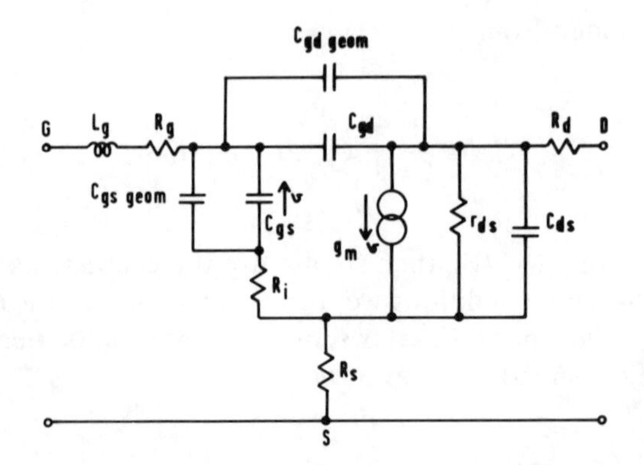


**Figure 1.10** Small-signal equivalent circuit of a GaAs FET.

$$i = \frac{d}{dt}(C'_{gs}V_{\text{dep}}) = C'_{gs}\frac{dV_{\text{dep}}}{dt} + V_{\text{dep}}\frac{dC'_{gs}}{dt} \tag{1.38}$$

Making the necessary substitutions from (1.34) and (1.35),

$$i = \omega(C_0 v'_{gs} + V_{gs}C_1)\cos(\omega t - \pi/2) + \text{components at } n\omega$$

Thus the current flowing through the capacitor consists of the normal component in-phase quadrature with the driving voltage at the driving frequency plus harmonic components, which must be allowed to flow in the input matching circuitry. Obviously, the harmonics generated by $g'_m$ are far more significant because they appear at the amplifier's output. The variable $C_0$, which is determined by (1.37), depends on the amplitude of the applied RF voltage and hence the fundamental current component is nonlinearly related to the applied voltage. The significance of this effect can be estimated by considering an idealized GaAs FET with a uniform doping profile in which case $C'_{gs}$ is given by [16]

$$C'_{gs} = l_g w_g \sqrt{\frac{qN_D\epsilon}{2\left[V_\phi - V_{gs}\left(1 + \frac{v'_{gs}}{V_{gs}}\cos\omega t\right)\right]}} \tag{1.39}$$

Substituting (1.39) into (1.36), $C_0$ can be calculated as a function of $v'_{gs}/V_{gs}$. It is assumed in the following that $V_{gs} = -|V_P|/2$ and without loss of generality it can be assumed that $V_\phi = 0$ because a nonzero value of $V_\phi$ merely requires a slight shift in bias point. Unfortunately, the integration required in (1.36) cannot be undertaken analytically, but it is easily calculated numerically and the results are shown in Figure 1.11. It can be seen that the variation in $C_0$ is small except when $v'_{gs}/V_{gs}$ approaches unity, which causes the capacitance to approach infinity over a portion of the RF cycle. A similar analysis can be undertaken for $C'_{gd}$, $g'_m$, and $R'_i$ but the conclusion is the same—namely, that none of the elements has a time-averaged value that is a strong function of the RF signal until $v'_{gs}/V_{gs}$ approaches unity. This conclusion is supported by the work of Tajima et al. [17] who present graphs of $C_{gs}$, $g_m$, and $R_i$ as a function of the input RF signal level (they did not consider $C_{gd}$ to be a nonlinear element). This conclusion is the theoretical justification for the widely used Cripps design technique to be discussed in Section 1.6.

While the above analysis gives some insight into the large-signal operation of a GaAs FET for signal levels below which gate current can flow, it is too simplistic to provide accurate predictions of large-signal performance because it neglects, among other things, the dependence of the nonlinear circuit elements on the drain-

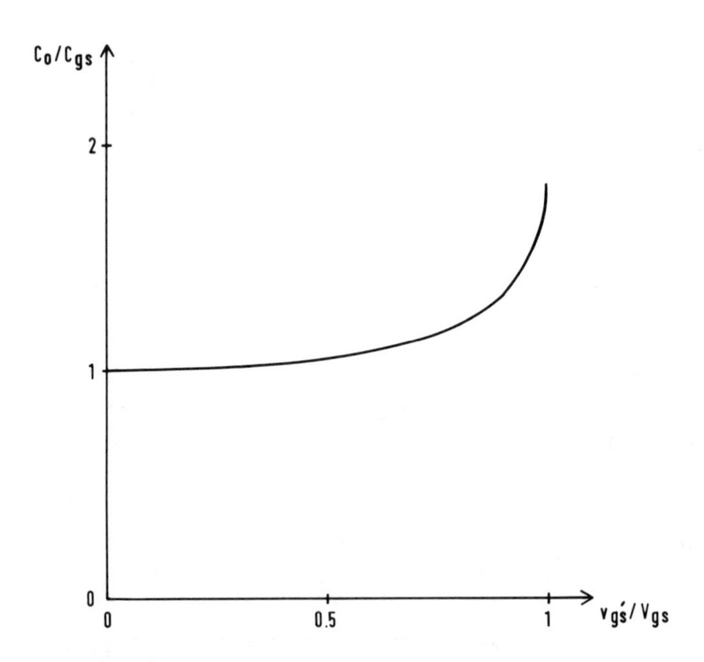


**Figure 1.11** Variation of gate-source capacitance with input signal level.

source voltage. A technique originating with Willing et al. [18] is gaining wide acceptance for nonlinear modeling and design of GaAs FET power amplifiers. The technique consists of measuring the small-signal *S* parameters over a wide range of gate and drain bias voltages and then fitting the standard FET equivalent circuit shown in Figure 1.10 at each bias point to the measured *S* parameters. In this way the complete bias dependence of the nonlinear circuit elements can be determined. Because small-signal measurements are made, this technique in effect determines the instantaneous values of the parameters as a function of bias voltage. Thus, the time-averaging process has been removed and the nonlinear parameters determined by this technique can show large variations with bias voltages. The nonlinear circuit element bias-dependent data are then used in a nonlinear simulation program (see Chapter 3 for a thorough discussion of this topic).

A complete large-signal equivalent circuit for a GaAs FET is shown in Figure 1.12. The diodes are included to account for the effect of forward gate-source current and gate-drain avalanching, which must be considered if we are to obtain realistic simulations. Depending on the bias point and the signal level, it is possible in some instances to ignore some of the nonlinearities without degrading the simulation accuracy significantly.

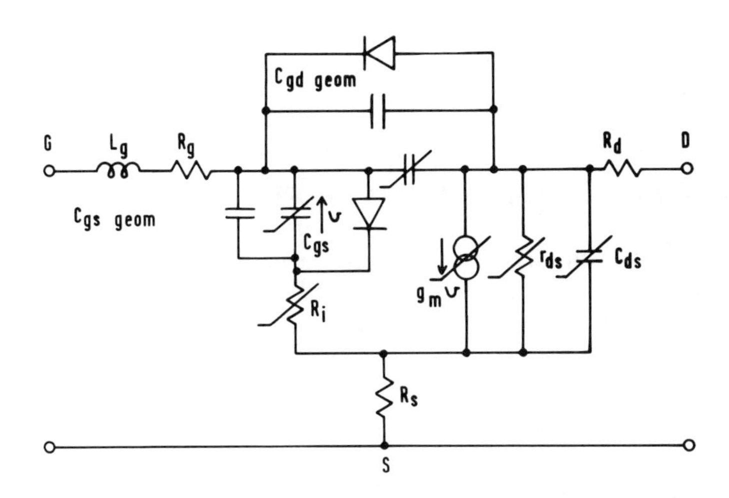


**Figure 1.12** Large-signal equivalent circuit of a GaAs FET.

Large-signal operation is inevitably associated with such undesirable attributes as gain compression, the generation of harmonics and intermodulation products, and *amplitude modulation to phase modulation* (AM-PM) conversion. That gain compression should occur can be deduced directly from the principle of conservation of energy since the RF output power can never exceed the sum of the dc and RF input powers; thus, as the RF input power is increased the gain must decrease if the dc input power remains fixed. The generation of harmonics and intermodulation products is a direct consequence of having to retain the higher order terms in the Fourier series of (1.35) and power series expansion of (1.37) for the circuit elements. AM-PM conversion refers to the fact that under large-signal conditions the circuit elements $C'_{gs}$, $C'_{gd}$, $R'_i$, and $g'_m$ depend on the signal level and hence the insertion phase of the FET is a function of the input signal level. Thus a pure amplitude-modulated signal applied to an FET will emerge with some added phase modulation.

Finally, it is important to distinguish between the terms *high power* and *large signal*. *Large signal* refers to a situation in which the RF voltage across the depletion capacitance is a significant fraction of the applied dc gate-source voltage and, as such, it has no relationship to the amount of output power produced from the FET. For example, large-signal effects occur in low output power FETs when driven hard. *High power* refers simply to the amount of output power produced and it is completely arbitrary what power level is defined as being "high power," but high power is not necessarily synonymous with large-signal operation because a given output power can be obtained either from an even higher power FET operated under small-signal conditions or from a lower power FET operated under large-signal conditions.

## 1.5 OVERVIEW OF DESIGN TECHNIQUES

It is instructive to consider how to design a small-signal amplifier for maximum gain in order to see where large-signal designs deviate from the small-signal case. The small-signal design process can be summarized as follows:

1. Select the transistor.
2. Select the bias point.
3. Measure the $S$ parameters at the intended bias point if not already known.
4. Assuming the transistor is unconditionally stable, compute the *maximum available gain* (MAG) from the $S$ parameters and compute the source and load impedances required to achieve this gain.
5. Synthesize the input and output matching networks to provide the required source and load impedances. In the case of broadband amplifiers it is necessary to compensate for the inherent 6 dB/octave gain roll-off of the GaAs FET, which can be done using either the input or output matching networks or by other techniques [19].

Figure 1.13 shows a schematic illustration of the resulting single-stage amplifier. As a result of the finite reverse gain $S_{12}$ of a transistor the input reflection coefficient $\Gamma_{in}$ depends on the reflection coefficient of the load and is determined by

$$\Gamma_{in} = S_{11} - \frac{S_{12}S_{21}\Gamma_L}{1 - S_{22}\Gamma_L} \tag{1.40}$$

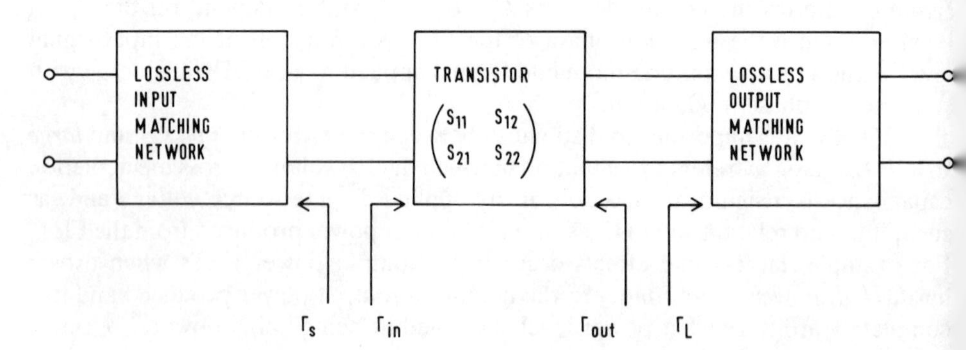


**Figure 1.13** Block diagram of a single-stage amplifier.

Similarly, the output reflection coefficient depends on the reflection coefficient of the source and is determined by

$$\Gamma_{\text{out}} = S_{22} - \frac{S_{12}S_{21}\Gamma_S}{1 - S_{11}\Gamma_S} \tag{1.41}$$

When the matching networks present to the FET the impedances required for maximum gain, this also results in a perfect match at the input and output of the amplifier. When this occurs the transistor is conjugately matched

$$\begin{aligned} \Gamma_{\text{in}}^{*} &= \Gamma_S \\ \Gamma_{\text{out}}^{*} &= \Gamma_L \end{aligned} \tag{1.42}$$

where $\Gamma_S$ and $\Gamma_L$ are determined from simultaneous solution of (1.40), (1.41), and (1.42). Note that only one value of $\Gamma_L$ and $\Gamma_S$ results in a perfect match and maximum gain. Any other value of $\Gamma_L$ and $\Gamma_S$ results in reduced gain and finite input and output VSWRs. Consequently, if the input and output matching networks are used to compensate for the 6 dB/octave gain roll-off by deliberately presenting non-optimum impedances to the FET, then this inherently results in large and unacceptable input and/or output VSWRs in broadband single-stage amplifiers. For example, an octave bandwidth amplifier would have an input VSWR of 14:1 at the low-frequency end if all the compensation occurred in the lossless input matching network or 3.5:1 if the compensation was split equally between the input and output lossless matching networks. To overcome the high VSWR problem, two such amplifiers are often connected in a balanced configuration [20] as shown in Figure 1.14 where the quadrature couplers prevent these high VSWRs from being seen at the terminals of the amplifier even though they are presented to the coupler internally.

There are other ways of overcoming the problem. Feedback or lossy matching networks or a distributed structure can be used, as can a two-stage amplifier with all the compensation occurring in the interstage network. These structures are shown in Figure 1.15 and a comparison of their relative merits for small-signal amplification can be found in [19]. However, the use of feedback or lossy matching networks in a power amplifier stage is generally precluded because of their deleterious effects on power output and gain. Distributed amplifiers are considered in further detail in Chapter 6. If a two-stage amplifier is used with the interstage network providing all the gain compensation, then for broad bandwidths one ends up requiring the input transistor to have a higher output power capability than the output device! The solution to this problem is to restrict the amount of gain compensation provided by the interstage network to, say, 6 dB/octave and to use this

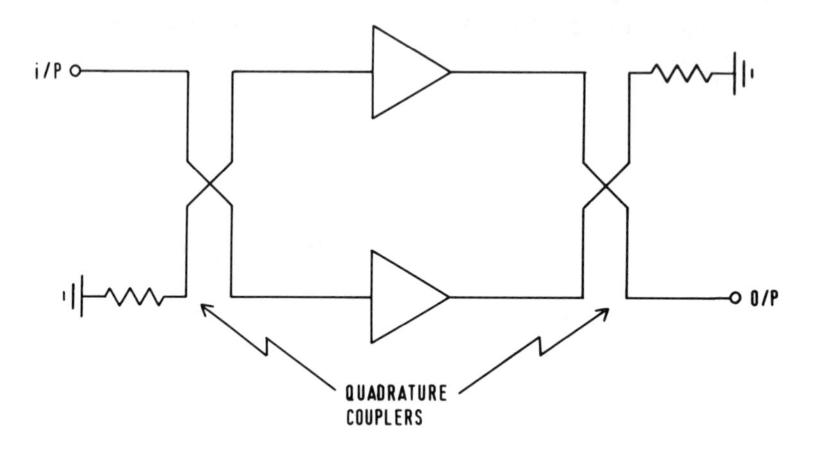


**Figure 1.14** Block diagram of balanced amplifier.

technique in conjunction with feedback around the first transistor or with a lossy input matching network.

The large-signal design process can be summarized as follows:

1. Select the transistor on the basis of its ability to provide the required output power with an adequate gain, efficiency, etc.
2. Select the bias point as discussed in Section 1.2.
3. Determine the load impedance at the fundamental (and harmonics if not a pure Class A amplifier) required for maximum output power.
4. Synthesize the lossless output matching network.
5. Determine the source impedance required to maximize the gain.
6. Synthesize the input matching network to achieve the required gain and gain flatness.

Considerable differences occur in the processes of designing small- and large-signal amplifiers. First, the load impedance required for maximum output power will not be the same as that required for maximum gain. Thus even narrowband power amplifiers always have a higher output VSWR than small-signal ones. They also have a lower gain, not only because they have to be mismatched at the output but also because the bias point for maximum output results in a lower $g_m$, and hence less gain, than the bias point for maximum gain.

Second, in theory, the lossless output matching network cannot provide any gain slope compensation since the network is constrained to provide the optimum load impedance to the FET for maximum power output. Thus, only the input matching network can be used for gain compensation.

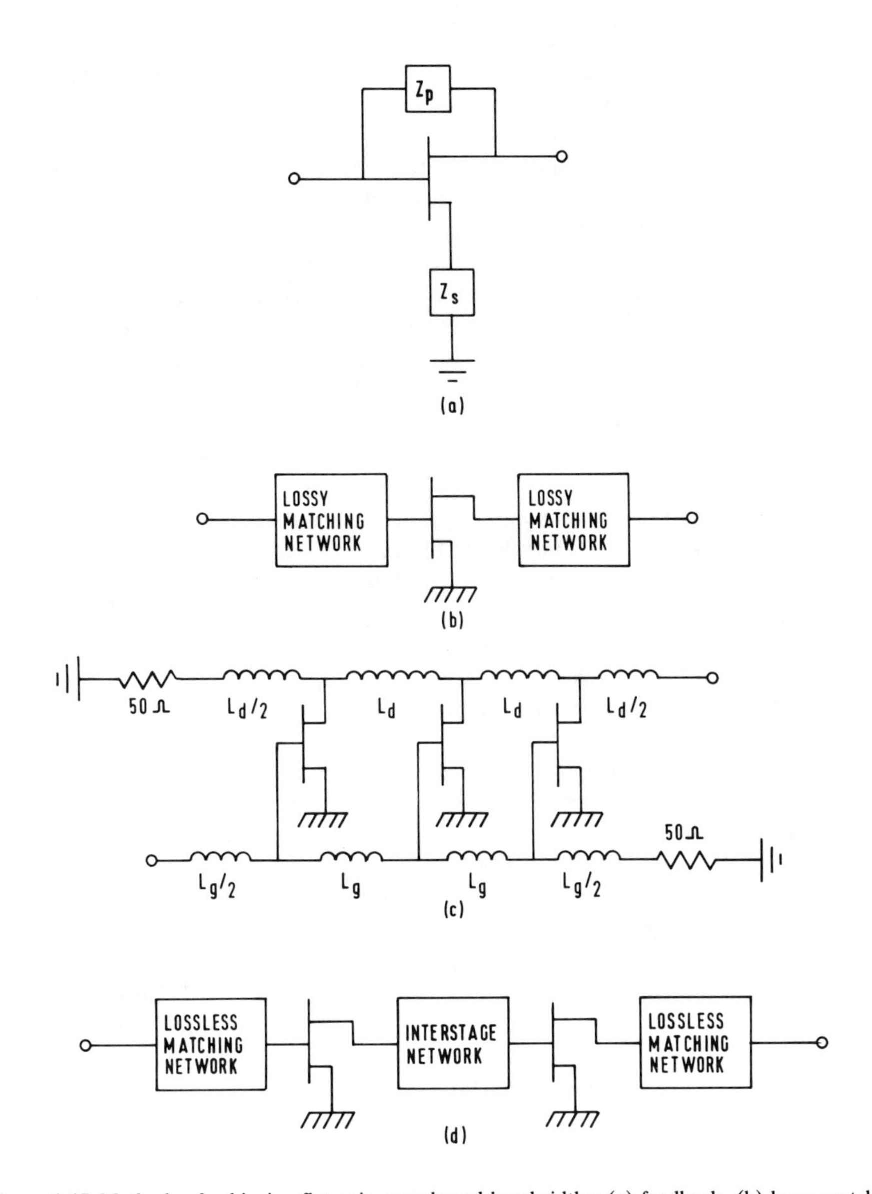


**Figure 1.15** Methods of achieving flat gain over broad bandwidths: (a) feedback, (b) lossy matching, (c) distributed amplifier, and (d) multistage amplifier.

Third, while a small-signal amplifier is uniquely characterized for all input signal levels by its $S$ parameters provided it is operated in the linear region, the large-signal amplifier has to be characterized at all possible input signal levels. In particular, the small-signal amplifier's stability is easily calculated from the $S$ parameters and its stability is independent of input signal level. The large-signal amplifier's stability, however, is a function of the input signal level and so one can end up with a power amplifier being unconditionally stable under large-signal conditions but unstable under small-signal conditions or vice versa.

Finally, how do you determine the optimum load impedance for maximum output power? In the small-signal case the load impedance for maximum gain can be determined directly from the measured $S$ parameters by numerical calculation, but the load impedance for maximum power output is completely unrelated to the $S$ parameters of the FET. The three most popular techniques for designing power amplifiers are the Cripps' method [21], load-pull, and nonlinear simulation. Each of these techniques is now reviewed briefly.

The Cripps technique is the simplest of the three methods and is widely used, particularly in the hybrid industry. The Cripps technique assumes that the optimum load resistance is identical to the values given in Table 1.1. and that the optimum load reactance is the complex conjugate of the FET's small-signal output reactance. The theoretical justification for this latter step is the observation in Section 1.4 that the output capacitance is not a strong function of the input signal level. While this technique is acceptable for hybrid power amplifiers in which the final amplifier can be tuned to achieve the desired response, it is clearly of limited use in monolithic power amplifier design where one cannot tune and where there is a very strong desire to have a "right first time" design philosophy because of the very large expense involved in a redesign.

In the load-pull technique, mechanized stub tuners are connected to the input and output of the FET and then adjusted until maximum output power is achieved. The tuners are then removed and the impedance that they present to the FET is measured on a vector network analyzer. This process is repeated for each frequency of interest. However, the technique has a number of limitations. For example, the loss of the tuner varies with tuner setting, thus making it difficult to determine when maximum power is achieved, the impedance seen by the FET is not the same as that presented by the tuner because the FET has to be measured in a test fixture and thus one has to correct for all the transitions involved, some tuner settings may cause the device to oscillate or to be destroyed, and one cannot independently control the impedance presented by the tuner at both the fundamental and harmonic frequencies. This latter problem is particularly serious for Class B design because it is virtually impossible for the tuner to present the optimum impedance to the FET. An alternative to the use of tuners for load-pull measurements is to use the active load-pull technique [22], but many of the above limitations still apply.

The last technique is based on nonlinear simulation, which is covered in depth in Chapter 3. This latter technique, although in its infancy at present, will without doubt become the industry standard method for the design of power amplifiers.

The technique has only become a practical reality due to the enormous advances that have been made in computer hardware.

## 1.6 BANDWIDTH LIMITATIONS OF REACTIVELY MATCHED AMPLIFIERS

### 1.6.1 Output Matching Network

In the design of a power amplifier the output matching network is required to transform the 50-$\Omega$ load impedance to the optimum value for maximum output power (i.e., $R_L$ as given in Table 1.1) and it simultaneously has to present the optimum reactance to the FET, which, although not exact, is assumed here for simplicity to be a shunt reactance equal to the complex conjugate of $C_{ds}$. Many studies, for example [23], have shown this to be a good approximation, particularly at relatively low frequencies where parasitic drain bond wire inductive reactances are negligible.

The above is a statement of the matching problem as seen by the FET, but for the study of matching networks it is conceptually simpler to invert the problem and express it in an entirely equivalent way as shown in Figure 1.16, namely, that the matching network is required to transform the complex load of $R_L$ in parallel with $C_{ds}$ to 50 $\Omega$ over a given bandwidth. A perfect match can be achieved at a spot frequency, but is impossible over a finite bandwidth, and Bode [24] showed that the reflection coefficient at the output is governed by

$$\int_0^\infty \ln \frac{1}{|S_{22}(\omega)|} d\omega \leq \frac{\pi}{R_L C_{ds}} \tag{1.43}$$

The minimum worst case reflection coefficient will occur when $|S_{22}(\omega)|$ is a constant over the band of interest, $\Delta f$, and unity elsewhere as shown in Figure 1.17, in which case the inequality of (1.43) can be simplified to

$$|S_{22}(\omega)| \geq \exp(-1/(2\Delta f R_L C_{ds})) \tag{1.44}$$

This is the theoretical best value using a matching network with an infinite number of elements. Practical networks with only a few elements will produce results considerably worse than this [25]. For example, suppose we want to match a load of 20$\Omega$ in parallel with 0.9 pF over 6 to 18 GHz. The very best match that can be achieved with an infinite number of elements is 1.23:1 VSWR, while a network with only one element will result in a 3.3:1 VSWR, and a two-element network will give 1.7:1 VSWR.

Since the matching network is lossless, the magnitude of the reflection coefficient seen by $R_{ds}$ is identical to the magnitude of the reflection coefficient seen by the 50-$\Omega$ load, thus the worst case values of load resistance seen by the FET, $R'_L$, are

$$R'_L = R_L \left( \frac{1 \pm |S_{22}|}{1 \pm |S_{22}|} \right) \tag{1.45}$$

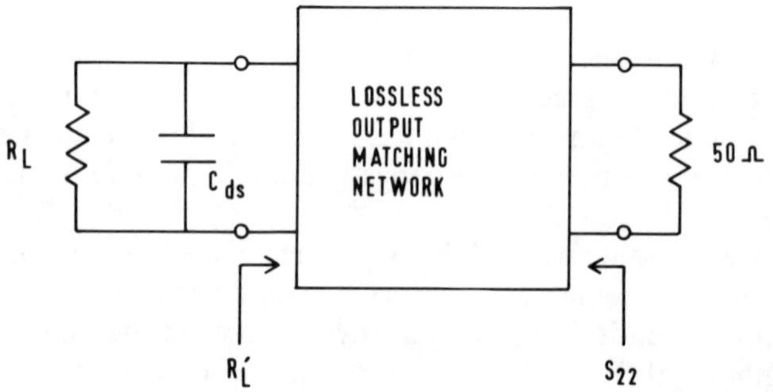

**Figure 1.16** Statement of output matching network problem.

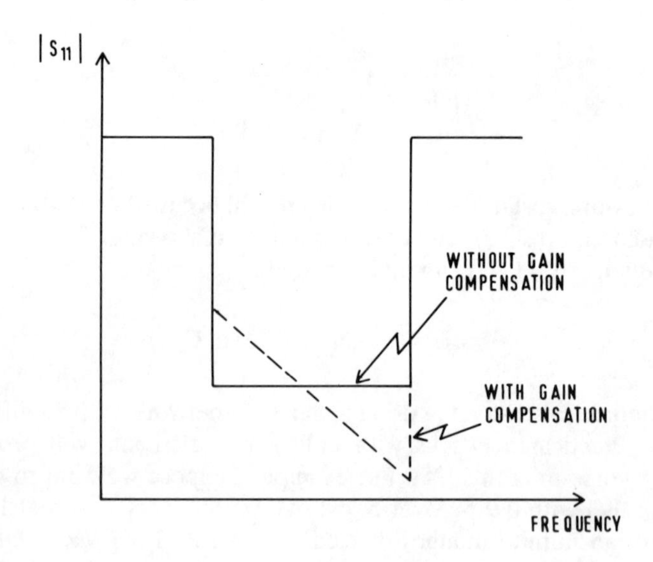


**Figure 1.17** Gain-bandwidth limitation of matching networks.

Figure 1.18 shows the IV characteristics of an ideal FET together with the optimum load line for Class A operation as given in Table 1.1. At some frequencies within the power amplifier's passband the load resistance presented by the matching network as given by (1.45) will be greater than $R_L$, in which case the drain current cannot be fully modulated as can be seen from Figure 1.18. As shown in Section 4.3.3. this causes a reduction in output power in the ratio [21]

$$\frac{P_{\text{out}}}{P_{\text{RFmax}}} = \frac{R_L'}{R_L} \tag{1.46}$$

Similarly, at some frequencies the load resistance will be less than $R_L$, in which case the drain-source voltage cannot be fully modulated. As shown in Section 4.3.3, this also causes a reduction in output power in the ratio

$$\frac{P_{\text{out}}}{P_{\text{RFmax}}} = \frac{R_L}{R_L'} \tag{1.47}$$

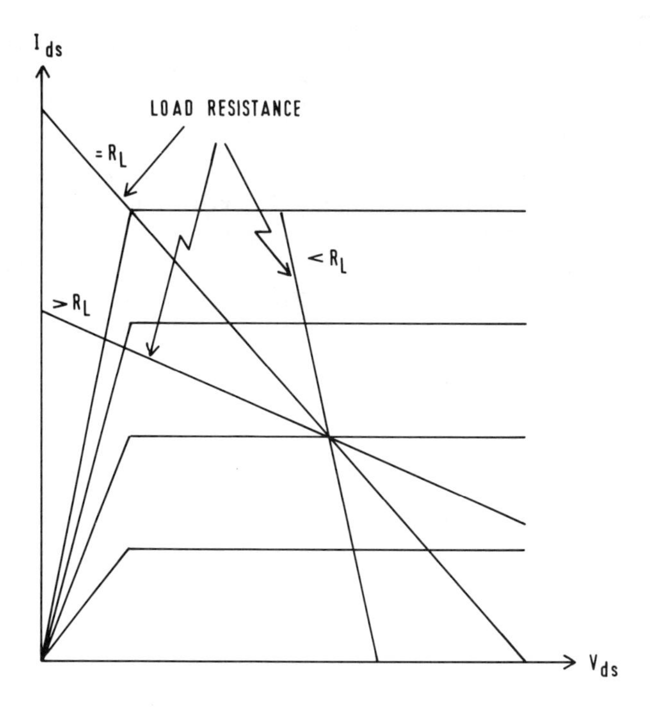


**Figure 1.18** Effect of using nonoptimum load impedances on a Class A amplifier load line.

*Example*

Suppose we want to design a 1-W, 6- to 18-GHz amplifier using a Toshiba JS 8855-AS power GaAs FET. From the data sheet $C_{ds} = 0.9$ pF and the calculated optimum load resistance is $R_L = 20\Omega$. Thus $|S_{22}| \leq 0.106$ and hence the extreme values of load resistance that the FET might see as given by (1.45) are 16.2 and 24.8$\Omega$ hence, the output power can vary between 1 and 0.8W—a 0.9-dB variation.

Note that if we simply needed to match a passive load of 20$\Omega$ in parallel with 0.9 pF, then $|S_{22}| = 0.106$ corresponds to a VSWR of 1.23:1 and a mismatch loss of 0.5 dB, figures that would be considered quite acceptable in most situations. However, when the load is an FET and one is trying to extract the maximum output power from it, then the matching network causes a power variation of 0.9 dB in the ideal case of an infinite lossless network. Thus, this example demonstrates how a variation in load resistance has a far more significant effect on power output than it does on VSWR or insertion loss.

If the gatewidth of an FET was increased tenfold, then in theory $R_L$ would decrease tenfold and $C_{ds}$ would increase tenfold with the result that the product $R_L C_{ds}$ would remain invariant, so it would be no more difficult to match a device having 10 times the output power. The fallacy in this argument is that it ignores practical reality. For example, if applied to the Toshiba power FET, then the matching network is required to transform 50 down to 2$\Omega$ rather than 20$\Omega$, which requires the use of very low and, therefore, very wide transmission lines. Such wide lines are a poor realization of low-impedance transmission lines. Also, circuit losses play a very significant role. For example, if the total microwave resistive loss in the matching network is 0.5$\Omega$, then one-quarter of the available output power is dissipated in the matching network rather than in the load. Since the loss of the network is proportional to its physical length, a technique sometimes adopted when using distributed matching networks is to use a very high dielectric constant substrate for the matching network. This also reduces the width of the transmission lines and makes the interface with the FET more compatible.

## 1.6.2 Input Matching Network

In the design of a power amplifier the input matching network is required to transform the 50-$\Omega$ source impedance to the complex conjugate of the FET's input impedance when the FET is loaded with the optimum load impedance for maximum output power. Although not exact, we assume here for simplicity that the input impedance of the FET under these circumstances can be modeled by a series $RLC$ network with $R = R_S + R_g + R_i$, $L = L_g + L_s$, and $C = C_{gs}$. Depending on the frequency and the relative values of $L_g$ and $C_{gs}$, the input reactance can be either inductive or capacitive, for example, the Toshiba JS 8855-AS power FET becomes nductive at 10 GHz. Obviously, this crossover frequency is reduced with higher power and, therefore, higher $C_{gs}$ devices.

The fundamental limit on the input reflection coefficient for a series *RC* network without gain compensation is given by

$$\int_0^\infty \frac{1}{\omega^2} \ln \frac{1}{|S_{11}(\omega)|} d\omega \leq \pi RC \tag{1.48}$$

and hence the minimum worst case reflection coefficient will occur when $S_{11}$ is a constant over the band of interest $\Delta f = f_H - f_L$, in which case the inequality of (1.48) can be simplified to

$$|S_{11}(\omega)| \geq \exp(-2\pi^2 RCf_L f_H/\Delta f) \tag{1.49}$$

Similarly, the fundamental limit on the input reflection coefficient for a series *RL* network without gain compensation is given by

$$\int_0^\infty \ln \frac{1}{|S_{11}(\omega)|} d\omega \leq \frac{\pi R}{L} \tag{1.50}$$

and hence the minimum worst case reflection coefficient is governed by

$$|S_{11}(\omega)| \geq \exp(-R/2\Delta f L) \tag{1.51}$$

Once again, these limits are the theoretical best values using a network with an infinite number of elements. As observed earlier, for high-power FETs the input impedance is modeled as a series *RL* network. However, while *R* decreases as $C_{gs}$ increases, *L* remains invariant leading to a higher input reflection coefficient as determined by inequality (1.51) (i.e., the input reflection coefficient does not remain invariant when the gate width is changed).

*Example*

Consider once more the problem of designing a 6- to 18-GHz amplifier using a Toshiba JS 8855-AS power FET. From the data sheet $R = 0.92\Omega$, $L = 0.9$ nH, and $C = 2.6$ pF. Thus, the crossover frequency is 10.4 GHz, which happens to be identical to the geometric mean frequency. In this case, therefore, the problem of matching a series *RLC* load over 6 to 18 GHz can be solved by transforming to the equivalent low-pass case of requiring a match over dc to 12 GHz of a series

$RL$ network with $R = 0.92\Omega$ and $L = 0.92$ nH, in which case inequality (1.51) shows that $|S_{11}(\omega)| \geq 0.65$, that is, a VSWR of 4.7:1 minimum. The mismatch loss is thus 2.4 dB; that is, the input matching network reduces the gain of the amplifier when terminated in the optimum load impedance for maximum output power by a minimum of 2.4 dB at all frequencies. However, because the input matching network does not provide any gain compensation, the gain of the amplifier will also have a 6 dB/octave slope.

The bandwidth limitations associated with matching networks designed to provide gain compensation have been considered by Ku and Peterson [26]. A lossless input matching network achieves its gain compensation by reflection and hence it is immediately evident from the integral inequalities (1.43), (1.48), and (1.50) that a higher reflection coefficient at the low-frequency end means that a lower reflection coefficient can be achieved at the upper frequency end as shown in Figure 1.17. Hence, the matching problem is eased. In fact, for the specific example above Ku and Peterson's work shows that it is theoretically possible to design a lossless matching network that has a 6 dB/octave attenuation slope and provides a perfect match at the upper cutoff frequency and hence no loss of gain would occur. However, the practical problem of transforming a 50-$\Omega$ source impedance to 1$\Omega$ using realizable elements is severe.

## ACKNOWLEDGMENT

The author gratefully acknowledges useful conversations with Dr. Steve Cripps of Hywave Associates concerning some aspects of Class A and B amplifiers.

## REFERENCES


1. Anadigics Inc. AWA 20611 Data Sheet.
2. Komiak, J. J., "Octave band 11W Power Amplifier MMIC," *Proc. IEEE Microwave and Millimeter Wave Monolithic Circuits Symp*. Dallas, TX, May 7–8, 1990, pp. 35–38.
3. Bahl, I. J., E. L. Griffin, and C. Andricos, "A 14W C Band Power Amplifier Employing MMIC Chips," *Microwave J.*, Vol. 32, No. 5, May 1989, pp. 295–303.
4. Basset, J. R., M. Avasarala, E. Chase, B. Kraener, and D. S. Day, "High Efficiency, High Power GaAs FET Amplifiers," *Proc. European Microwave Conf. Workshop*, Wembley, England, September 5–8, 1989, pp. 137–142.
5. Texas Instruments TGA 8220 Data Sheet.
6. Williams, R. E., and W. Show, "Graded Channel FETs: Improved Linearity and Noise Figure," *IEEE Trans. on Electron Devices*, Vol. ED-25, June 1978, pp. 600–605.
7. Kushner, L. J., "Output Performance of Idealised Microwave Power Amplifiers," *Microwave J.*, Vol. 32, No. 10, October 1989, pp. 103–116.
8. Cripps, S. C., "GaAs FET Power Amplifier Design," Technical Note 3.2, Matcom Inc.

9. Snider, D. M., "A Theoretical Analysis and Experimental Confirmation of the Optimally Loaded and Overdriven RF Power Amplifier," *IEEE Trans. on Electron Devices*, Vol. ED-14, December 1967, pp. 851–857.
10. Nishiki, S., and T. Nojima, "Harmonic Reaction Amplifier—A Novel High-Efficiency and High-Power Microwave Amplifier," *Proc. IEEE Int. Microwave Symp.*, Las Vegas, NV, June 9–11, 1987, pp. 963–966.
11. Geller, B. D., and P. E. Goettle, "Quasi-Monolithic 4GHz Power Amplifiers with 65% Power Added Efficiency," *Proc. IEEE MTT-S Int. Microwave Symp.*, New York, NY, May 24–25, 1988, pp. 835–838.
12. Bahl, I. J., E. L. Griffin, A. E. Geissberger, C. Andricos, and T. F. Brukiewa, "Class-B Power MMIC Amplifiers with 70 Percent Power-Added Efficiency," *IEEE Trans. on Microwave Theory and Techniques*, Vol. MTT-37, September 1989, pp. 1315–1320.
13. Walker, J. L. B., "Improving Operation of Classic Broadband, Balanced Amplifiers," *Microwaves & RF*, Vol. 26, September 1987, pp. 175–182.
14. Lane, J. R., R. G. Freitag, H. K. Hahn, J. E. Degenford, and M. Cohn, "High-Efficiency 1-, 2-, and 4-W Class-B FET Power Amplifiers," *IEEE Trans. on Microwave Theory and Techniques*, Vol. MTT-34, December 1986, pp. 1318–1326.
15. Pavio, A. M., and A. Kikel, "A Monolithic or Hybrid Broadband Compensated Balun," *Proc. IEEE MTT-S Int. Microwave Symp.*, Dallas, TX, May 8–11, 1990, pp. 483–486.
16. Ladbroke, P. H., *MMIC Design: GaAs FETs and HEMTs*, Norwood, MA: Artech House, 1989.
17. Tajima, Y., B. Wrona, and K. Mishima, "GaAs FET Large-Signal Model and Its Application to Circuit Design," *IEEE Trans. on Electron Devices*, Vol. Ed-28, February 1981, pp. 171–175.
18. Willing, H. A., C. Rauscher, and P. de Santis, "A Technique for Predicting Large-Signal Performance of a GaAs MESFET," *IEEE Trans. on Microwave Theory and Techniques*, Vol. MTT-26, December 1978, pp. 1017–1023.
19. Walker, J. L. B., "Designing GaAs MMIC Transistor Amplifiers," *J. Monolithic Technology*, Vol. 1, No. 1, April 1988, pp. 18–21 and 44.
20. Engelbrecht, R. S., and K. Kurokawa, "A Wideband Low-Noise L band Balanced Transistor Amplifier," *Proc. IEEE.*, Vol. 53, March 1965, pp. 237–246.
21. Cripps, S. C., "A Theory for the Prediction of GaAs FET Load-Pull Power Contours," *Proc. IEEE MTT-S Int. Microwave Symp.*, Boston, MA, May 31–June 3, 1983, pp. 221–223.
22. Actis, R., and R. A. McMorran, "Millimetre Load Pull Measurements," *Applied Microwave*, Vol. 1, No. 3, November/December 1989, pp. 91–102.
23. Harris Microwave Semiconductor, Application Note 2, April 1986.
24. Bode, H. W., *Network Analysis and Feedback Amplifier Design*, New York: Van Nostrand Reinhold, 1945.
25. Levy, R., "Explicit Formulas for Chebychev Impedance Matching Networks, Filters and Interstages," *Proc. IEE*, Vol. 111, No. 6, June 1964, pp. 1099–1106.
26. Ku, W. H., and W. C. Peterson, "Optimum Gain—Bandwidth Limitations of Transistor Amplifiers as Reactively Constrained Active Two-Port Networks," *IEEE Trans. on Circuits and Systems*, Vol. CAS-22, June 1975, pp. 523–533.

# Chapter 2
# High-Power GaAs FETs

**Y. Aoki and Y. Hirano**
*Fujitsu Compound Semiconductor Division*

## 2.1 INTRODUCTION

### 2.1.1 The Development of the High-Power GaAs FET

The first reported development of a prototype *gallium arsenide field effect transistor* (GaAs FET) was by Mead [1] in 1966. Many reports on the development of the quality of GaAs materials and basic FET prototype technology followed, until Baechtold et al. [2] and Liechti et al. [3] in 1972 announced a GaAs FET with a maximum oscillation frequency of more than 50 GHz. Research then moved rapidly in the direction of practical applications. The first announcements of a high-power GaAs FET were made simultaneously by Fukuta et al. [4] and Napoli et al. [5] in 1973. The development of the high-power GaAs FET had to surmount at least three major obstacles.

1. *Heat dissipation*: Gallium arsenide has only about one-third of the thermal conductivity of silicon (0.46 versus 1.5 W/cm °C for silicon), so that the heat created in high-power operation must be more efficiently dissipated.
2. *Breakdown voltage*: The FET structure must have high drain and gate breakdown voltages to withstand high RF voltages.
3. *Electrode layout*: Limited space on the chip face requires efficiency in channel area layout and source, drain, and gate electrode configuration technology.

Fukuta's group solved the first problem by thinning the GaAs substrate to 80 $\mu$m, the second by offsetting the gate relative to the source and drain, and the third by crossing the drain over the source and the gate to create a mesh source pattern. As a result, they achieved a gate-drain breakdown voltage of 10V, and a high output power of 1.6W at 2 GHz with $V_{ds}$ = 8V using a 1-mm-gatewidth FET.

Napoli's group used a multigate construction to overcome the second problem, and solved the third by using an $n - n^+$ continuous-growth GaAs wafer and utilizing recess etching with metal source and drain electrodes as a mask. They achieved a drain breakdown voltage of 11.5V and an output power of 0.8W at 4 GHz. The conquest of these three problems had a major influence on the subsequent development and preproduction of the high-power FET.

The latter half of the 1970s and the first half of the 1980s saw intense competition to develop the high-power FET. A 4W FET at 4 GHz [6] was announced in 1976, followed by 14W at 4 GHz in 1978, and 25W at 6 GHz [7] in 1980. Figure 2.1 charts the annual increase in the reported output power of high-power GaAs FETs in relation to the operating frequency. Many of these were internally matched FETs, that is, they contained impedance matching circuits on the input and output sides of the FET chip. The technologies developed through this process of competition are discussed in the following sections, but many GaAs material improvements and production process refinements not discussed here have been made that have contributed to the enhancement of the capabilities of the FET. In the 1980s the focus of development shifted to higher power-added efficiency and improved distortion qualities as well as to operating frequency and output power. In 1984, Fukaya et al. [8] reported the development of a C-band FET with 43% power-added efficiency and 10W of output power.

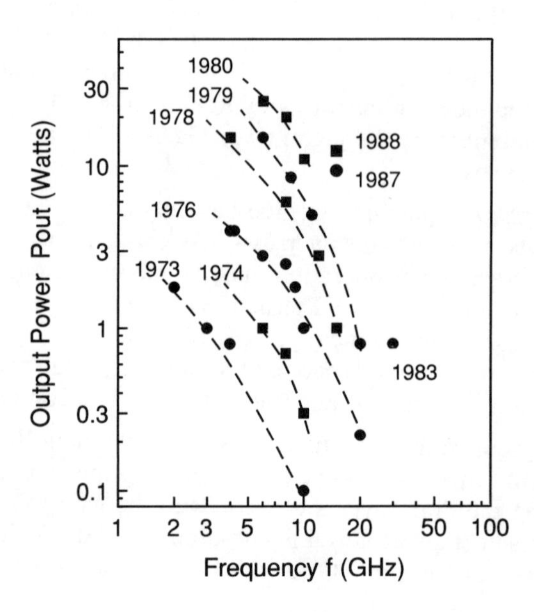


**Figure 2.1** Annual development of the high-power FET: $P_{out}$ versus frequency chart.

In another development, Mimura et al. [9] in 1980 invented a FET with a new structure, called the *high electron mobility transistor* (HEMT). High-power devices using the HEMT structure have been reported since 1984. Demonstrations have shown that the HEMT has high-frequency characteristics superior to the GaAs FET and can be used as an active device at millimeterwave frequencies. Other devices using semiconductor heterojunctions, such as the *heterojunction bipolar transistor* (HBT) and the heterojunction *metal-insulator-semiconductor FET* (MISFET), were reported with superior microwave characteristics. Considerable development activity focused on these HEMT and heterojunction devices. HEMT and heterojunction MISFET technologies were not developed independently of the GaAs FET. Rather, they were the product of a fusion of process technology and crystal growth technology from the development of the FET. The HEMT and other heterojunction elements were superior (in terms of noise figure and gain) to the GaAs FET and replaced it for small-signal applications, but they have never replaced the GaAs FET for high-power applications. At the same time, GaAs FET performance has been enhanced by incorporating heterojunctions in some architectural areas.

As a post-GaAs material, indium gallium arsenide/gallium arsenide (InGaAs/GaAs) heterojunctions, indium aluminum arsenide/gallium arsenide (InAlAs/GaAs) heterojunctions, indium phosphide (InP), and other materials have been proposed. Because, in general, high electron mobility in a material means a small bandgap, high-mobility materials are poorly suited for high-power use, but they are useful for small-signal applications. The low electric-field mobility of InP is less than that of GaAs; however, its greater saturated electron velocity has made it attractive as a material for high-power devices. The appearance of heterojunction elements and new materials seems likely to push the FET industry from microwave into millimeterwave frequencies.

This chapter will address the design and manufacturing technology of the GaAs MESFET.

### 2.1.2 Current Technology for the High-Power GaAs FET

In contrast to the emphasis on improved material qualities and precision fabrication techniques in the development of the small-signal FET, high-power FET development has concentrated on improvements in channel and chip construction. The figures of merit of the small-signal FET are noise figure and gain, so that improvement efforts concentrate on material qualities such as enhanced electron mobility and electron velocity for greater transconductance, reduced gate capacitance and parasitic resistance, and improvements in channel microstructure under the gate. In the high-power FET, there are a number of critical properties including output power, gain, efficiency, and distortion for which it is necessary to increase breakdown voltage, decrease thermal resistance, improve pattern layout and source-drain cross-section structure, and maintain a constant transconductance over the entire operating area of $V_{ds}$ and $I_{ds}$. Materials and construction must also be rugged enough to operate continuously for long periods, and to withstand excessive input signal levels. In other words, high-power FET improvements must encompass the entire chip. Therefore, although it is true that the development of the high-power FET has benefited from the many material manufacturing and precision fabrication technologies provided by the small-signal FET development process, improvement of the high-power FET requires a focus on diametrically opposite characteristics. The technologies used in the high-power FET are classified in Table 2.1.

Each technology is outlined briefly here, then discussed in detail later in the chapter. To lower thermal resistance, the thermal conduction paths of the poorly conductive GaAs must be made as short as possible and a thick layer of a highly conductive metal (usually gold) deposited to aid heat diffusion. Techniques that reduce chip thickness also lower the thermal resistance by creating shorter heat conduction paths. It is now possible to thin GaAs chips to a uniform thickness of 25 $\mu$m. *Plated heat sink* (PHS) technology plates the chip with metal on the back and side surfaces. In addition to lowering thermal resistance, this relieves stress during chip mounting. *Flip-chip* technology involves depositing a thick plating over the electrodes on the chip surface, then connecting the chip in an upside-down position. This reduces the thermal resistance and eliminates the variability in inductance due to wire bonding.

In the high-power FET, the design must efficiently lay out a channel area and connect electrodes to each other within the limited confines of a chip surface.

**Table 2.1**
Technologies Used in High-Power FETs

| *Item* | *Technology* |
|---|---|
| Thermal resistance | Thin chip |
| | Plated heat sink |
| | Flip chip |
| Pattern | Dielectric crossover |
| | Air-bridge crossover |
| | Via-hole connection |
| | Interdigital electrode layout |
| | Fishbone gate layout |
| Breakdown voltage | Recessed channel |
| | $n^+$ implantation |
| Gain, distortion | Heterobuffer |
| | Hi-lo profile |
| Reliability | Refractory metal |
| | Passivation |

This requires gate layout patterns and electrode connection techniques that are not used in the small-signal FET. The *fishbone gate*, as its name implies, looks like the bones of a fish with the bus line as the backbone and an array of figures projecting to either side. In the *interdigital* layout, source and drain electrodes are interwoven like clasped fingers. When these patterns are used, a technique is necessary for connecting the various electrodes. Two commonly used connecting techniques are the *air-bridge crossover* and the *dielectric crossover*. The air-bridge crossover builds connections of overlay metal-like bridges in the air and has the advantage of reducing parasitic capacitance, but it requires a complex manufacturing process. A technique called *via-hole connection* reaches electrodes by making holes in the surface of the GaAs chip, then fabricating an electrode from the back (or front) through the hole to a connection on the front (or back). This technique is limited in that it can only be used with ground connections, but it greatly increases the freedom of design in pattern layouts and it reduces grounding inductance.

To increase gate breakdown voltage, structural solutions to prevent the formation of locally concentrated electric fields are required. In a planar FET, with source, gate, and drain lying in the same plane, electric fields normally concentrate at the metal edge of the gate nearest the drain, so that the gate breakdown voltage tends to be low. In *recess construction* technology a recessed area is created in the GaAs material at the location of the gate, narrowing the active area below the gate. This has the effect of preventing local concentration of the electric field at the edge of the gate, as well as increasing the transconductance. An $n^+$ layer is formed beneath the source and drain electrodes using ion implantation to create an upside-down recess construction, which reduces the source resistance and increases the gate breakdown voltage.

High gain and low distortion are also important characteristics for a high-power FET. Because these factors depend on material quality, channel microstructure, and the profile of the active area, little of this technology has been made public. In *heterojunction buffer* construction, in which AlGaAs is normally used as a buffer layer, the heterojunction energy gap at the GaAs and AlGaAs interface prevents electrons from entering the buffer layer and the substrate. The increased output resistance provides greater gain. "*Hi-lo*" *profiling* reduces the donor concentration of the channel near the surface relative to deeper areas (or makes the surface an insulator), thereby increasing the gate breakdown voltage, improving the linearity of transconductance, and reducing distortion.

The reliability of FET amplifiers is many times higher than that of the traveling-wave tube amplifier, the preceding generation of device, because the FET is a solid-state element. However, the use of microfabrication processes also means that the variability in materials and structure can adversely affect reliability. To prevent electromigration under high electric fields over periods of continuous operation, electrodes are made from refractory metals, with a protective layer of passivation.

Not shown on the list in Table 2.1 are the less visible but important advances in substrate materials, epitaxial growth, and ion implantation technologies.

The GaAs FET exists today because of the combined effects of the technologies listed in Table 2.1. The remainder of this chapter presents the various stages in the process of designing and manufacturing FETs and points out the various ways in which these technologies have been adapted and utilized.

## 2.2 HIGH-POWER FET DESIGN: FET CHANNEL CROSS-SECTION DESIGN

The design of high-power FETs is much more complex than that of small-signal and other low-noise FETs. In the low-noise FET, major design concerns virtually start and end with the channel cross-section structure, whereas high-power FET design must also consider significant issues in chip pattern design. High-power FET design can therefore be considered in two stages:

1. Channel cross-section design;
2. Chip pattern design.

Channel cross-section design involves designing the small FET units that function as elements of the high-power FET. At this level, performance indicators such as gain, gate breakdown voltage, drain breakdown voltage, drain current, output power, and efficiency are designed. Here it is important to understand the relationship of the performance indicators to structural qualities such as the doping density profile of the active area, gate length, and recess structure. The relationships among performance indicators and design considerations are summarized in Table 2.2.

Chip pattern design involves the design of the entire FET, incorporating the FET units described previously. Performance indicators designed at this level include total FET output power, efficiency, gain, and thermal resistance. Here it is important to (1) design gate finger length and power supply pad location without losing high-frequency FET unit characteristics, (2) consider the operating frequency, and (3) design the gate spacing and chip thickness to achieve an acceptable thermal resistance. The relationships among performance indicators and design considerations are summarized in Table 2.3.

### 2.2.1 The Flow of the Design Process

In this section, we will discuss the process of designing the channel, the heart of the FET. Narrowly defined, the channel is the semiconductor layer below the gate through which the current passes, but here we will use it to mean all electrode and semiconductor components between the source electrode and the drain electrode.

**Table 2.2**
High-Power FET Channel Cross-Section Design Objectives

| *Design Item* | *Characteristic* | *Performance Indicator* |
|---|---|---|
| Epitaxial layer density | Drain current | Output power |
| Epitaxial layer thickness | Transconductance | Gain |
| Epitaxial layer structure | Gate breakdown voltage | Efficiency |
| Gate length | Output resistance | Distortion |
| Recess structure | Gate capacitance | Cutoff frequency |
| Offset gate | Parasitic resistance | Reliability |
| Source-drain spacing | | |
| Electrode metal | | |
| Passivation film | | |

**Table 2.3**
High-Power FET Pattern Design Objectives

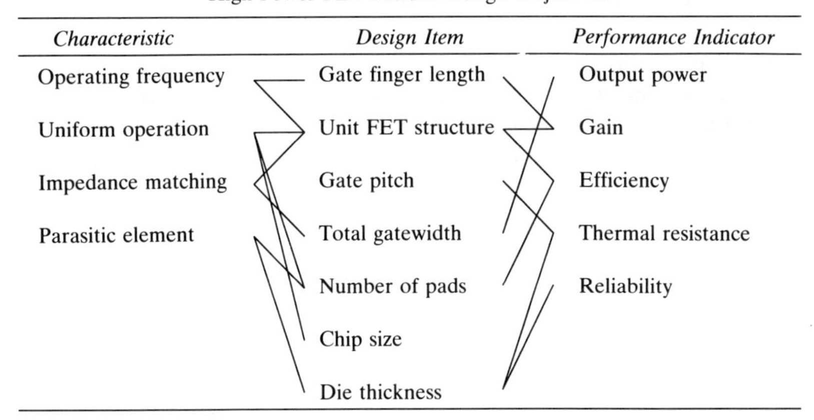

| *Characteristic* | *Design Item* | *Performance Indicator* |
|---|---|---|
| Operating frequency | Gate finger length | Output power |
| Uniform operation | Unit FET structure | Gain |
| Impedance matching | Gate pitch | Efficiency |
| Parasitic element | Total gatewidth | Thermal resistance |
| | Number of pads | Reliability |
| | Chip size | |
| | Die thickness | |

The design objective is to optimize the small FET unit. The entire FET will be designed using gatewidth computations based on characteristics of the FET unit. Therefore, the design of the FET unit is the basis of the design of the entire FET chip. FET unit design is basically the same as small-signal FET design, except that where the small-signal FET must maximize the gain and minimize the noise figure within a limited operating range, the high-power FET unit must have as wide a operating range as possible (the range of available drain current and voltage), an it must operate as uniformly as possible within that range. Therefore, the FET uni must have a high breakdown voltage and drain current and a high gain. For thi reason, the objectives of channel design are in some respects the opposite of thos in the small-signal FET. The design objectives are shown in Table 2.2. The desig processes linking design objectives and associated system characteristics are th subject of this chapter.

### 2.2.2 Designing the Epitaxial Wafer Structure

The most important point of FET design is the epitaxial structure of the wafer which determines how the active area of the FET is formed. Here we discuss dono density and active layer thickness and their relation to FET characteristics. Epitaxia layer growth can now be controlled at the atomic level; thus, it is no longer necessar for the active layer to be a single layer of uniform density. In addition, ion implan tation now makes possible the creation of three-dimensional active layers of th desired depth and density wherever needed on the wafer. These development have increased the freedom of design, and the design of active layer structures no must go beyond output power and gain to include efficiency and distortion char acteristics, even as the manufacturing techniques are being developed to mak them a reality.

We will focus on the area just below the FET gate and compute the drai current and transconductance, assuming a simple profile. We will assume that th epilayer profile has density steps as shown in Figure 2.2 [10]. To compute the drai current we need to make the following three assumptions (approximations):

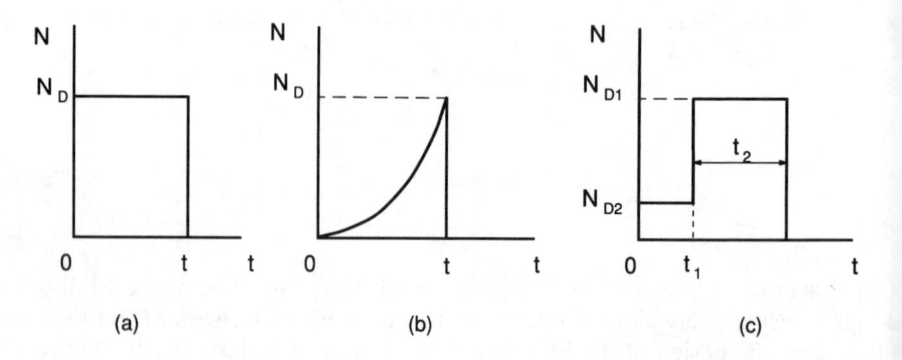


**Figure 2.2** Various doping profiles: (a) uniform doping, (b) power law doping (hip-up doping), ar (c) step doping (hi-lo doping).

1. Gradual channel;
2. Long gate length;
3. Electron velocity saturation.

hese assumptions are actually more appropriate for a "fat gate" FET with gate engths of 2 μm or more, but here they will do as a first approximation for design urposes. We will also assume that all of the voltage applied at the gate is applied o the depletion layer. The interface between the Schottky junction gate metal and he GaAs layer will be the origin, oriented in the direction of the depth of the emiconductor layer. Assumptions 1 and 2 assure that the electric field is in the epletion layer, also oriented in the direction of its depth. When current flows hrough the channel, the positive donor charge is cancelled by the negative charge f electrons, so there is no net charge. Assume that the field is oriented parallel o the gate electrode in the channel. The width of the channel can be computed y solving Poisson's equation in one dimension, namely the direction of the depth f the GaAs layer.

### *Jniform Doping*

Jniform doping is the most traditional and simple profile. The fully open channel urrent $I_F$, pinch-off voltage $V_P$, drain current $I_{ds}$, and transconductance $g_m$ as a unction of the gate-source voltage are given, respectively, by

$$I_F = qv_{\text{sat}} w_g N_D t \tag{2.1}$$

$$V_P = \frac{qN_D t^2}{2\epsilon} \tag{2.2}$$

$$I_{ds} = I_F \left[ 1 - \left( \frac{V_{gs}}{V_P} \right)^{1/2} \right] \tag{2.3}$$

$$g_m = \frac{I_F}{2\sqrt{V_P V_{gs}}} \tag{2.4}$$

vhere $v_{\text{sat}}$ is the saturated electron drift velocity and $w_g$ is the gatewidth.

### *Power Law Doping*

Power law doping is rather fictitious. However, it can be approximated by gradual ontrol of the quantity of donor impurity during epitaxial wafer growth. The "hip-up'' profile is considered to be approximately described by this doping profile. The resulting electrical characteristics of the FET are given by

$$N(d) = N_D d^n \qquad \text{for } 0 \leq d \leq t \tag{2.5}$$

$$I_F = q v_{\text{sat}} w_g N_D \frac{t^{n+1}}{n+1} \tag{2.6}$$

$$V_P = \frac{qN_D}{\epsilon} \frac{t^{n+2}}{n+2} \tag{2.7}$$

$$I_{ds} = I_F \left[ 1 - \left( \frac{V_{gs}}{V_P} \right)^{(n+1)/(n+2)} \right] \tag{2.8}$$

$$g_m = \frac{I_F}{V_P^{(N+1)/(N+2)}} \frac{n+1}{n+2} V_{gs}^{-1/(N+2)} \tag{2.9}$$

In the special case of $n = 0$, (2.6) to (2.9) reduce to those in the uniform doping situation, namely, (2.1) to (2.4).

*Step Doping*

Let the donor density on the GaAs surface region be lower (in this case, zero) than in the interior. We assume the surface layer doping density to be zero to avoid anomalous properties near $V_{gs} = 0$. Let the donor density be represented by $N_D$ and let the thickness of the surface layer and donor-doped layers be $t_1$ and $t_2$, respectively. From the one-dimensional Poisson equation, the gate pinch-off voltage $V_P$ (including the built-in Schottky barrier voltage) at which the donor-doped layer is completely depleted (pinched-off) is given by

$$V_P = \frac{qN_D}{2\epsilon} t_1^2 \left[ \left( 1 + \frac{t_2}{t_1} \right)^2 - 1 \right] \tag{2.10}$$

Also, the fully open channel current is given by

$$I_F = q v_{\text{sat}} w_g N_D t_2 \tag{2.11}$$

etting the depletion depth within the donor-doped layer be $d$, then $d$, $I_{ds}$, and $g_m$ re given by

$$d = t_1 \left( \left\{ \frac{V_{gs}}{V_P} \left[ \left(1 + \frac{t_2}{t_1}\right)^2 - 1 \right] + 1 \right\}^{1/2} - 1 \right) \tag{2.12}$$

$$I_{ds} = I_F \left(1 - \frac{d}{t_2}\right) \tag{2.13}$$

$$g_m = \frac{I_F \left(\frac{t_1}{t_2}\right) \left[ \left(1 + \frac{t_2}{t_1}\right)^2 - 1 \right]}{2V_P \left\{ \frac{V_{gs}}{V_P} \left[ \left(1 + \frac{t_2}{t_1}\right)^2 - 1 \right] + 1 \right\}^{1/2}} \tag{2.14}$$

ample computations are shown in Figure 2.3. From (2.13) and (2.14), we can

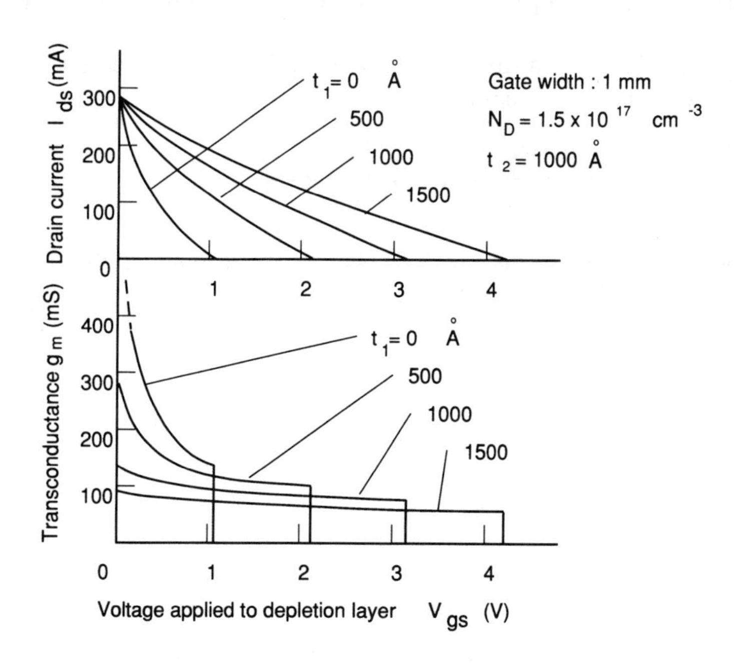


gure 2.3 Graph of $I_{ds}$ versus $V_{gs}$ and $g_m$ versus $V_{gs}$ characteristics for uniform and step doping profiles.

deduce the following: When we look at the transconductance $g_m$ in terms of th
relation between $t_1$ and $t_2$, we see that if $N_D t_2$ is constant, then $g_m$ is inverse
proportional to $t_1$. Also, if $N_D t_2$ and $t_1$ are held constant, then $g_m$ is inverse
proportional to $t_2$ and its dependency on $V_{gs}$ is reduced. If we hold $N_D t_2$ consta
and take the extreme case in which $t_2 \rightarrow 0$, then we get

$$V_P = \frac{qN_D t_1 t_2}{\epsilon} \tag{2.15}$$

$$g_m = \frac{\epsilon V_{\text{sat}} w_g}{t_1} \tag{2.16}$$

and $g_m$ remains constant independent of the value of $V_{gs}$ and the FET has ou
standing low-distortion characteristics (see Sec. 1.2.1). The case in which $t_2 = 0$
only theoretical; in practice, a donor-doped region must have a finite thickne
since $v_{\text{sat}}$ decreases as the donor density is increased. However, in devices such
HEMTs where the donor supply layer and electron transit layer are separated,
is possible to create a thin, extremely highly doped donor layer. HEMTs usi
planar doping produce a high-power output in the millimeterwave range [11].
the special case of $t_1 = 0$, (2.10) to (2.14) reduce to those for the uniform do
ng case.

Next, let us look at distortion. Consider a dc gate voltage of $V_{gs} = V_P/2$,
which we add a high-frequency sinewave:

$$V_{gs} = \frac{V_P}{2} + v_{gs}(t) = \frac{V_P}{2}(1 + p \sin \omega t) \qquad (2.1$$

where $p$ is a parameter in the range (0, 1) expressing the amplitude of the inp
RF signal. Expanding (2.13) in powers of $v_{gs}(t)$, we get

$$\frac{I_{ds}}{I_F} = a_0 + a_1 v_{gs}(t) + a_2 v_{gs}^2(t) + a_3 v_{gs}^3(t) + \cdots \qquad (2.1$$

where

$$a_0 = 1 - \frac{t_1}{t_2}\left\{\frac{1}{\sqrt{2}}\left[\left(1 + \frac{t_2}{t_1}\right)^2 + 1\right]^{1/2} - 1\right\} \qquad (2.1$$

$$a_n = (-1)^n \frac{(2n - 3)!!(t_1/t_2)\{[1 + (t_2/t_1)]^2 - 1\}^n}{\sqrt{2}\, n! V_P \{[1 + (t_2/t_1)]^2 + 1\}^{n-(1/2)}} \qquad (2.2$$

vhere

$$N!! \equiv N(N-2)(N-4)\cdots 1, \qquad \text{and } (-1)!! \equiv 1$$

f $N_D t_2$ is finite and $t_2 \to 0$, none of the high-order harmonic components will be above $a_2$, which is ideal from the standpoint of distortion. Although this is impos-ible to achieve, the lower the ratio of $t_2$ to $t_1$, the fewer high-order terms above $a_2$, and the better the distortion characteristics will be. Also, in this example, we et the surface layer density to zero to avoid anomalous results near $V_{gs} = 0$, but n practice the surface layer density must be finite. However, even though the urface layer density is finite, its presence still improves the distortion performance. As will be discussed later, this can also produce a higher gate breakdown voltage han can be achieved by pure uniform doping. Similar results can be achieved using on implantation methods by setting the acceleration voltage used to implant the lonor so as to reduce the donor density at the surface.

Other factors exist that are hard to build into a computational model, but vhich must nevertheless be included in a theoretical discussion. These include the quality of the Schottky gate, defects and deterioration in the GaAs layer from gate ormation, dielectric polarization at the edges of the metal gate due to stress against he semiconductor layer, surface states in the interface between the gate metal and he semiconductor and at the sides of the gate, and many others. These are problems hat process technology has not overcome, which, therefore, must be addressed at he design stage.

As we have seen from this simplified model, the following points must be onsidered in designing an epitaxial structure:

1. Determine the product of the doping density $N_D$ and thickness of the donor-doped layer $t$ in relation to the desired value of drain current per unit gate-width. More precisely, consider the dependence of saturated velocity on doping density in (2.1). To design to a level of precision that permits quantitative comparison with the manufactured result, it is necessary to use numerical analysis of the two-dimensional Poisson equation and other more advanced methods. To determine a saturated drain current value for design purposes, it is first necessary to consider breakdown voltage, efficiency, and gain. However, from experimentation we know that a value of 280 to 320 mA/mm of gatewidth should be satisfactory.
2. In a step profile epitaxial structure, determine the ratio $t_2/t_1$ so as to meet the desired distortion level. Our illustration used a simplified step profile, but there is also a "hip-up" profile in which the density of the deeper part of the donor-doped layer is greater than that of the surface part. In a hip-up profile, the ratio of high-density to low-density areas will affect distortion and must be chosen accordingly.

3. Determine $t_2$ so as to obtain the desired value of transconductance, $g_m$. Th greater the value of $g_m$, the better. However, care must be taken to conside its effect on other parameters. If $t_2$ is too small, $N_D$ must be large, which wi reduce the gate breakdown voltage and $v_{sat}$, adversely affecting performance In designing the epitaxial structure, therefore, the first step is to select th values of the functional indicators (saturated drain current, gate breakdow voltage, transconductance) that will satisfy or approach the desired perfor mance indicator values (output power, gain, distortion), then determine th doping density and thickness to best satisfy the functional indicator targets In a high-power FET, the design of the buffer layers must be considered i addition to the active areas. It is known that when operating close to pinch off, applying a high voltage to the drain causes the drain current to increase This is thought to be due to electrons escaping into the buffer layer whe there is a high field in the channel. This type of drain current flow is prevente by improving the buffer layer quality or creating a heterojunction in som part of the buffer layer. A GaAs/GaAlAs superlattice buffer layer is efficie in masking the effect of the GaAs substrate.

### 2.2.3 Determination of the Gate Length

Determination of the gate length is the most important design factor in a microwav FET. Gate length has the greatest effect on cutoff frequency and small-signal ga by directly determining the gate-source capacitance. The essential importance short-gate technology in the small-signal FET is due to the fact that gate lengt directly determines the gain and noise figure. Its importance in high-power FE design also cannot be overstated due to its effect on gain and efficiency. In th section we examine the relationship between gate length and gate-source capac tance (hereafter called *gate capacitance*), and discuss the design of gate length.

In an FET a built-in depletion layer exists due to the Schottky barrier betwee the GaAs semiconductor layer and the gate metal. In the depletion layer, the don atoms are positively ionized, having released their electrons to a very thin she at the interface between the GaAs and the metal. In the channel layer, the Ferr level is virtually at the donor level and the bottom of the conduction band is lowere Within the channel, the positive charge of the ionized donor atoms cancels th negative charge of the electrons at the gate metal-GaAs interface, so that there no net charge present. This dipole layer is the physical origin of the gate-sour capacitance. This stored charge, and consequently the depletion depth, varies sympathy with the applied gate voltage. Let us compute the gate capacitance the step profile example presented in the previous section. Let gate length ar gatewidth be represented by $l_g$ and $w_g$, respectively. The epilayer profile is th same as that shown in Figure 2.2. All approximations are the same as previousl

Fringing capacitance at the sides of the gate is ignored. With a gate-source voltage of $V_{gs}$ applied (including the Schottky built-in voltage $V_\phi$), the thickness of the depletion layer $d$ is given by (2.12). The total ionized donor charge $Q$ is given by

$$Q = qN_D d l_g w_g$$
$$= qN_D l_g w_g t_1 \left( \left\{ \frac{V_{gs}}{V_P} \left[ \left(1 + \frac{t_2}{t_1}\right)^2 - 1 \right] + 1 \right\}^{1/2} - 1 \right) \quad (2.21)$$

Differentiating $Q$ with respect to $V_{gs}$, we obtain the gate capacitance $C_{gs}$. This could be computed directly, but by simply comparing

$$g_m = \frac{\partial I_{ds}}{\partial V_{gs}} = qN_D V_{\text{sat}} w_g \frac{\partial d}{\partial V_{gs}} \quad (2.22)$$

and

$$C_{gs} = \frac{\partial Q}{\partial V_{gs}} = qN_D l_g w_g \frac{\partial d}{\partial V_{gs}} \quad (2.23)$$

we obtain the relational equations

$$\omega_T = 2\pi f_T = \frac{g_m}{C_{gs}} = \frac{v_{\text{sat}}}{l_g} \quad (2.24)$$

Thus, for calculations with this model, $g_m$ has the same reliance on $V_{gs}$ as $C_{gs}$ has, whereas $f_T$ is a constant with no reliance on $V_{gs}$. Gate capacitance $C_{gs}$ is proportional to gate length, from which it can be seen that the most effective way to reduce gate capacitance is to reduce gate length. The proportional relationship of gate length and gate capacitance (ignoring fringing capacitance) can also be verified experimentally. Also, the right side of (2.24) demonstrates that $l_g/v_{\text{sat}}$, the time required for an electron to pass under the gate, is the inverse of the angular cutoff frequency. Does this mean that the gate length should be shortened indefinitely? The answer is no. Equation (2.24) holds for relatively long gate lengths, but as the gate length becomes shorter, the longitudinal (parallel to the channel) component of the electric field in the depletion region below the gate becomes significant. With gate lengths of approximately 0.5 $\mu$m or less, the gate length becomes limited by the epilayer structure. With extremely short gate lengths, the output resistance becomes small, and the output power and gain decrease. For a high-power FET, the answer is expressed in terms of the aspect ratio—the ratio of gate length to active layer thickness (i.e., the distance from the deepest part of the channel to the gate metal). The aspect ratio normally has a value of at least 5. As we will explain in the next section, even though shortening gates may be effective in increasing the gain and the high-frequency performance, it is not as effective as a means of increasing output power. Therefore, the best solution is to make gates short, but not excessively, so that the aspect ratio and gate breakdown voltage are degraded.

### 2.2.4 The Scaling Law

The scaling law is a central principle in determining channel design. This law has wide application, well beyond the world of the FET. Aircraft, ship, and automobile designs are subjected to fluid dynamic testing of scale models made from design drawings. The relation of the model's air resistance, lift, and buoyancy to that of the full-scale item is determined by the equations of fluid dynamics. Dimensional analysis of these equations leads to the derivation of the scaling law. Thus the scaling law allows us to understand properties of a full-scale body without actually constructing it, by experimenting with models. In this section we will discuss the optimization of FET structural factors (gate length, epitaxial layer structure and thickness, bias conditions, termination impedance) together with dimensional analysis and scaling rules based on Poisson's equation. The same scaling law is applicable to FETs of any construction, as long as their operation can be described by Poisson's equation. Our analysis is based on the following assumptions and limitations:

1. Carrier density, electric field, and potential in the semiconductor (GaAs) can be described in terms of Poisson's equation. Here we use a two-dimensional model of a cross section perpendicular to the gate electrode. Operation parallel to the gate electrode are assumed to be homogeneous and scaled in terms of gatewidth only.
2. The conduction of holes (i.e., the minority carriers in GaAs) is neglected.
3. The ambient temperature of the FET (as well as channel temperature when operating) is significantly higher than that of the donor impurities. In other words,

$$kT \gg E_{\text{conduction band}} - E_{\text{donor}} \quad (2.25)$$

   where $k$ is Boltzmann's constant. Deep trap levels and deep impurity levels are not considered. Donor impurities are fully activated, and the depletion layer is completely depleted.
4. Quantum effects are not considered. For example, our model does not include discrete electron levels in the semiconductor heterojunction interface, tunne

effects due to spreading or overlapping of electron wave functions, electron velocity overshoot in high fields, anisotropy of electron velocity or diffusion constant due to crystal orientation, and the like.

5. Electron velocity is a monotonic, continuous, arbitrary function of the absolute value of the electric field:

$$\mathbf{V} = \mu(E)\mathbf{E}, \qquad E = |\mathbf{E}| \tag{2.26}$$

where $\mu(E)$ is the generalized mobility relating electron drift velocity and electric field.

Based on the preceding assumptions and limitations, the operation of the FET can be determined from the two-dimensional Poisson equation

$$\nabla \cdot \mathbf{E}(x, y) = -\frac{\rho(x, y)}{\epsilon(x, y)} \tag{2.27}$$

where

$$\mathbf{E}(x, y) = \nabla\phi(x, y) \tag{2.28}$$

and

$$\rho(x, y) = q[N(x, y) - N_D(x, y)] \tag{2.29}$$

and where $\phi$ means electric potential, $N$ is the electron concentration, and $\rho$ is the charge density. The current density is given by

$$\mathbf{j}(x, y) = qN(x, y)\mathbf{v}(x, y) \tag{2.30}$$

The equation for current continuity is

$$q\frac{\partial N(x, y)}{\partial t} = \nabla \cdot \mathbf{j}(x, y) \tag{2.31}$$

In the preceding equations the unknown quantities are

1. Potential distribution;
2. Electric field distribution;
3. Electron density distribution;
4. Current distribution.

Each is related to the others, and all must be determined self-consistently. Of these, the electric field distribution is the item that most closely represents the operation of the FET because electron drift velocity, as shown in (2.26), is normally a function of the electric field. From a microscopic point of view, electrons are accelerated by the Coulomb force of the field and forced to move. Since the electric field governs the state of the electrons, two FETs with identical field distributions can be considered equivalent despite differences in potential, current, or size. This is the fundamental concept of scaling. So let us consider two FETs, which we will call A and B, with the geometrically similar cross sections shown in Figure 2.4. Any location $(x, y)$ in FET A corresponds to an equivalent location $(x', y')$ in FET B. Let $M$ be the ratio of the size of A to B. Symbolically, this relationship can be expressed as

$$(x, y) \Leftrightarrow (x', y') = M(x, y) \tag{2.32}$$

where $M$ must be real and positive. Thus any values for the dimensions of the cross section of either FET can be converted using (2.32), such as gate length $l_g$,

$$l_g \Leftrightarrow l_g' = Ml_g \tag{2.33}$$

or channel thickness $a$,

$$a \Leftrightarrow a' = Ma \tag{2.34}$$

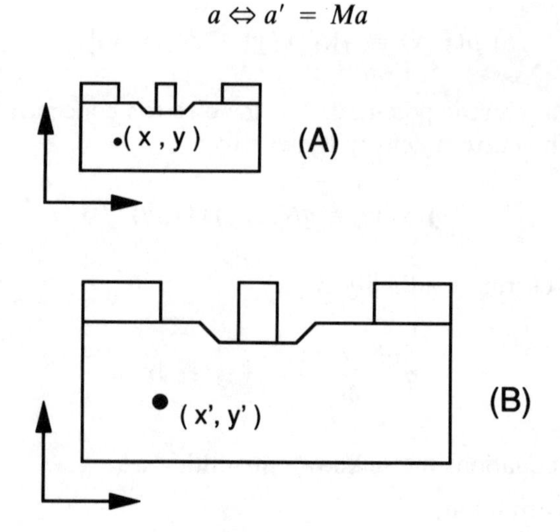


**Figure 2.4** Channel cross sections of two equivalent FETs.

Let us consider FETs A and B equivalent; that is, A and B have identical electric field distributions that can be expressed by the same function, namely:

$$\mathbf{E}(x, y) \Leftrightarrow \mathbf{E}(x', y') = \mathbf{E}(x, y) \tag{2.35}$$

Now, using (2.32) and (2.35), let us investigate the remaining variables:

1. *Semiconductor potential distribution* $\phi(x, y)$: The distribution $\phi(x, y)$ can be computed as a path integral of $\mathbf{E}(x, y)$. The potential difference between two points $(x_1, y_1)$ and $(x_2, y_2)$ is

$$\begin{aligned}\phi(x_2, y_2) - \phi(x_1, y_1) &= \int_{(x_1, y_1)}^{(x_2, y_2)} \mathbf{E}(x, y) \cdot d\mathbf{s} \\ &= M^{-1} \int_{(x_1', y_1')}^{(x_2', y_2')} \mathbf{E}'(x', y') \cdot d\mathbf{s} \\ &= M^{-1}[\phi(x_2', y_2') - \phi(x_1', y_1')]\end{aligned} \tag{2.36}$$

Thus, if we assume equal potential at our reference points [i.e., $(x, y) = (x', y') = (0, 0)$], then the following relationship holds in FETs A and B:

$$\phi(x, y) \Leftrightarrow \phi(x', y') = M\phi(x, y) \tag{2.37}$$

2. *Electron density distribution* $N(x, y)$: From Poisson's equation, (2.27),

$$\begin{aligned}\epsilon^{-1}(x,y)q[N(x,y) - N_D(x,y)] &= \nabla \cdot \mathbf{E}(x,y) \\ &= M\nabla' \cdot \mathbf{E}(x',y') \\ &= \epsilon^{-1}(x',y')q[N(x',y') - N_D(x',y')]\end{aligned} \tag{2.38}$$

As long as the same semiconductor material is used in both FETs, $\epsilon(x, y) = \epsilon(x', y')$. Because $N$ and $N_D$ are independently determined quantities, we may assume "detailed balance" in (2.38). The following relationships therefore hold for $N$ and $N_D$:

$$N(x, y) \Leftrightarrow N(x', y') = M^{-1}N(x, y) \tag{2.39}$$

$$N_D(x, y) \Leftrightarrow N_D(x', y') = M^{-1}N_D(x, y) \tag{2.40}$$

3. *Current density distribution*: As shown in (2.30), current density is the product of $N$ and $v$. From (2.26) and (2.35), we know that

$$v(x, y) \Leftrightarrow v(x', y') = v(x, y) \tag{2.41}$$

so that

$$\mathbf{j}(x, y) \Leftrightarrow \mathbf{j}(x', y') = M^{-1}\mathbf{j}(x, y) \tag{2.42}$$

Next, let us see how FET structural factors and performance characteristics are affected by scaling. In principle, the scaling property of each quantity is determined by its dimension. Let us introduce a new scaling factor $N$, which is the ratio of the gatewidth of FET A to B.

*Structural Factors*

Quantities relating to length:

$$\text{Gate length } l_g \Leftrightarrow l'_g = Ml_g \tag{2.43a}$$

$$\text{Source-gate spacing } l_{gs} \Leftrightarrow l'_{gs} = Ml_{gs} \tag{2.43b}$$

$$\text{Source-drain spacing } l_{ds} \Leftrightarrow l'_{ds} = Ml_{ds} \tag{2.43c}$$

Active layer dimensions:

$$\text{Active layer thickness } a \Leftrightarrow a' = Ma \tag{2.43d}$$

$$\text{Donor density } N_D \Leftrightarrow N'_D = M^{-1}N_D \tag{2.43e}$$

If the active layer is formed by ion implantation processes, then,

$$\text{Peak donor density location } t_p \Leftrightarrow t'_p = Mt_p \tag{2.43f}$$

$$\text{Donor density half-bandwidth } a_p \Leftrightarrow a'_p = Ma_p \tag{2.43g}$$

$$\text{Peak donor density } N_{dp} \Leftrightarrow N'_{dp} = M^{-1}N_{dp} \tag{2.43h}$$

Gatewidth:

$$\text{Gatewidth } w_g \Leftrightarrow w'_g = Nw_g \tag{2.43i}$$

Applied voltage is scaled in the same way as potential:

$$\text{Source electrode potential } V_s \Leftrightarrow V_s' = MV_s \tag{2.43j}$$

$$\text{Drain-source potential } V_{ds} \Leftrightarrow V_{ds}' = MV_{ds} \tag{2.43k}$$

$$\text{Gate-source potential } V_{gs} \Leftrightarrow V_{gs}' = MV_{gs} + (M - 1)V_\phi \tag{2.43l}$$

In scaling the gate-source potential, $V_\phi$ is considered a physical constant not subject to scaling.

*Operating Characteristics*

The dc performance characteristics can be converted as follows:

$$\text{Drain current } I_{ds} \Leftrightarrow I_{ds}' = NI_{ds} \tag{2.44a}$$

$$\text{Pinch-off voltage } V_P \Leftrightarrow V_P' = MV_P + (M - 1)V_\phi \tag{2.44b}$$

$$\text{Transconductance } g_m \Leftrightarrow g_m' = M^{-1}Ng_m \tag{2.44c}$$

$$\text{Drain conductance } g_{ds} \Leftrightarrow g_{ds}' = M^{-1}Ng_{ds} \tag{2.44d}$$

$$\text{Stored charge under gate } Q_g \Leftrightarrow Q_g' = MNQ_g \tag{2.44e}$$

$$\text{Gate-source capacitance } C_{gs} \Leftrightarrow C_{gs}' = NC_{gs} \tag{2.44f}$$

$$\text{Feedback capacitance } C_{gd} \Leftrightarrow C_{gd}' = NC_{gd} \tag{2.44g}$$

$$\text{Delay time } \tau \Leftrightarrow \tau' = M\tau \tag{2.44h}$$

$$\text{Channel resistance } R_i \Leftrightarrow R_i' = MN^{-1}R_i \tag{2.44i}$$

None of these conversions appears to need specific explanation.

Next, let us consider the conversion of $Y$ parameters, one of the RF characteristics. The $Y$ parameters have the dimensions of conductance. The four parameters of $\{Y\}$ are converted in the same fashion and can also be expressed in terms of the elements of the FET equivalent circuit model, which include $g_m$, $g_d$, $C_{gs}$, $C_{gd}$, $R_i$, and $\tau$. This means that whatever type of equivalent circuit model we use, all elements have the dimensions of conductance, that is, $g_m$, $g_d$, $\omega C_{gs}$, $\omega C_{gd}$, and $R_i^{-1}$ are converted in the same way. Thus in order to convert

$$\omega C_{gs} \Leftrightarrow (\omega C_{gs})' = M^{-1}N\omega C_{gs} \tag{2.45}$$

$$\omega C_{gd} \Leftrightarrow (\omega C_{gd})' = M^{-1}N\omega C_{gd} \tag{2.46}$$

we must add the conditions

$$\omega \Leftrightarrow \omega' = M^{-1}\omega \quad \text{or} \quad f \Leftrightarrow f' = M^{-1}f \tag{2.47}$$

Note that this means that $\omega\tau$ is invariant. Under the conditions of (2.47), the $Y$ parameters become

$$\{Y\}_{@f} \Leftrightarrow \{Y'\}_{@f} = M^{-1}N\{Y\}_{@f} \tag{2.48}$$

Of course, impedance is the reverse:

$$\{Z\}_{@f} \Leftrightarrow \{Z'\}_{@f} = MN^{-1}\{Z\}_{@f} \tag{2.49}$$

The significance of (2.48) is that the $Y$ parameters $\{Y\}$ of FET A at frequency $f$ are related to the $Y$ parameters $\{Y'\}$ of FET B at frequency $f' = M^{-1}f$ by the function $\{Y'\}_{@f} = M^{-1}N\{Y\}_{@f}$. A similar relationship applies for the $Z$ parameters as (2.49) shows.

*Performance Criteria*

First let us discuss $S$ parameters. The $S$ parameters are dimensionless values pertaining to signal ratios at the input and output ports of an element. The $S$ parameters are also characterized by the characteristic impedance $Z_0$ of the system in which the element is placed. Thus if we allow conversion of $Z_0$ as follows:

$$Z_0 \Leftrightarrow Z_0 = MN^{-1}Z_0 \tag{2.50}$$

then the $S$ parameters remain unchanged:

$$\{S\}_{@f} \Leftrightarrow \{S'\}_{@f} = \{S\}_{@f} \tag{2.51}$$

However, in the microwave and millimeterwave frequency ranges, the characteristic impedance is set at $50\Omega$ and cannot be scaled. If we narrow our discussion to focus only on $S$ parameters in 50-$\Omega$ systems, then the scaling of $S$ parameters can take the same format as any conversion from $Z_0 = 50\Omega$ to $Z_0 = MN^{-1} \times 50\Omega$ [12]:

$$\begin{aligned}
S_{11@f} \Leftrightarrow S'_{11@f} &= AD_s^{-1}[(1 - \Gamma S_{22})(S_{11} - \Gamma^*) + \Gamma S_{12}S_{21}]_{@f} \\
S_{21@f} \Leftrightarrow S'_{21@f} &= AD_s^{-1}S_{21}(1 - |\Gamma|^2)_{@f} \\
S_{12@f} \Leftrightarrow S'_{12@f} &= AD_s^{-1}S_{12}(1 - |\Gamma|^2)_{@f} \\
S_{22@f} \Leftrightarrow S'_{22@f} &= AD_s^{-1}[(1 - |\Gamma|S_{11})(S_{22} - \Gamma^*) + \Gamma S_{12}S_{21}]_{@f}
\end{aligned} \tag{2.52}$$

where

$$\Gamma = (Z_0' - Z_0)/)(Z_0' + Z_0)$$
$$D_S = (1 - \Gamma S_{11})(1 - \Gamma S_{22}) - \Gamma^2 S_{12} S_{21}$$
$$A = (1 - \Gamma)/(1 - \Gamma^*)$$

Let's look at gain next. We have shown that $S$ parameters are invariable if we allow conversion of the characteristic impedance. Because gain is independent of the characteristic impedance, it also should not change. To put it more correctly, if FET A in a system with a source impedance of $Z_S$ and load impedance $Z_L$ has a gain of $G$ at frequency $f$, then FET B in a system with source impedance $Z_S = MN^{-1}Z_S$ and output impedance $Z_L = MN^{-1}Z_L$ will also have a gain of $G$ at a frequency of $f' = M^{-1}f$, that is,

$$G_{@f} \Leftrightarrow G_{@f'} = G_{@f}, \qquad Z_S' = MN^{-1}Z_s, \qquad Z_L' = MN^{-1}Z_L \tag{2.53}$$

Thus far we have been speaking of gain in general, but the relationships in (2.53) also hold for maximum available power gain, the *figure of merit*.

Now we investigate output power. Because the fully open channel current is the product of the donor density and thickness of the active region, the scaling relation is given by

$$I_F \Leftrightarrow I_F' = NI_F \tag{2.54}$$

Drain-source breakdown voltage obeys the rule from (2.43):

$$V_{dsB} \Leftrightarrow V_{dsB}' = MV_{dsB} \tag{2.55}$$

Since output power is proportional to the product of these two values,

$$P_{\text{out}} \Leftrightarrow P_{\text{out}}' = MNP_{\text{out}} \tag{2.56}$$

In reality, however, the breakdown voltage does not behave according to (2.55) due to the significant effects of the recess structure, as discussed in the next section. It is basically determined by the product of the density and thickness of the active area, so that (2.56) is not necessarily meaningful in practice. It is true, however, that as an FET is reduced in size by direct application of the scaling laws, the donor density of the active area increases, which lowers the gate breakdown voltage and reduces the output power. Thus caution is needed in applying the scaling law in high-frequency, high-power applications, so that the result does not lower the breakdown voltage.

It is worth mentioning that the noise figure $F$ scales in the same way as gain. The noise parameters, namely, $F_{min}$, $R_n$, and $Y_{opt}$ are scaled the same as noise figure, resistance, and $Y$ parameters, respectively. Associated gain can also be converted using (2.53).

The significance of the scaling law can be summarized as follows. Suppose FET A has the following dimensions:

- Gate length: $l_g = 2\ \mu m$
- Donor density: $N_D = 1 \times 10^{17}\ cm^{-3}$
- Active layer thickness: $a = 0.2\ \mu m$
- Gatewidth: $w_g = 1$ mm

Also suppose that this FET has the following characteristics: $G = 10$ db at $V_{ds} =$ 8V, $I_{ds} = 100$ mA, and $f = 4$ GHz. The corresponding FET B with dimensions

- Gate length: $l_g = 1\ \mu m$
- Donor density: $N_D = 2 \times 10^{17}\ cm^{-3}$
- Active layer thickness: $a = 0.1\ \mu m$
- Gatewidth: $w_g = 2$ mm (i.e., $M = 0.5$, $N = 2$)

should have the characteristics $G = 10$ db at $V_{ds} = 4V$, $I_{ds} = 100$ mA, and $f =$ 8 GHz with the same matching circuit ($MN = 1$).

Thus it is possible to use the scaling law to design reduced-size FETs with improved high-frequency characteristics. Note that the scaling illustration used here is based on Poisson's equation and is applicable to FETs with a gate length large than 1 $\mu$m. When technologies such as submicron gate length, thin epilayer structure, heterojunctions, and others are introduced, quantum mechanical effects must be taken into account for quantitative analysis. Such subjects are beyond the scope of this book. However, for determining the direction of FET design efforts, scaling principles based on Poisson's equation as presented here are extremely effective

### 2.2.5 Breakdown Voltage and Recess Structure

Breakdown voltage, one of the key criteria in FET design, consists of drain breakdown voltage and gate breakdown voltage. In this section, we present the mechanism of breakdown voltage and discuss recess structure, one of the techniques used to increase breakdown voltage.

#### *Drain Breakdown Voltage*

When the FET channel is open, allowing current to flow between the drain and source terminals, the level of drain-source voltage at which the FET is destroyed is called *drain breakdown voltage*. Because the drain-source current is ohmic, thi s not, as the name might imply, the breakdown. What actually happens is that, ong before any type of heat-caused destruction occurs, the FET fails at a certain evel of drain voltage. Mimura et al. [13] discovered that high drain voltage levels aused the edges of the drain electrodes to emit light and related the effect to drain oltage. Yamamoto et al. [14] and Higashisaka [15] explained the dependency of ailure on gate bias and drain bias, the phenomenon of light emission and the nechanism of destruction. From this work, we understand that the reverse gate urrent consists of two types: the breakdown current observed when the gate bias s deep and the excess gate current observed when gate bias is shallow. The excess ate current is closely related to the amount of light emitted by the FET, and xperiments have verified that the more light an FET emits, the higher the excess ate current. The light emitted by the FET is nearly white, and it is visible under n ordinary optical microscope. The light emitted by the semiconductor is generally aused by radiation energy from the recombination of electron-hole pairs. Because his light emission occurs in the range of visible light, it is considered to indicate he existence of high-energy electron-hole pairs. The excess gate current resulting rom high drain voltage is thought to be caused by holes flowing to the gate.

In $n$-type GaAs, electrons are the majority carriers. Holes, the minority arriers, do not exist under normal bias conditions. The existence of holes is sugested by the occurrence of avalanche-like hole formation near the surface of the ET operating layer under high electric fields. By measuring the location and ntensity of light emission, we can pinpoint the location and strength of field conentration in the FET. In reality, FETs destroyed due to drain voltage (as well as ETs on the threshold of destruction) show tiny pits on the GaAs surface in ocations that coincide exactly with those emitting light. Figure 2.5(a) and (b) show ET structures and the observed locations of failure. Failure occurred in the GaAs ear the edge of the drain electrode in the planar FET shown in (a), and in the -$n^+$ interface in the FET shown in (b). In structure (a), the failure is believed to e caused by avalanche breakdown due to field concentration near the edge of the rain electrode, and there are problems arising from incomplete ohmic contact in he drain electrode. In structure (b), field concentration appears to have occurred ecause of discontinuity and incompleteness in the crystals in the $n$-$n^+$ interface, nd there are structural problems related to the steep horizontal changes in donor ensity. To increase drain breakdown voltage, therefore, there must be complete hmic contact with the electrodes, and a structure must be developed that eliminates teep changes in donor density along the channel.

Comparing Figure 2.5(a) and (b) with respect to drain breakdown voltage, he FET with the $n^+$ area (b) has a drain breakdown voltage 1.4 times that of (a) 6, 16]. Therefore, we conclude that an $n^+$ area formation is good for obtaining igher drain breakdown voltages.

Another way of increasing drain breakdown voltage is a recessed gate strucure. Figure 2.6 shows various recess structures and drain breakdown voltages in

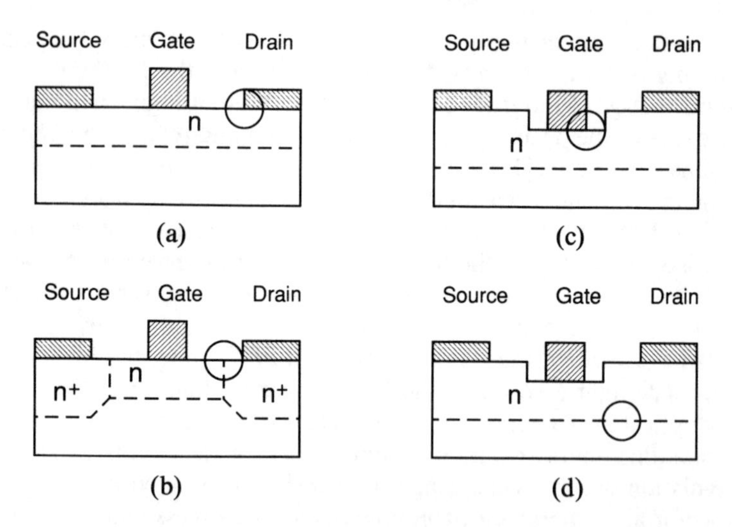


**Figure 2.5** Channel cross sections of several types of FET and location of the failure point [15]: (a edge of the drain electrode in a planar FET, (b) $n - n^+$ boundary in an ion-implante FET, (c) edge of the gate electrode in a steeply recessed FET, and (d) interface betwee the buffer and active layers beneath the recess edge.

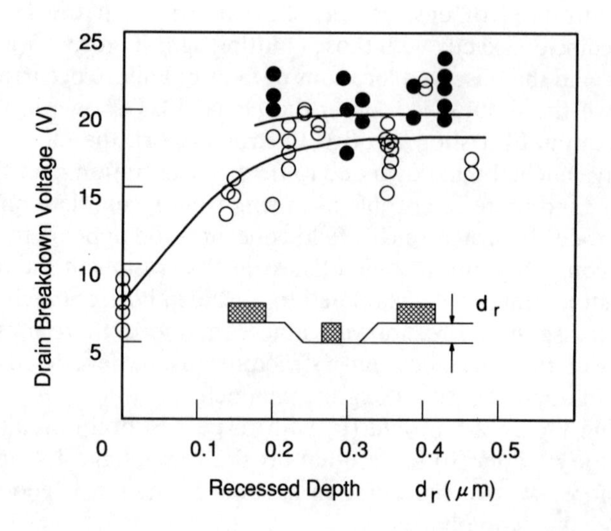


**Figure 2.6** Drain breakdown voltage versus recess depth at $V_{gs} = 0$ [15]. Open circles: abrupt reces structure; solid circles: graded recess structure.

relation to recess depth [15]. A graded recess is more effective than an abrupt recess in reducing field concentration near the recess edges, and it also provides higher drain breakdown voltage. Recessing also increases $g_m$, and is widely used in FETs. Achieving the desired recess structure, however, often creates problems in process technology, so that it is important to consider process limitations in the design stage.

### *Gate Breakdown Voltage*

Figure 2.5(c) and (d) show examples of FET failure under deep gate bias. There appear to be two modes of failure. Failures occurring deep in the activated area of the GaAs wafer appear to be failures of the GaAs wafer buffer layer and $n^-$ layer [15]. Failures at the edge of the gate electrode show leakage of current accompanied by emission of light. The gate current has been shown to be due to avalanching at the point of Schottky contact. This is the most common mode of failure due to gate breakdown.

Gate breakdown voltage was estimated by Wemple et al. [17] and later expanded by Hikosaka et al. [18] to include FETs with a recessed structure. Here we will introduce the calculation of gate breakdown voltage in Wemple's FET model, a cross section of which is shown in Figure 2.7(a). Consider an FET to which the maximum gate-drain voltage is applied. The electric field lines emanating from the depletion layer terminate at the end of the gate electrode nearest the drain. As the gate-drain voltage increases and the field grows in strength, it will eventually reach the level of critical field strength at this point. In Figure 2.7, $l_{\text{eff}}$ is the effective gate length, the length of the metallic gate electrode at which the field lines from the depletion layer terminate. According to Wemple, $l_{\text{eff}}$ is 0.5 $\mu$m, and for any gate shorter than this, $l_{\text{eff}} = l_g$. Let the donor density of the active layer be $N_D$, active layer thickness be $a$, thickness of the surface depletion layer be $d_s$, and the lateral length of the depletion layer toward the drain be $D$. The critical field strength of GaAs is represented by $E_a$. The field at the depletion layer just below the gate electrode is $E_1$, and the field at the end of the depletion layer toward the drain is $E_2$, represented according to Gauss's theorem as

$$E_1 = \frac{qN_D a}{\epsilon} \tag{2.57}$$

$$E_2 = \frac{qN_D(a - d_s)D}{\epsilon l_{\text{eff}}} \tag{2.58}$$

When the sum of $E_1 + E_2$ reaches $E_a$, breakdown occurs. The critical value of $E_2$ is thus

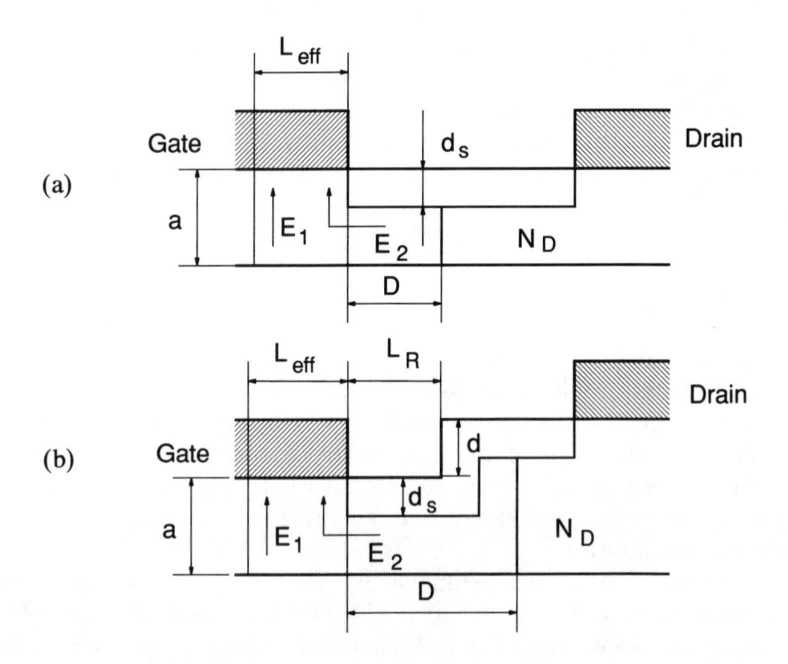


**Figure 2.7** Schematic cross-sectional sketch of the gate-to-drain region [17, 18]: (a) planar structure and (b) recessed structure. Shaded areas indicate donor-doped regions, which electrically interact with charge on the gate electrode.

$$E_2 = E_a - E_1 = E_a\left(1 - \frac{qN_D a}{\epsilon E_a}\right) \tag{2.59}$$

At breakdown, the value of $D$ is

$$D = \frac{\epsilon l_{\text{eff}} E_a}{qN_d(a - d_s)}\left(1 - \frac{qN_D a}{\epsilon E_a}\right) \tag{2.60}$$

Thus, gate breakdown voltage $V_{dgB}$ can be written as

$$V_{dgB} = \frac{DE_2}{2} = \frac{\epsilon l_{\text{eff}} E_a^2}{2qN_D(a - d_s)}\left(1 - \frac{qN_D a}{\epsilon E_a}\right)^2 \tag{2.61}$$

Furthermore, if we use the recessed structure shown in Figure 2.7(b), the gate breakdown voltage is calculated as

$$V_{dgB} = \frac{\epsilon l_{\text{eff}} E_a^2}{2qN_D(a - d_s)} \left(1 - \frac{qN_D a}{\epsilon E_a}\right)^2 + \left(1 - \frac{a - d_s}{a + d - d_s}\right)\left[E_a l_R\left(1 - \frac{qN_D a}{\epsilon E_a}\right) - \frac{qN_D(a - t_s) l_R^2}{\epsilon l_{\text{eff}}}\right] \tag{2.62}$$

The critical field strength $E_a$ is dependent on donor density. When $N_D = 1.5 \times 10^{17}$ cm$^{-3}$, $E_a = 7 \times 10^5$ V/cm [19]. Figure 2.8 presents computed and observed gate breakdown voltage values as a function of recess length. Breakdown voltage is greatly affected by the existence (or lack) of a surface depletion layer. In reality, FETs are given various types of surface treatments, so that observed values will fall between the two values. As the illustrated data show, the thickness of the surface depletion layer has a great effect on gate breakdown voltage. The thickness of the surface depletion layer varies according to the surface state density of GaAs; the greater the surface state density, the deeper the surface depletion layer. Because the donor charge in the surface depletion layer interacts with electrons and fields captured at the surface state, it loses its function as a channel. Therefore, it is important to apply surface treatments that make the surface depletion layer as thin

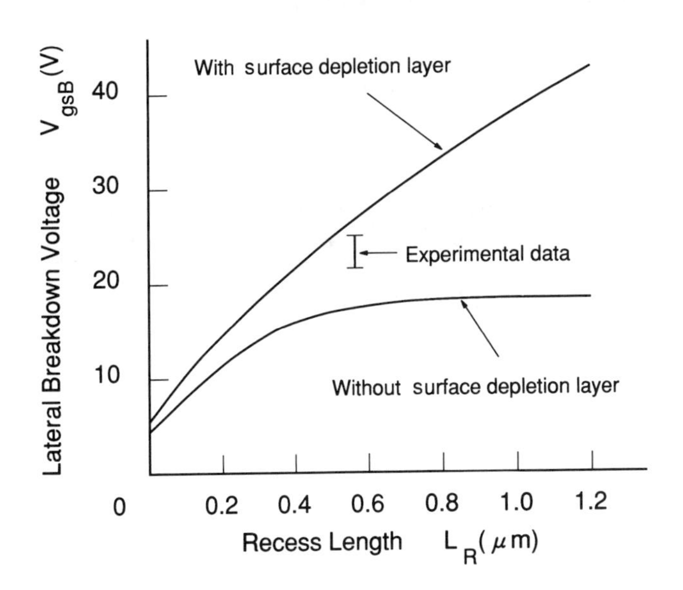


**Figure 2.8** Calculated and measured lateral gate breakdown voltage versus recess length [30, 31] with the following conditions: $N_D = 1.5 \times 10^{17}$ cm$^{-3}$, $a = 0.12$ $\mu$m, $d = 0.38$ $\mu$m, $l_g = 0.7$ $\mu$m, $L_{\text{eff}} = 0.5$ $\mu$m, $I_F = 300$ mA/mm, $E_a = 7 \times 10^5$ V/cm, and $d_s = 0.085$ $\mu$m.

as possible. Gate breakdown voltage can therefore be increased by adjusting donor density and thickness as well as recess shape.

Another important characteristic is that the distance between the gate and drain electrodes must be greater than the horizontal spread of the depletion layer. If an $n^+$ layer is used, this also holds true for the distance between the gate electrode and the $n^+$ layer. Note that the light-emission phenomenon discussed in the preceding part is an extremely useful method of directly measuring for uniform FET operation and matching performance [15].

### 2.2.6 Parasitic Resistance

Besides the structure below the gate and gate recess discussed in the previous two sections, the critical elements of FET design include the reduction of parasitic elements not directly related to the active operation of the FET. Parasitic elements affecting channel cross-section design include gate resistance, ohmic contact resistance, source resistance, drain resistance, and parasitic gate capacitance. Ohmic contact resistance and parasitic resistance relate primarily to process technology and material properties and will not be discussed here. Gate resistance $R_g$ is the resistance in the metal of the gate finger, as shown in Figure 2.9. Fukui [20] empirically expressed gate resistance with the following simple formula:

$$R_g \approx \frac{\rho w_{gu}^2}{3 l_g h w_g} \tag{2.63}$$

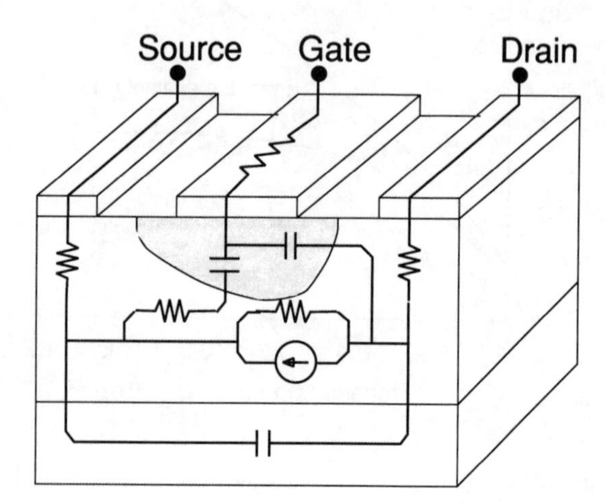


**Figure 2.9** Cross-sectional view of an FET channel and part of the small-signal equivalent circuit of the device.

vhere $\rho$ is the resistivity of the gate metal, $w_{gu}$ is the unit gatewidth, and $h$ is the ;ate metal thickness. The gate is assumed to be rectangular. A factor of one-third s used as a first approximation of the effective ratio of $R_g$ to the total dc resistance f the gate finger. Assuming an aluminum gate, for example, with resistivity $\rho =$ .5 $\times 10^{-6}\ \Omega$ cm, total gatewidth $w_g = 500\ \mu$m, unit gatewidth $w_{gu} = 250\ \mu$m, ate length $l_g = 0.8\ \mu$m, and gate metal thickness $h = 0.65\ \mu$m, the above formula ives a gate resistance of 2.0 $\Omega$. As (2.63) suggests, gate resistance can be reduced y decreasing the unit gatewidth. However, it is more efficient to increase the cross ection of the gate metal $l_g h$. In practice, the gate length, where the metal forms Schottky junction with the semiconductor, must be kept short, so that T-shaped r "mushroom" gate metal cross sections are often used to achieve both shorter ates and lower gate resistance. Equation (2.63) expresses gate resistance in terms f dc current. Resistance in high-frequency signals will be greater than this value ue to the skin effect. Because of attenuation, high-frequency waves will not pen-trate into the metal beyond the skin depth. As a first approximation, $h$ in (2.63) an be replaced by the skin depth $\delta$. The skin depth is given by

$$\delta = \sqrt{\frac{2}{\omega\mu\sigma}} \tag{2.64}$$

here $\mu$ is permeability and $\sigma$ is conductivity. With copper, for example (conduc-vity $\sigma = 5.8 \times 10^7$ S/m), $\delta = 0.66\ \mu$m. At X-band and higher frequencies, the kin depth approaches the dimensions of actual gate length. Therefore, in the high-ower FET, even at very low levels of dc gate current, high-frequency gate current ill flow, meaning that the gate length and cross section must be designed with onsideration for high-frequency current capacity and gate resistance.

Source resistance is the resistance in the GaAs from the edge of the source lectrode nearest the gate to the edge of the gate electrode nearest the source. ormally high fields do not exist between the source and gate, so that the source esistance $R_S$ can be estimated using the formula for low-field resistance,

$$R_S = \frac{\rho l_{sg}}{w_g a} \tag{2.65}$$

here $l_{sg}$ is the distance between source and gate electrodes and $a$ is the channel ickness. The resistivity $\rho$ of GaAs is dependent on doping density and, according Fukui [20], can be estimated by

$$\rho \approx 0.11\ N_D^{0.82} \tag{2.66}$$

here $N_D$ is expressed in units of $10^{16}$ cm$^{-3}$. This expression is valid when $N_D$ is in e range of $10^{16}$ to $10^{19}$ cm$^{-3}$. For example, assuming that the donor density

$N_D = 1 \times 10^{17}$ cm$^{-3}$, source-gate spacing $l_{sg} = 2$ μm, gatewidth $w_g = 1$ mm, and channel thickness $a = 0.25$ μm, then $\rho = 0.017$ Ω cm, and $R_s = 1.33$ Ω.

As the equivalent circuit diagram in Figure 2.9 shows, source resistance and gate resistance lie in series at the input of the FET. Because source resistance acts as an input-output feedback element, it has a major effect on high-frequency gain. In terms of dc operating characteristics, source resistance causes an apparent decrease in transconductance. When a voltage $V_{gs}$ is applied between gate and source, the actual voltage $V'_{gs}$ applied to the gate capacitance is reduced by the quantity $R_s I_{ds}$, that is, $V'_{gs} = V_{gs} - R_s I_{ds}$. Because $I_{ds}$ is equal to $g_m V'_{gs}$, $V'_{gs}$ and $V_{gs}$ are related by $V'_{gs} = V_{gs}/(1 + R_s g_m)$. Therefore, the relation between the intrinsic FET transconductance $g_m$ $(= \partial I_{ds}/\partial V'_{gs})$ and measurable transconductance $g'_m$ $(= \partial I_{ds}/\partial V_{gs})$ is given by

$$g'_m = \frac{g_m}{1 + R_s g_m} \tag{2.67}$$

Source resistance $R_s$ thus causes an apparent reduction in $g_m$. Source resistance may be reduced by three methods: (1) creating an $n^+$ layer below the source electrode by ion implantation, (2) reducing the source-gate spacing (called *offset gate design*), or (3) creating recesses.

Drain resistance is the resistance in the GaAs between the gate and the drain. Under weak fields, it could be estimated by the same formula used for source resistance. But because high fields exist between the gate and drain, estimating drain resistance is not a matter of applying a simple formula. It can be estimated using the equivalent circuit analysis presented in the next section. Drain resistance is dependant on bias, and may become negative under high drain voltage. This is because as electrons transfer from a state near the $\Gamma$ point to a state near the $L$ point or $X$ point, a region of negative differential resistance is created (i.e., decreasing velocity as field strength increases).

### 2.2.7 Equivalent Circuits

In developing a high-power FET, the first step is to construct FET units using the design steps we have discussed. The next step is to use the characteristics of the FET unit as the foundation for designing the complete FET pattern. In this section we examine the determination of equivalent circuit parameters, the most basic method of evaluating FET unit performance. The equivalent circuit in Figure 2.1[illegible] is applicable to an FET with a comparatively small gatewidth. [In cases where forward or reverse gate breakdown has occurred, it is necessary to add a resistor (or diode element) in parallel with the capacitor.] Equivalent circuit elements ma[illegible]

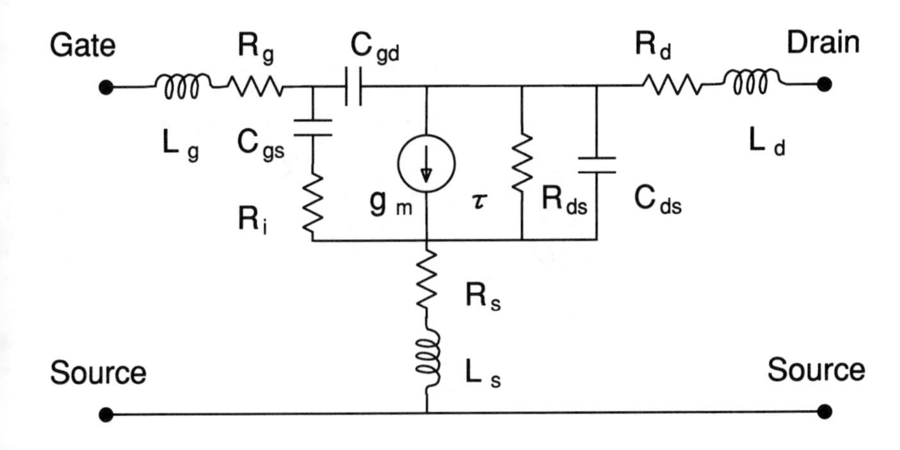


igure 2.10 Complete small-signal equivalent circuit of an FET.

e estimated either by Fukui's method [20] of determining all values other than apacitance by dc measurement or by measuring *S* parameters and then adjusting hem to the equivalent circuit elements. The best method is, as a first step, to etermine gate, source, and channel resistance by Fukui's method, and then, using hese values as initial values, to adjust the equivalent circuit elements to realize he measured *S* parameters. Equivalent circuit fitting can be done easily with ommercially available software. Figure 2.11 shows an example of an equivalent ircuit and the *S* parameters as a function of frequency. The arrows indicate how parameters change when circuit element values are increased.

High-power FETs operate over a wide range of drain currents and voltages; herefore, the unit FET must be evaluated at several bias points. Gain and output ower can be generally estimated by *S* parameter measurement and equivalent ircuit derivation, using at least three points on the load line. The *Y* parameters f the intrinsic FET shown in Figure 2.10 can be expressed as follows:

$$
\begin{aligned}
Y_{11} &= \frac{j\omega C_{gs}}{1 + j\omega C_{gs} R_i} + j\omega C_{gd} \\
Y_{12} &= -\,j\omega C_{gd} \\
Y_{21} &= \frac{g_m \exp(-j\omega\tau)}{1 + j\omega C_{gs} R_i} - j\omega C_{gd} \\
Y_{22} &= \frac{1}{R_{ds}} + j\omega(C_{gd} + C_{ds})
\end{aligned}
\tag{2.68}
$$

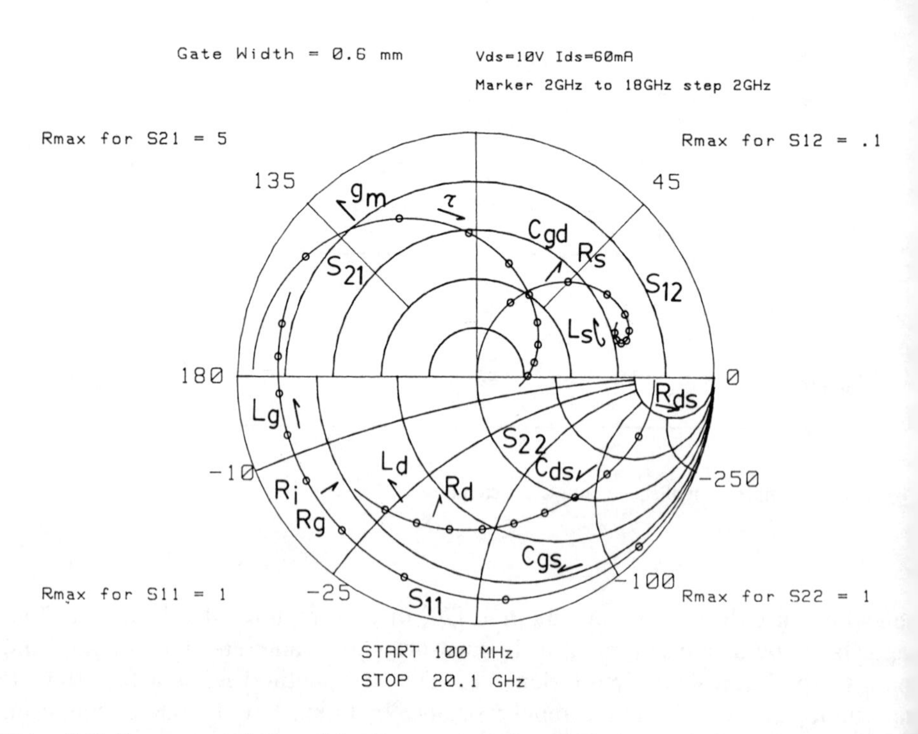


**Figure 2.11** Frequency variation of the $S$ parameters of an FET. Arrows indicate the main chang when the value of each parameter increases. Circuit parameters: $g_m$ = 53 mS; $\tau$ = 4.5 $C_{gs}$ = 0.58 pF; $C_{gd}$ = 0.031 pF; $C_{ds}$ = 0.17 pF; $R_i$ = 2.2Ω; $R_{ds}^{-1}$ = 3.1 mS; $R_g$ = 0.2Ω; = 1.35Ω; $R_d$ = 1.0Ω; $l_g$ = 0.17 nH; $L_s$ = 0.01 nH; $L_d$ = 0.15 nH; $V_{ds}$ = 8V; $I_{ds}$ = mA; and $W_g$ = 600 μm.

The $Y$ parameters of several of these FET units connected in parallel are obtaine simply by multiplying the FET unit $Y$ parameters by the number of elements connection:

$$\{Y_t\} = t\{Y\} \tag{2.6}$$

where $t$, a positive real number, not necessarily an integer, is the number of FE units connected in parallel. When $\{Y_t\}$ is converted into $S$ parameters, it represen the $S$ parameters of the entire FET. The $S$ parameters derived in this way a tentative approximations and ignore the effects of FET pattern layout, but th can be used to predict the $S$ parameters of the entire FET. The formulas to conve $Y$ parameters to $S$ parameters are:

$$
\begin{aligned}
S_{11} &= \frac{(1 - y_{11})(1 + y_{22})y_{12}y_{21}}{(1 + y_{11})(1 + y_{22}) - y_{12}y_{21}} \\
S_{12} &= \frac{-2y_{12}}{(1 + y_{11})(1 + y_{22}) - y_{12}y_{21}} \\
S_{21} &= \frac{-2y_{21}}{(1 + y_{11})(1 + y_{22}) - y_{12}y_{21}} \\
S_{22} &= \frac{(1 - y_{22})(1 + y_{11}) + y_{12}y_{21}}{(1 + y_{11})(1 + y_{22}) - y_{12}y_{21}}
\end{aligned}
\tag{2.70}
$$

ote that $y_{ij} = Y_{ij}Z_0$ for $i, j = 1, 2$.

## .3 HIGH-POWER FET DESIGN: FET PATTERN DESIGN

### .3.1 The Flow of the Design Process

high-power FET is not simply an agglomeration of FET units. To satisfy a set f desired performance characteristics for output power, gain, efficiency, and disrtion, the designer must determine an appropriate total gatewidth, unit gatewidth, hip size, and the number of bonding pads. In this section, we present methods of etermining these parameters, and discuss related aspects of FET structure. The verall flow of FET pattern design is shown in Figure 2.12.

### .3.2 Output Power and Total Gatewidth

nce the input and output characteristics of the FET unit have been determined, is a relatively easy matter to determine the total gatewidth of the completed ET. Let us represent the FET unit gatewidth by $w_{gu}$, and the related output power t the 1-dB gain compression point by $P_{1dBu}$ watts. If the target value for the 1-dB ompression point for the total chip is $P_{1dBt}$ watts, then all we need to do to arrive t the total FET gatewidth $w_{gt}$ is to calculate $w_{gu}\, P_{1dBt}/P_{1dBu}$. Of course, in practice, utput power is not simply scaled to gatewidth. One reason for this is the decrease f drain current because of the increase in operating channel temperature in large ETs. Also, as we will discuss later, another reason is the loss in power and the duction of gain due to the combining of multiple cells.

At present, the saturated drain current of a high-power FET per millimeter f gatewidth is of the order of 300 mA/mm. The gate breakdown voltage is around 0V, and the knee voltage is around 2V. Therefore, the maximum linear output ower per millimeter of gatewidth is about $0.3 \cdot (20 - 2)/8 = 0.675$ W/mm in the eal case. In practical FETs, however, nonuniformities of operation reduce the

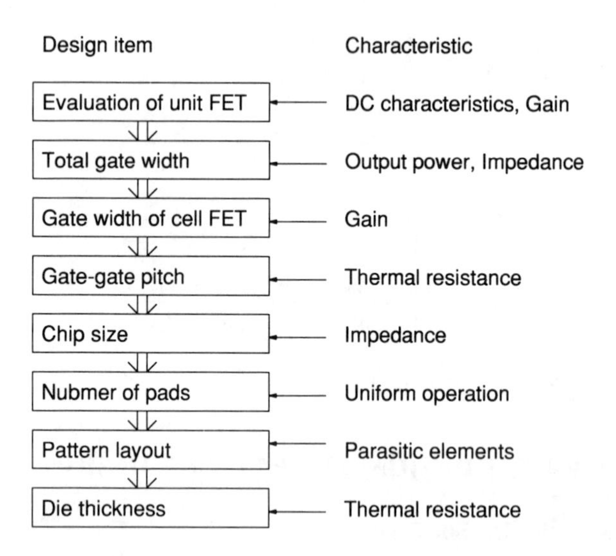


**Figure 2.12** Overall flow of FET pattern design.

output power. A typical value for $P_{1dB}$ is 0.50 W/mm in our experience. As a rul of thumb, to obtain an output power of $P_{1dB}$ watts, a gatewidth of $P_{1dB} \cdot 2$ mm needed. This rough estimate is surprisingly accurate and holds true for practical any active layer density or thickness. The reason for this is that the determinan of output power, namely, the product of fully open channel current and drain-ga breakdown voltage, do not depend on active layer density or thickness, at least the level of first approximation. If we let the fully open channel current be $I_F$ ar drain-gate breakdown voltage be $V_{dgB}$, then the maximum linear output powe $P_{\text{RFmax}}$, can be approximated by

$$P_{\text{RFmax}} \approx \frac{I_F V_{dgB}}{8} \quad (2.7$$

Assuming uniform doping, $I_F$ is given by [see (2.1)]

$$I_F = qN_D a v_{\text{sat}} w_g \quad (2.7$$

According to Wemple's model, defined by (2.61), $V_{dgB}$ can be approximated by

$$V_{dgB} = \frac{\epsilon L_{\text{eff}} E_a^2}{2qN_D a} \quad (2.7$$

Here we ignore the surface depletion layer and vertically oriented fields. Hence the maximum linear output power is given approximately by

$$P_{\text{RFmax}} \approx \frac{\epsilon w_g l_{\text{eff}} v_{\text{sat}} E_a^2}{16} \tag{2.74}$$

Ignoring the dependence of electron saturation velocity and critical electrical field on donor density, $P_{\text{RFmax}}$ per unit gatewidth depends only on the physical constants of the semiconductor material.

As the gate length decreases to the order of 0.5 $\mu$m, we can set the effective gate length equal to the actual gate length, that is, $l_{\text{eff}} = l_g$ and, after substituting into (2.74), obtain

$$P_{\text{RFmax}} f_T = \frac{\epsilon w_g v_{\text{sat}}^2 E_a^2}{32\pi} \tag{2.75}$$

In terms of the unit gatewidth of a high-power FET, the product of the performance indicators $P_{\text{RFmax}}$ and $f_T$ can be described in terms of the physical constants of the semiconductor material. This formula describes the limits of FET capability. Based on the FET data developed thus far and (2.75), it can be said that

$$P_{\text{RFmax}} f_t \approx 9 \text{ W Ghz/mm} \tag{2.76}$$

If $f_T$ is 20 GHz, $P_{\text{RFmax}}$ per 1 mm of gatewidth is thus 0.45W. If we shorten the gate length and set $f_T$ at 40 GHz, then $P_{\text{RFmax}}$ per 1 mm of gatewidth becomes 0.225W. These limits can only be exceeded by introducing new materials or concepts.

Determination of the total gatewidth also involves the impedance of an FET. For a unit gatewidth of 1 mm, saturated drain current of 300 mA/mm, and an effective drain voltage of 18V (i.e., 20V less 2V), then maximum output power will occur with a load impedance of 18/0.3 or 60 $\Omega$. In this case, $P_{1\text{dB}}$ is of the order of 0.5W. Now if 10 such FET units are placed together in parallel to form an FET with a total gatewidth of 10 mm, then the optimum load impedance will be around 6$\Omega$, and $P_{1\text{dB}}$ about 5W. Put 20 units together for an FET with a total gatewidth of 20 mm and the optimum load impedance becomes 3$\Omega$, and $P_{1\text{dB}}$ about 10W. Now if we want to use these FET chips in an amplifier application with an output of, for example, 8 to 9W, we will face a choice between one chip with a total gatewidth of 20 mm or two of the 10-mm chips. The choice will actually involve circuit design information outside the scope of our discussion here, but will include allowing for the frequency range and comparative bandwidth of the amplifier, as well as the impedance conversion ratio. Normally, an FET with a large gatewidth will have a low impedance, so that the conversion ratio for 50$\Omega$ will be greater, making matching more difficult over a wide bandwidth. In other words, in wide bandwidth amplifiers it is better to use a greater number of FETs of relatively smal gatewidth. FETs of larger gatewidth are also harder to use for high-frequenc applications due to their size. FET gatewidth should be appropriate for the intende application.

Figure 2.13 shows equivalent circuit $S$ parameters for an FET with a gatewidt of 1 mm, and for several such FETs connected together in parallel. Using th theory of impedance scaling, the $S_{11}$ and $S_{22}$ parameters of the 1-mm FET (char acteristic impedance 50$\Omega$) and the same parameters of an FET with a gatewidt of $n$ mm (characteristic impedance 50/n$\Omega$) are geometrically congruent and math ematically equivalent. (The reason that $S$ parameters appear different for eac gatewidth is that the characteristic impedance is not scaled but fixed at 50$\Omega$. Impedance-matching circuits at any given frequency and bandwidth can therefor be designed and created using the same factors for a 1-mm-gatewidth FET at 50$\Omega$ as for an $n$-mm-gatewidth FET at 50/n$\Omega$. The only problem is that the latter FE must require an impedance converter to transform the 50/n$\Omega$ system to a 50-$\Omega$ system, a ratio which increases and which restricts the comparative bandwidth a $n$ increases. This is the same as saying that based on an equal 50-$\Omega$ impedance the FET impedance decreases as the FET size increases, requiring multiple imped ance conversion steps to convert to 50$\Omega$.

### 2.3.3 Determination of the Unit Gatewidth

The high-power FET normally uses a comb-shaped gate electrode alignment. Th length of one gate finger is therefore called the *width* of the unit gate. Gates ar usually ultrafine lines of 1 $\mu$m or less in length, with considerable signal loss a microwave frequencies, and from the standpoint of signal loss it would seem pref erable to keep the unit gatewidth as short as possible. The shorter the fingers however, the more of them that are required, increasing the width of the chip (th length of the "comb") and skewing the signal phase delay. To select a unit gatewidt for an FET design, therefore, the first steps are to compute the signal loss in th gate fingers and to determine the allowable range of widths.

Considering the gate fingers as a distributed circuit system, we calculate th attenuation and phase rotation of the input signal in the direction of the gatewidt [15, 21, 22]. Resistance and inductance in the source and drain electrodes are smal in comparison to the gate electrode and, hence, are ignored here. Similarly, drain gate feedback capacitance $C_{gd}$ is nearly one order of magnitude less than the gate source capacitance and is also ignored here. Letting the high-frequency resistance inductance, and gate-source capacitance per unit gatewidth be, respectively, $r_g$, $l_g$ and $c_{gs}$, we will use $y_{in}$ to designate the FET input admittance per unit gatewidth

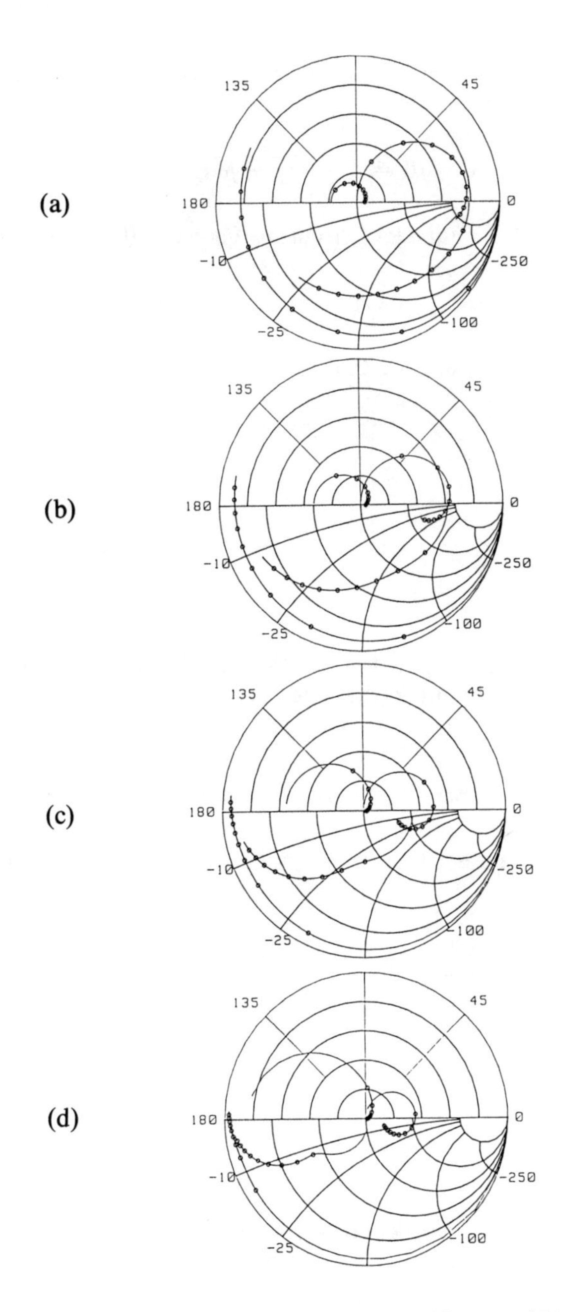


**Figure 2.13** Equivalent circuit $S$ parameters for an FET with a gatewidth of 1 mm, and for several such FETs connected together in parallel: (a) 0.5 ($w_g$ = 0.5 mm), (b) 1 ($w_g$ = 1 mm), (c) 2 ($w_g$ = 2 mm), and (d) 4 ($w_g$ = 4 mm).

The propagation constant, $\gamma = \alpha + j\beta$, of the gate transmission line can be computed using the formula

$$\gamma = \alpha + j\beta = \sqrt{y_{\text{in}}(r_g + j\omega l_g)} \quad (2.77)$$

For the input side of the FET equivalent circuit in Figure 2.10, $y_{\text{in}}$ is given by

$$y_{\text{in}} = (r_{\text{in}}^{-1} + j\omega c_{gs}), \qquad r_{\text{in}} = r_s + r_i \quad (2.78)$$

From (2.77) and (2.78), we can deduce that

$$\alpha = \omega c_{gs} \sqrt{-\frac{l_g}{2c_{gs}} + \frac{1}{2}\left[\left(\frac{l_g}{c_{gs}}\right)^2 + \left(r_{\text{in}}\omega l_g + \frac{r_g}{\omega c_{gs}}\right)\right]^{1/2}} \quad (2.79)$$

$$\beta = \omega c_{gs} \sqrt{\frac{l_g}{2c_{gs}} + \frac{1}{2}\left[\left(\frac{l_g}{c_{gs}}\right)^2 + \left(r_{\text{in}}\omega l_g + \frac{r_g}{\omega c_{gs}}\right)\right]^{1/2}} \quad (2.80)$$

Note that $r_{\text{in}}^{-1} \gg c_{gs}$, and $l_g/r_{\text{in}} \gg r_g c_{gs}$. Letting the signal voltage at the gate feed point be $v_0$, the signal voltage input to the gate can be expressed as a function of the gate position $w$ as follows:

$$v(w) = v_0 e^{-\alpha w}(\cos \beta w + j \sin \beta w) \quad (2.81)$$

If we let the unit gatewidth be $w_u$, then the signal attenuation and phase rotation are

$$\left|\frac{v(w)}{v_0}\right| = e^{-\alpha w_u}, \qquad \Delta\phi = \beta w_u \quad (2.82)$$

The effective microwave power dissipation per unit gatewidth, $p_{\text{in}}$, is

$$p_{\text{in}} = \frac{v_0^2 e^{-2\alpha w}(\cos \beta w)^2}{r_{\text{in}}} \quad (2.83)$$

The FET gain will be constant regardless of location on the gate electrode, so that at position $w$ the output power per unit gate width $p_{\text{out}}$ will be

$$p_{\text{out}} = Gp_{\text{in}} = \frac{Gv_0^2 e^{-2\alpha w}(\cos \beta w)^2}{r_{\text{in}}} \quad (2.84)$$

The total output power $P_{out}$ can be obtained by integration of $p_{out}$ over the unit gatewidth from $w = 0$ to $w = w_u$:

$$P_{out} = \frac{Gv_0^2}{r_{in}} \int_0^{w_u} e^{-2\alpha w} (\cos \beta w)^2 \, dw \tag{2.85}$$

Of course, if there is no propagation loss or phase rotation in the gate electrode, then the total output power $P_{out0}$ is simply

$$P_{out0} = \frac{w_u G v_0^2}{r_{in}} \tag{2.86}$$

The difference between (2.85) and (2.86) is the effect propagation loss and phase rotation in the gate electrode have on output power:

$$\begin{aligned} \frac{P_{out}}{P_{out0}} &= \frac{1}{w_u} \int_0^{w_u} e^{-2\alpha w} (\cos \beta w)^2 \, dw \\ &= \frac{1}{w_u} \left[ \frac{2\alpha^2 + \beta^2}{4\alpha(\alpha^2 + \beta^2)} - \frac{\mathrm{e}^{-2\alpha w_u}}{4\alpha} - \frac{e^{-2\alpha w_u}}{4(\alpha^2 + \beta^2)} (\alpha \cos 2\beta w_u - \beta \sin 2\beta w_u) \right] \end{aligned} \tag{2.87}$$

Figure 2.14 shows a calculation of the dependence of output power reduction on unit gatewidth. The parameters used are listed in the figure. Because short gate lengths have greater inductance per unit gatewidth, they cause greater phase rotation. Also because short gate lengths have greater resistance per unit gatewidth, they have greater output power loss.

For design purposes, the maximum phase rotation should be no more than $\pi/16$ radians [23], corresponding to being able to represent the FET as a lumped element. The allowable range of output power loss should be in the region of 0.5 dB. For the example in Figure 2.14, the result is a C-band FET of 150 $\mu$m, an X-band FET of 100 $\mu$m, and a Ku-band FET of 75 $\mu$m.

### 2.3.4 Chip Size

In general, the wider an FET chip (horizontally, with respect to vertically oriented input/output terminals), the lower the gain. This tendency is especially noticeable in a high-power FET composed of a number of cells. Four reasons can be given for this reduction of gain:

1. Phase rotation at the gate input feed point due to longer gate bus paths;
2. Matching loss due to higher impedance ratios;
3. Nonuniform channel temperature;

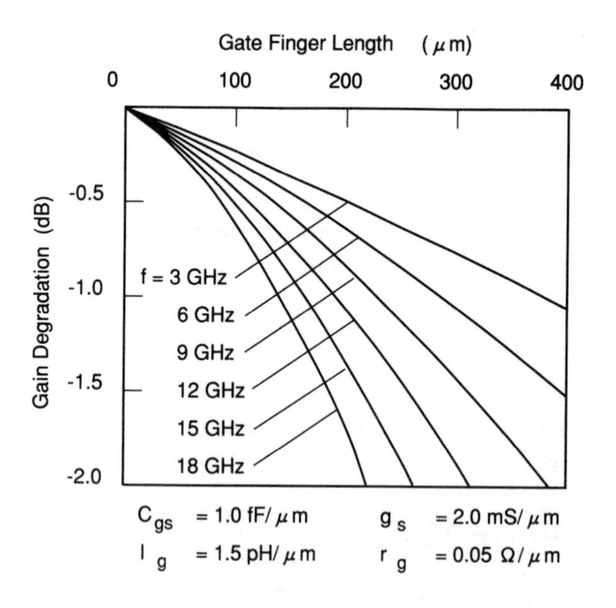


**Figure 2.14** Calculated gain degradation versus gate finger length.

4. Nonuniform operation from cell to cell due to variations in materials and processes.

Of these, item 2 can be avoided by MMIC technology. Item 3 can be controlled to some extent by using a lower thermal resistance construction. To deal with items 1 and 4, however, it is necessary to reduce the chip size.

Figure 2.15 shows the relation between the maximum width of the active area and the design frequency [23]. Lines A and B represent a width of $\lambda/8$ and $\lambda/16$ respectively, in GaAs at the design frequency. A width of $\lambda/16$ represents a level at which phase rotation is definitely not observable in experiments. When $L$ is in region 1 below B, parallel operation is maintained throughout the active area. In region 2, between A and B, there is some reduction in gain, but the FET still operates efficiently. In region 3 above A, the FET cannot be considered as a lumped element, and external circuitry is required to control the phase rotation in power input and output. According to Figure 2.15, for example, the maximum value of $L$ at 8 GHz is 0.65 mm. This is the limiting value of the horizontal dimension of the active layer to achieve in-phase operation of all unit cells in the FET.

### 2.3.5 Determination of the Number of Pads

The high-power FET is composed of a number of small FETs, and for a high-power FET to function satisfactorily, it is important that all of the small FET

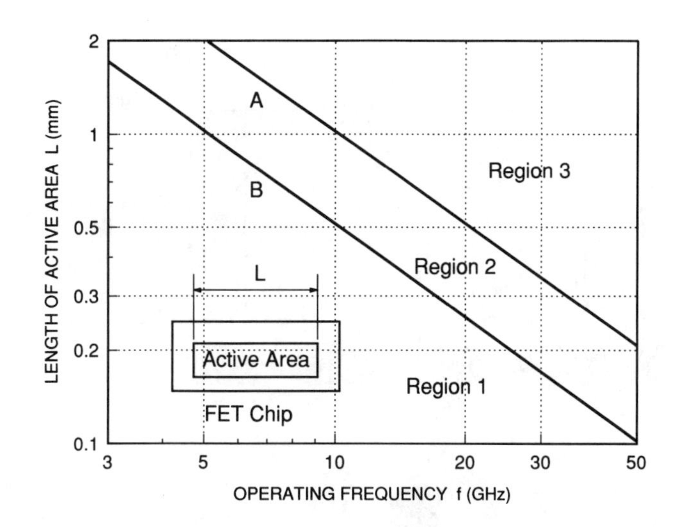


**gure 2.15** Relationship between the lateral length of the active region of an FET and the design frequency. Line A corresponds to $\lambda/8$ and line B to $\lambda/16$ of the frequency.

perate uniformly. In the preceding section the length of individual gate fingers as discussed. In a high-power FET it is not practical to create a power supply id for each finger. Instead, multiple bus lines carry signals from each power supply id to the individual fingers. Next, we discuss how to determine the proper number power pads, or the number of gate fingers per pad. Increasing the number of ids will reduce the phase differential in high-frequency signals reaching each nger, but it also increases the parasitic capacitance of each electrode. Reducing e number of pads will have the reverse effect. Consider the input signal phase fferential in an FET with the interdigital (comb-type gate) structure shown in gure 2.16 [24].

Let the total gatewidth of the FET chip be $w_g$, and the number of gate pads $N_{\text{pad}}$. (The number of drain pads is fundamentally equal to the number of gate ids. However, as Figure 2.16 shows, the drain electrodes are all linked together, aking a large bonding area, and do not necessarily constitute pads, so in reality is more common to determine the number of drain power supply wires from the ain current requirement or impedance matching.) Let us call the FETs grouped ound one gate pad an *FET cell unit*. If the total gatewidth of the FET cell unit $w_{gc} = w_g/N_{\text{pad}}$, gate-to-gate spacing is $L_{gg}$, and the number of gate fingers per ET cell is $n$, then the longest distance from the gate power supply to gate finger $L_p$ is represented by [25]

$$L_p = \frac{w_{gc}}{n} + \frac{(n-1)L_{gg}}{2} \tag{2.88}$$

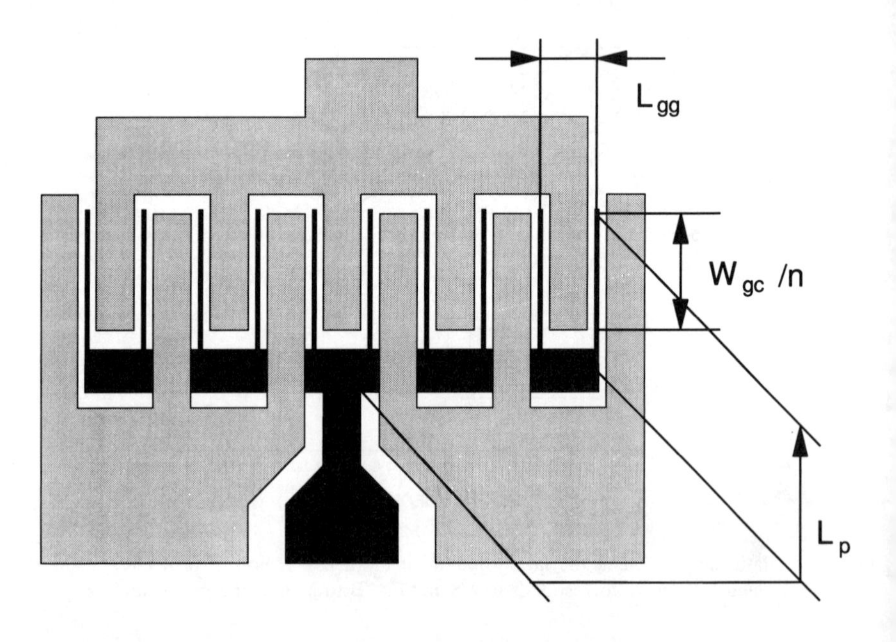


**Figure 2.16** Typical pattern layout of an interdigital FET and key dimensions.

Differentiating (2.88) with respect to $n$, we can determine the optimum numbe of $n$ for which $L_p$ is minimized. Thus,

$$\frac{\partial L_p}{\partial n} = -\frac{w_{gc}}{n^2} + \frac{L_{gg}}{2} = 0 \quad (2.8$$

and, hence,

$$n = \sqrt{\frac{2w_{gc}}{L_{gg}}} \quad (2.9$$

The maximum allowable value for the gate finger length $w_{gu}$, as discussed in th previous section, must be such that

$$w_{gu} > \frac{w_{gc}}{n} \quad (2.9$$

From this limiting condition, the number of gate pads must be

$$N_{\text{pad}} > \frac{w_g L_{gg}}{2 w_{gu}^2} \tag{2.92}$$

This gives the minimum number of gate pads. For example, with a total gatewidth of 10 mm, a maximum gate finger length of 100 $\mu$m, and a gate-to-gate spacing of 10 $\mu$m, then at least five gate pads are required. Gate-to-gate spacing is normally determined with respect to thermal resistance, which is discussed in detail in Chapter 5.

Pads also have electrostatic capacitance with respect to the ground, which affects the input and output ports in the form of a parallel parasitic capacitance. When this parasitic capacitance includes the fringing effect related to the length of the pad perimeter, it becomes non-negligible. Let the substrate thickness be $H$, electrode area be $S$, and electrode perimeter be $P$, with the dielectric constant of GaAs represented by $\epsilon (= \epsilon_0 \epsilon_r)$. By experiment, we know that the electrostatic capacitance $C$ of the electrode can be approximated by

$$C = 1.3 \frac{\epsilon S}{H} + 0.92 \frac{\epsilon P}{\log H} \tag{2.93}$$

In this expression, $S$ is expressed in square centimeters, $P$ and $H$ in centimeters, and $C$ in farads. The first term comes from the parallel plate capacitance of the pad, $C_0 = \epsilon S/H$, while the second term comes from the capacitance of a charged line at height $H$ above a ground plane, $C_1 \propto P/\log H$. The coefficients have been determined experimentally. The number of pads is thus a trade-off between the limit given by (2.92) and the need to minimize the electrostatic capacitance given by (2.93).

### 2.3.6 Pattern Layout

As mentioned, the high-power FET is constructed from a number of small FETs. When connecting a number of FET source, drain, and gate electrodes to power supply pads, it is impossible to avoid crossing some electrodes over others. This makes it necessary to develop crossover structures [16] for the electrodes. The first FET, designed by Fukuta et al. [4], was the mesh-source FET. Part of the source resembled a cutaway mesh, with many island-like drains and clover leaf-shaped gates connected through the insulating $SiO_2$ layer by crossover electrodes. This is derived from the mesh-emitter transistor, a high-power transistor form also devised by Fukuta. The structure with multiple basic FETs connected by crossover electrodes is common and inevitable in a high-power FET.

Another effective pattern for the high-power FET is the comb-type gate FET, or the interdigital source-drain FET, shown in Figure 2.16. The gate fingers are

all parallel and are connected by bus lines, which form the back of the comb. Th
gate bus line normally crosses over the source electrodes. Compared to the earlie
mesh-source construction, this has the advantage of lower parasitic inductance
and because the gate fingers are all oriented in the same direction, it is easier t
manufacture. Gate-source crossovers are of two types, with either the gate o
source uppermost, depending on the electrode material and the manufacturing pro
cess used.

In another technique, the air-bridge structure, the middle part of the gate
source crossover rises like a bridge above the chip. This structure is used in situation
where parasitic capacitance is a problem, particularly in high-frequency FETs. A
SEM photograph of such an FET is shown in Figure 2.17. A protective insulatin
film lies over the lower electrode, to guard against the potentially fatal breakdow
that would occur if the bridge collapsed and the electrodes came into contact wit
each other. The manufacturing processes involved are somewhat more complicate
than for the normal crossover, which passes through the insulating layer. The ai
bridge structure is normally used in FETs designed for X-band and higher fre
quencies.

The flip-chip structure [26], depicted in Figure 2.18, was designed to furthe
reduce parasitic capacitance. The source electrodes are covered with about 20 $\mu$
of plating, then three dimensionally separated from the gate and drain electrode
Gate and drain electrodes are beam leads, bonded directly to the respective packag
terminals. (In some cases, the gate and drain electrodes are also thickly plate

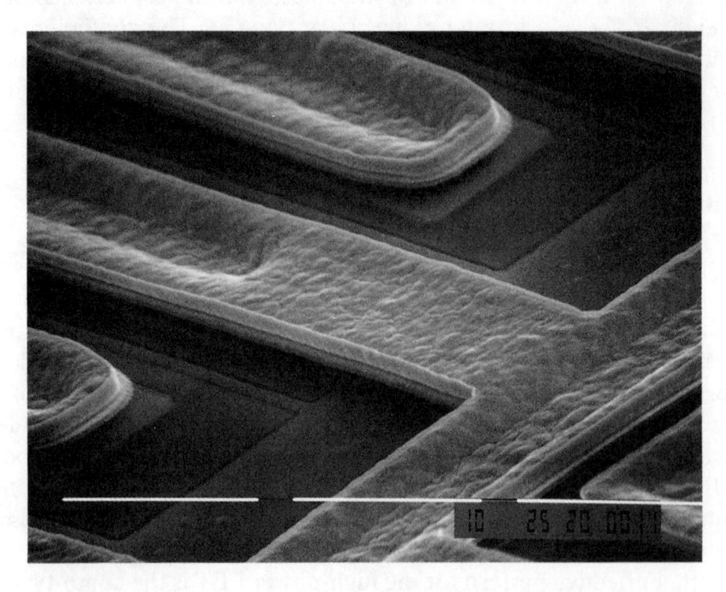

**Figure 2.17** Diagonal view of an air-bridge overlay structure.

nd the three terminal connections are made simply by placing the chip into its ackage.) Both parasitic capacitance and inductance are lower because no wires re used. The drawbacks are that visual inspection is not possible after assembly nd that extreme flatness is required in the package heat sink.

Several other types of pattern layouts are also in use. Other gate finger layouts, n addition to the interdigital (comb-type) layout, include those shown in Figure .19:

1. Multiple-stage comb-type structure (bypass-gate structure);
2. Fishbone construction;
3. Comb-type gates in facing rows.

ach pattern is used for its own particular advantages in layout efficiency or dis-ortion, efficiency, or frequency characteristics.

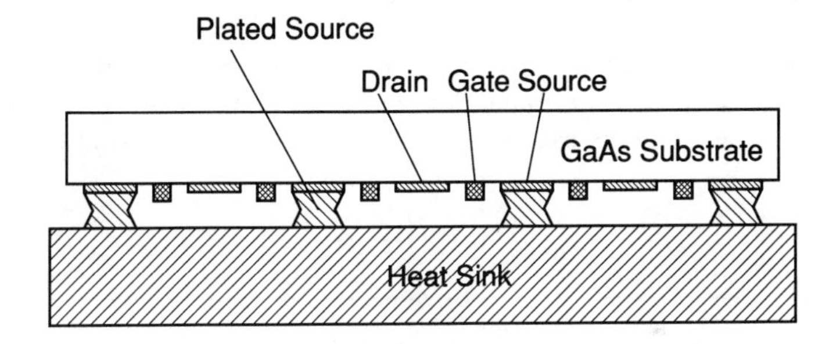


igure 2.18 Cross-sectional view of flip-chip structure.

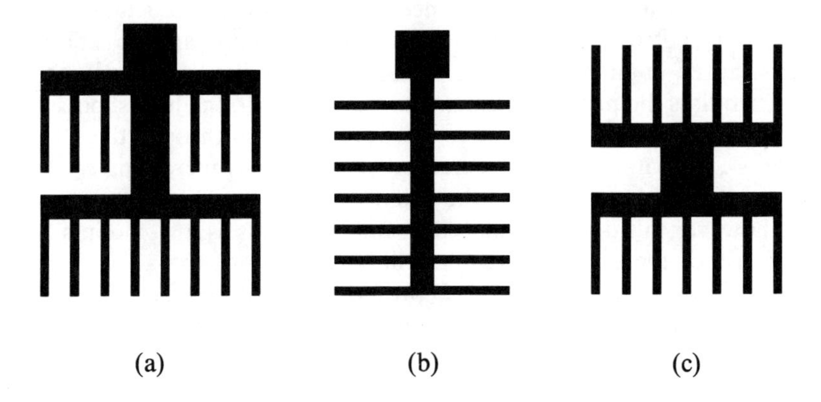


igure 2.19 Various FET gate pattern layouts: (a) multistage comb-type structure, (b) fishbone type structure, and (c) comb-type structure in facing rows.

In some instances, the gate power supply may use a *tournament* structure a shown in Figure 2.20. Because all distances from the gate pad to gate fingers ar equal, highly uniform operation results. In some large chips, dummy gate pads ar positioned at the edge of the chip as shown in Figure 2.21. Thus when several suc chips are placed in a row, these side gates can be wired together for uniforn operation and to suppress oscillations. For drain pads, a technique exists that allow wires to be positioned anywhere, depending on outside matching circuits. This typ of pad construction is effective in cases where drain wires must be more numerou than gate wires due to a high drain current. In interdigital FETs, drain pads an gate pads may be alternated rather than paired. This makes the shortest distanc from the gate pad to drain pad the same for all gate fingers and eliminates the los of efficiency due to signal phase differences.

### 2.3.7 Chip Backside Structure

Because high-power FETs are relatively large, parasitic elements can have a cor siderable effect on gain. The maximum available power gain of an FET, $G_{ama}$ including parasitic elements can be approximated by the following formula [6]:

$$G_{\text{amax}} \approx \frac{\left(\frac{f_T}{f}\right)^2}{4G_{ds}\left(R_i + R_s + R_g + \frac{\omega_T L_s}{2}\right) + 2\omega_T C_{gd}(R_i + R_s + 2R_g + \omega_T L_s)} \qquad (2.9$$

where $g_{ds}$ is the drain-source conductance, $L_s$ is the source inductance, and $C_{gd}$ the gate-drain feedback capacitance. Of these, $R_s$, $R_g$, and $L_s$ are parasitic element and $C_{gd}$ is caused in part by parasitic elements. Naturally, construction of a hig gain FET requires the parasitic elements to be reduced to the lowest possible leve In particular, $C_{gd}$ and $L_s$ contribute to a lowering of gain through the effect of th multiplier $\omega_T$, so that it is hard to increase $G_{\text{amax}}$ simply by increasing $\omega_T$. Feedbac capacitance $C_{gd}$ is dependent on the cross-sectional structure of the gate-drai channel, however this does not cause the same problems as it would in a low-nois FET, because of the comparatively low donor density of the active layer of th high-power FET. On the other hand, $L_s$ presents major problems in large-size hig power FETs. So as not to reduce $G_{\text{amax}}$, it follows from (2.94) that the conditio

$$\omega_T L_s \ll 2R_i \qquad (2.9$$

must be satisfied. This makes it desirable to reduce $L_s$. When source electrod

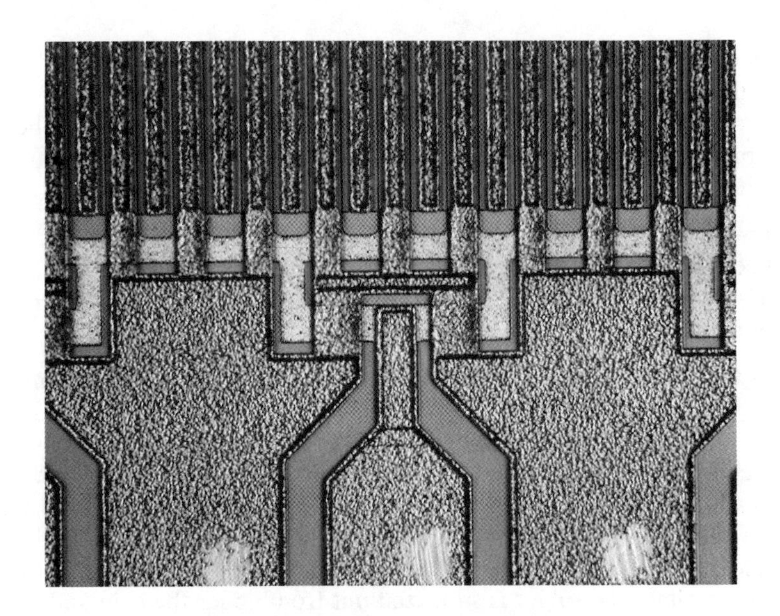

**igure 2.20** Overview of part of a tournament-type gate feed pattern.

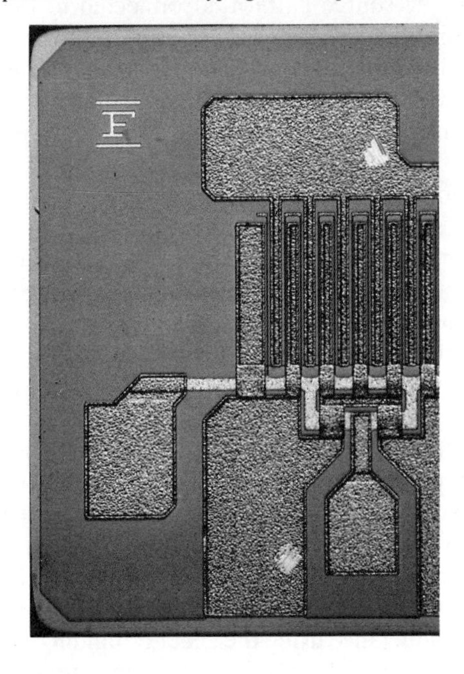

**gure 2.21** Photograph of a side gate pad.

are grounded by the use of wires, the position of the source electrode and the shape of the chip make it difficult to reduce $L_s$. If we assume values of $g_m$ = 10 mS, $C_{gs}$ = 1 pF, $R_i$ = 2.5$\Omega$, all per 1 mm of gatewidth, then $L_s$ must be far les than 50 pH, and the allowable limit of source inductance is reached at $L_s \approx$ 1 pH. Methods developed to reduce $L_s$ include:

1. Flip-chip;
2. Sheet grounding;
3. Via-holes;
4. Source island.

Flip-chip technology was discussed in the previous section. Sheet grounding is technique developed by Fukuta et al. [16] in which all source electrodes on th FET face are connected, and they reach to the edge of the chip. All source elec trodes are covered with an insulating film, except at the chip edge. A metal scree plate, of about the same thickness as the chip, is laid over the ground electrod plate on which the chip is mounted. When the chip is attached beside the scree with a solder material such as AuSn and AuSi, and pressure applied from the sid of the screen plate, the solder is squeezed out from under the chip and crawls up covering those areas of the source pattern that have not been protected by th insulating layer. Thus the source pattern is connected to the ground terminal o the back of the chip through the solder that has crept up around the screen plate Using this method, and assuming a typical number of source pads, source inductanc is approximately 50 pH per source pad. This is an order of magnitude less tha when normal wires are used. The complicated assembly process, however, is detracting point.

The most commonly used means of reducing source inductance is to emplo via-holes. The source electrodes are grounded through via-holes formed from th back of a thin FET chip. The via-hole concept was verified experimentally in 197 by Kohn [27]. In 1977 it was actually used in a 4-GHz FET application by D'Asar et al. [25, 28] with a linear gain improvement of some 2 dB over the ribbon-typ of source earthing technique. Figure 2.22 shows a cross-sectional drawing of th via-hole structure that was most widely used in high-power FETs in the early 1990 If the objective is to reduce source inductance, the via-hole connection should b immediately beneath the source electrode. To simplify the manufacturing proces it is often placed below the source electrode pad, as shown in the illustration. the GaAs chip is only about 25 $\mu$m thick, the via-hole connection will be only 3 to 40 $\mu$m in length. With advances in GaAs etching technology, it has becom possible to create via-hole connections in GaAs chips up to 100 $\mu$m thick. The chi in the drawing has a plated heat sink (PHS) for lower thermal resistance. Figu 2.23 compares source inductance using the sheet grounding and via-hole techniqu [29, 30]. Let us experiment with two FETs having a total gatewidth of 1.2 mn one constructed using sheet grounding (with a chip thickness of 70 to 80 $\mu$m) an

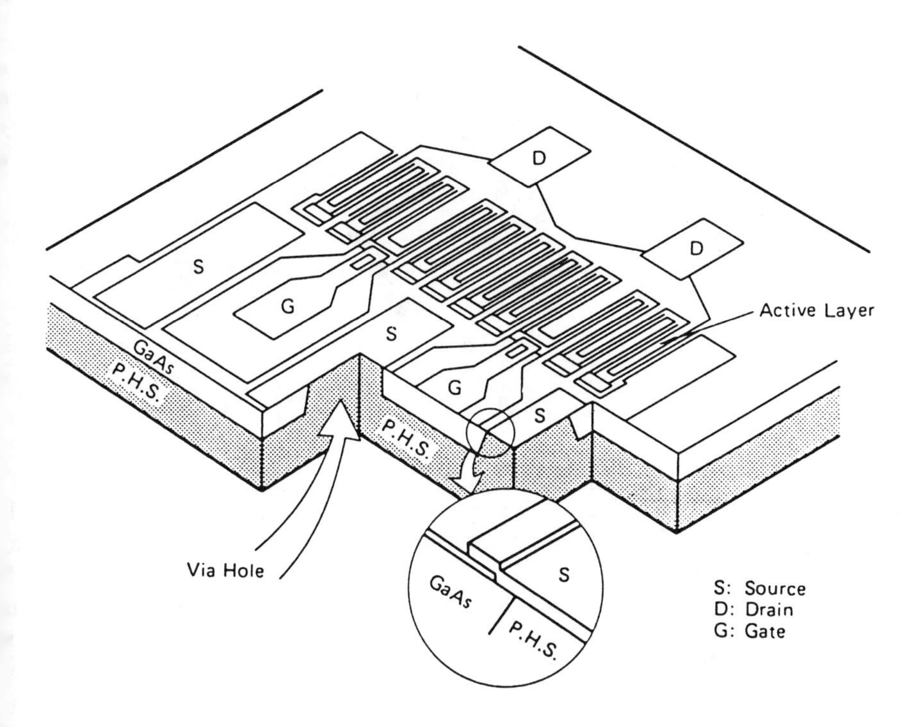


**igure 2.22** Cross-sectional drawing of a via-hole structure.

ne using via-hole connections, both with electrodes shorted by gold ribbon. Using he circuit shown in Figure 2.23, we will estimate inductance using measurements f the transmission parameter $S_{21}$. The $S_{21}$ and $L_s$ parameters are related by

$$S_{21} = \frac{2j\omega L_s}{Z_0} \tag{2.96}$$

leasurements show that source inductance using sheet grounding is about 40 pH, hile source inductance using via-holes is reduced by a factor of three and satisfies he target conditions in (2.95). Because via-hole technology allows the ground to e placed at any desired spot on the chip, it is widely used in MMIC applications.

A further advance in via-hole technology, in which each source electrode has via-hole connection and is constructed as an island, is called the *source-island tructure* [31]. A cross section is shown in Figure 2.24. This structure has advantages oth for lower inductance and for lower thermal resistance. It is used in millime- erwave FETs and has achieved outstanding results.

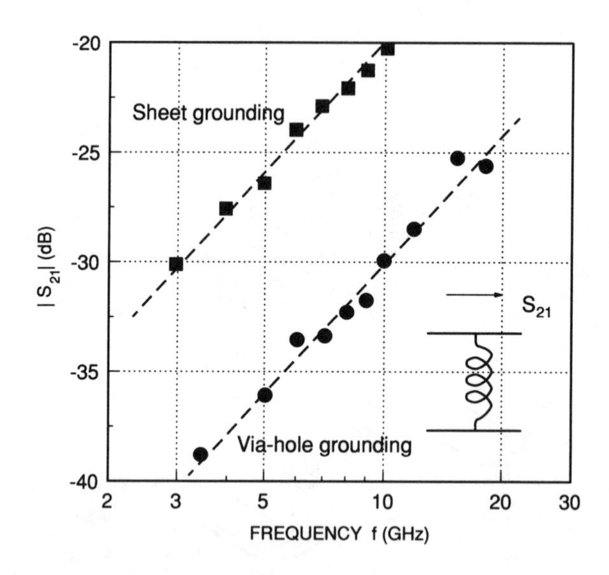


**Figure 2.23** Comparison of source inductance using sheet grounding and via-hole techniques. T gatewidth of the FET is 1.2 mm. Source inductance $L_s$ is obtained from the gradient $|S_{21}|$ with respect to frequency.

## 2.4 THERMAL PROPERTIES

The high-power FET, like any other high-power device, is constructed so as attain higher output power levels by consuming and dissipating more dc powe As long as the FET has a nonzero thermal resistance (see Chap. 5), this dissipat power in the form of Joule heat raises the temperature of the FET, thereby reduci the performance of the device (reduced electron mobility increases the series pa asitic resistance $R_s$ and decreases transconductance $g_m$) as well as accelerating th aging of the Schottky gate properties and ohmic contact resistances, thereby shor ening the life of the FET. In this section, we discuss the temperature dependen of electron drift velocity and field properties, and demonstrate that the dc ar microwave properties of the FET are also dependent on temperature.

Electrons in a GaAs FET collide repeatedly with donor impurities (impuri scattering) and with the lattice atoms (phonon scattering) as they move from sour to drain. In the area between the gate and drain where the field is stronges electrons undergo an intervalley transition from the lowest valley of the conductic band ($\Gamma$-point) to the upper valleys ($L$-point and $X$-point states). In the low valley, electrons have a small effective mass ($m^* = 0.068m_0$) and therefore hav high mobility. These "light" electrons are easily accelerated by the electric fiel Electrons in the upper valley, however, are "heavy," with a larger effective ma

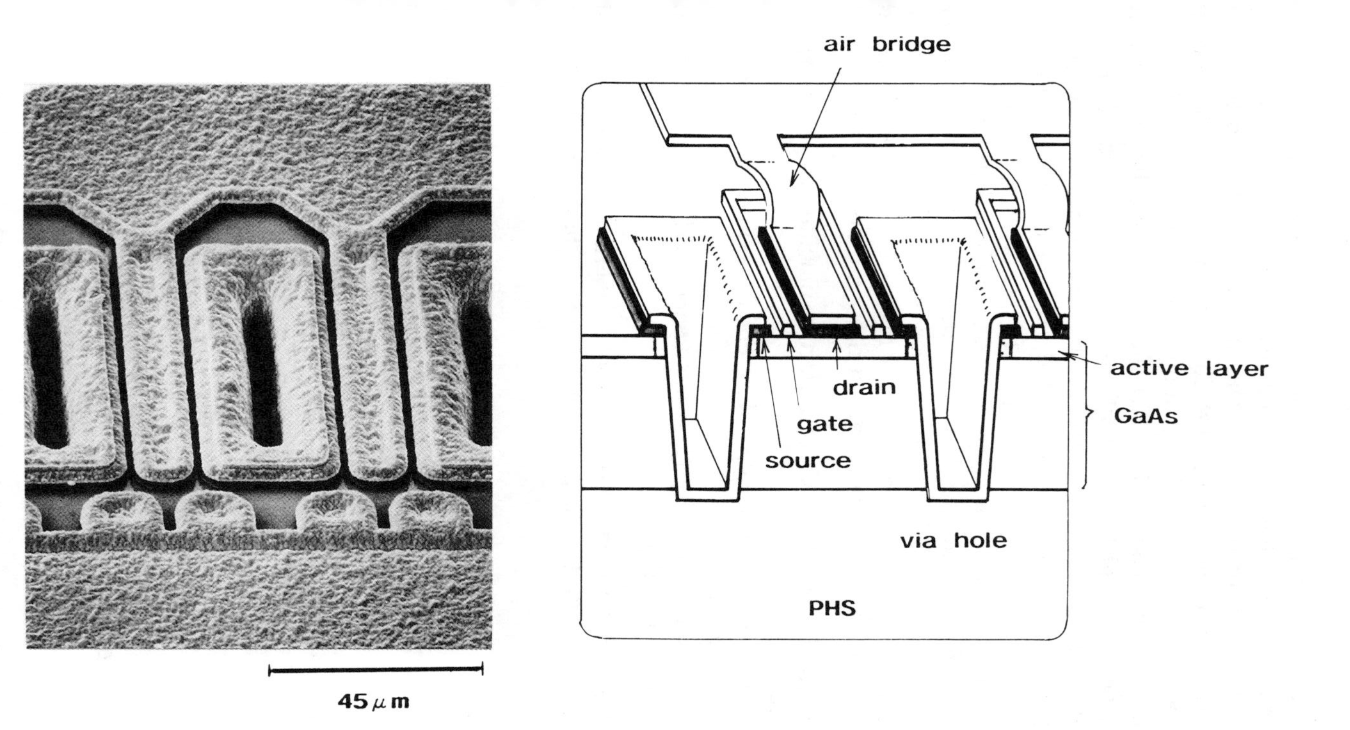


**Figure 2.24** SEM photograph and schematic cross section of a source-island via-hole structure. (Courtesy of Mitsubishi Electric Corporation)

($m^* = 1.2m_0$). Thus, in the relatively weak field of the source-gate region, almo all electrons are in the lower valley state, and they can accelerate without underg ing an intervalley transfer (or scattering). The intervalley transition that occu when the electrons traverse the gate-drain region is sufficiently complete that tl electron drift velocity reaches velocity saturation and therefore cannot increase.

Higher temperatures significantly affect the relationship between electrc velocity and electric field, which plays a vital role in the function of the Ga/ FET, and they adversely affect FET operation. The dependence of GaAs electrc velocity on temperature has been demonstrated experimentally using the Gur oscillation effect, as shown in Figure 2.25 [15]. The experiment used a Gunn devi with an epitaxial operating layer thickness of 28 $\mu$m and a carrier density of 1 $10^{15}$ cm$^{-3}$. It was thought that the velocity-field property would be less temperatu dependent than in a GaAs MESFET with a carrier density of $10^{17}$ cm$^{-3}$, howev no qualitative difference was discovered. From Figure 2.25 we can deduce tl following about the temperature dependence of the $v$-$E$ properties:

1. In low-field regions (less than 1 kV/cm), electron mobility is highly depende on temperature.
2. Threshold fields exhibiting negative differential mobility grow stronger temperature increases.
3. Saturated electron drift velocity $v_{sat}$ is less temperature-dependent than tl low-field electron mobility.

From these observations it can be inferred that the temperature dependence of tl parasitic resistance $R_s$ in the low-field source-gate region has a greater effect c

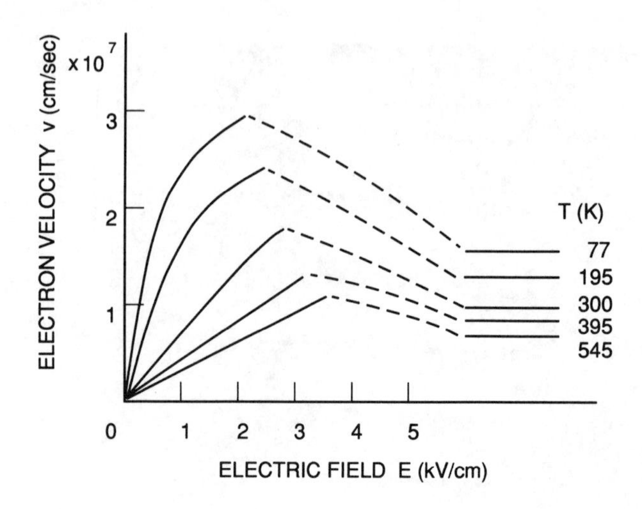


**Figure 2.25** Dependence of GaAs electron velocity on electric field and temperature [15].

he GaAs FET than the temperature dependence of saturated electron drift velocity n the higher fields of the FET intrinsic region. As the channel temperature rises, he saturated drain current $I_{dss}$ and transconductance $g_m$ decrease, and there is a light increase in pinch-off voltage $V_P$. There is virtually no change in the voltage t which the drain current becomes saturated, the knee voltage $V_k$. Intrinsic transconductance $g_m$ and measurable $g'_m$ are related by (2.67). As the temperature ncreases, the decrease in $v_{sat}$ causes $g_m$ to decrease. Furthermore, as $R_s$ increases ue to the decrease in low-field mobility, the combination of these effects causes $g'_m$ to decrease. Also, because $I_{dss}$ is the product of $V_P$ and $g_m$, it will also decrease t higher temperatures, reflecting the decrease in $v_{sat}$. These changes in dc properties re accompanied by a deterioration in high-frequency properties as well.

Experimentally obtained changes in output power, linear gain, and other FET parameters with respect to temperature change in an FET are shown in Figure .26. As the channel temperature increased from $-40°$ to $+60°$C, the linear gain decreased by 0.015 dB/°C, and the output power decreased by 0.008 dB/°C. Linear gain thus decreases by 1.5 dB over a 100°C increase in temperature. This underlines the importance of minimizing thermal resistance. Detailed models for calculating thermal resistance are given in Chapter 5.

## .5 MANUFACTURING

The steps used to manufacture an FET chip are called *wafer processes*. In this ection we will trace the flow of wafer processes, addressing the processes required o produce specific design elements and explaining how the finished results are evaluated. We will not get into the details of individual manufacturing processes.

### .5.1 Epitaxial Wafer Growth

Several methods are used to introduce donor impurities into a semi-insulating GaAs substrate in order to form the active layer:

1. Liquid phase epitaxy;
2. Vapor phase epitaxy;
3. Molecular beam epitaxy;
4. Metal-organic chemical vapor deposition;
5. Ion implantation.

Liquid phase epitaxy is now virtually obsolete. Vapor phase epitaxy is the most widely used method for crystal growth. The most commonly used source is $AsCl_3$/Ga/$H_2$. Hydrogen gas ($H_2$) is passed through liquid arsenic trichloride ($AsCl_3$) at room temperature, and a mixture of the two is carried into a reaction vessel. At the entrance to the vessel a gallium boat is placed that is heated to between 800°

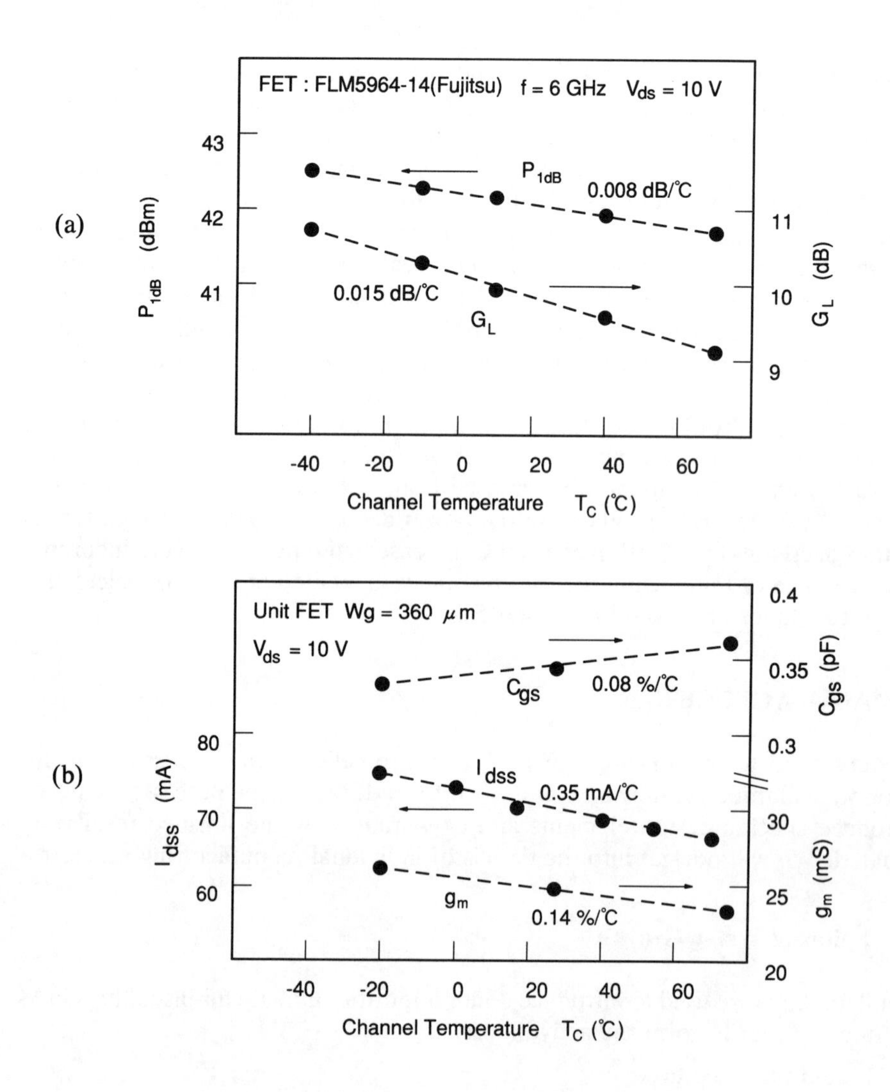


**Figure 2.26** (a) Variation of output power and gain with temperature. (b) Variation of $I_{dss}$, $g_m$, and C with temperature.

and 900°C. Here the $AsCl_3$ and $H_2$ react: The arsenic separates from the chlorine and combines with the vaporized gallium to form gallium arsenide (GaAs). A GaAs wafer at 700° to 800°C is placed in the flow below the gallium source. Epitaxial growth occurs as the wafer comes into contact with the GaAs vapor. Donor impurities are added by mixing dopant materials with the reagent gases, or by placing solid donor materials at controlled temperatures in the reaction vessel. With thi ethod, the carrier density profile can be lowered to the order of $1 \times 10^{17}$ cm$^{-3}$ ) $1 \times 10^{16}$ cm$^{-3}$ within a thickness of several hundred angstroms.

Metal-organic chemical vapor deposition (MOCVD) is one type of vapor hase epitaxy in which organic metals are used as sources. Gallium sources used iclude trimethyl gallium ($(CH_3)_3Ga$) and the usual arsenic source is arsine gas $AsH_3$). At room temperature, arsine is a gas and trimethyl gallium a liquid, which mixed with hydrogen gas as a carrier. Within the reaction vessel, the GaAs wafer placed on a carbon susceptor and heated by high-frequency induction to promote pitaxial growth. Donor impurities are added in the form of organic gases. The IOCVD method is able to create extremely steep changes in carrier density profile s well as excellent heterojunctions, and it is widely used in manufacturing het-rojunction devices and lasers. This method ranks alongside the molecular beam pitaxy (MBE) method in its ability to form steep interface profiles and outperforms IBE in creating epilayers containing phosphorus, such as InP.

MBE heats and vaporizes a substance in a vacuum, then causes it to accu-nulate on a semiconductor substrate by means of a molecular beam. With GaAs, gallium arsenide substrate is placed with effusion cells of gallium and arsenic in n ultrahigh vacuum. When heated, the gallium and arsenic break down into nolecular form inside the cells. In the vacuum, the mean free path of the molecules s of the order of 100 km so that the vaporized molecules travel straight toward he substrate. Donor impurities are added by molecular beams from cells containing he basic impurity substances. The growth rate and impurity density can be con-rolled by opening and closing shutters between the cells and the wafer, allowing he manufacturer to make steep changes in the doping profile and heterojunction tructure at the atomic level.

The epitaxial growth technologies discussed here are limited, however, to the nanufacture of uniform surface layers. To form specific two-dimensional surface reas of a desired depth and density, it is necessary to use ion implantation, in vhich electrically accelerated dopant ions are driven into a semi-insulating sub-trate. The level of accelerator energy can be varied between 30 and 300 keV in order to regulate the depth of the active area. The ion dose is normally between $0^{12}$ and $10^{14}$ atoms/cm$^2$. Common $n$-type dopants are silicon (Si) and selenium Se), and $p$-type dopants include beryllium (Be). The dopant ions strike the crystal tructure and in the resulting inelastic scattering process, are trapped (implanted) vithin it. Thus the profile $N(x)$ can be represented by a Gaussian distribution:

$$N(x) = N_0 \exp\left[-\frac{(x - R_p)^2}{2\sigma^2}\right] \tag{2.97}$$

vhere the peak value of the projection range $R_p$ and the width of the distribution

$\sigma$ are determined by the implantation energy and materials. The problem with ion implantation is that the donor (or acceptor) layers are not formed simply by implan tation, but must be activated by a heating process called *annealing*, which causes the dopant ions to rediffuse, lengthening the tails of the Gaussian distribution Annealing temperatures are high, around 850°C, and must be carefully matched to the requirements of other processes. Ion implantation has become an indispen sable part of integrated circuit manufacture. Although heterojunction structure and other aspects of FET technology are developing and changing rapidly, MBE and MOCVD technology represent the present mainstream of epitaxial wafe growth technologies. Ion implantation will be used together with MBE and MOCVD to reduce parasitic elements.

The most commonly used criteria for evaluating wafers grown by these meth ods are:

1. Sheet resistance;
2. Doping profile ($C - V$ measurement);
3. Mobility (Hall measurement).

Sheet resistance measurement probably requires no special explanation. Doping profile measurement is used to infer the density of the activated carrier (electrons in the case of $n$-type GaAs) with respect to depth. The usual method is to create a Schottky diode pattern, like the one in Figure 2.27, on the wafer surface. Th voltage applied to the Schottky terminal is then used to vary the depletion laye capacitance in the Schottky junction, and the carrier density is then computed a a function of the depth. If we assume that the Schottky junction is a one-sided abrupt junction, then the carrier density versus depth profile $N(x)$ (the density of the ionized donors) can be computed as a function of depletion layer capacitance $C$ by the formula

$$N(x) = \frac{2}{q\epsilon A^2}\left[\frac{d}{dV}\left(\frac{1}{C^2}\right)\right]^{-1} = -\frac{C^2}{q\epsilon A^2}\left(\frac{dC}{dV}\right)^{-1} \quad (2.98)$$

where $\epsilon$ is the dielectric constant of GaAs and $A$ is the surface area of the Schottk junction. The depth $x$ below the surface of the Schottky junction is

$$x = \frac{\epsilon A}{C} \quad (2.99)$$

Even with no voltage applied to the Schottky junction, its built-in potential create a depletion layer, which prevents direct profile measurement near the surface Conversely, applying a positive voltage to the Schottky junction causes forwar

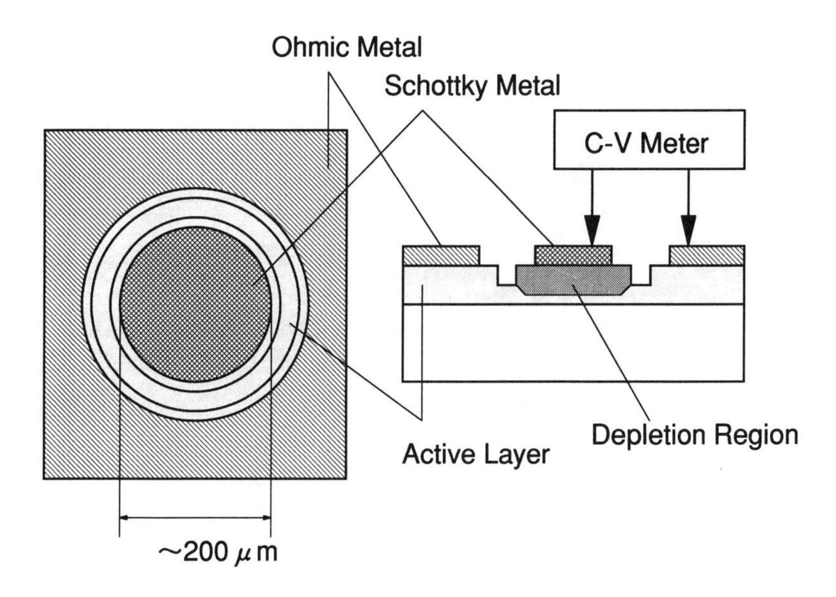


**Figure 2.27** Schematic view of Schottky diode and setup for C-V measurement.

current to flow as the junction approaches breakdown, and measurements of carrier density depth $N(x)$ become meaningless. To be effective, therefore, measurements must monitor conductance as well as capacitance. In the pattern shown in Figure 2.27, for example, $N(x)$ appears reliable at conductance values of $10^{-4}$ S and lower, while at $10^{-3}$ S or higher, $N(x)$ readings should be rejected. For deeper ranges, useful data on carrier density can be obtained by step-etching the wafer to several different depths, then placing $C - V$ measurement patterns at each respective depth. Figure 2.28 is an example of a $C - V$ profile obtained in this way.

Another important aspect of the wafer epilayer is mobility, both Hall mobility and drift mobility. Hall mobility quantifies the ease with which the carrier can be deflected by a magnetic field, and drift mobility quantifies the ease with which the carrier can be accelerated by an electric field. The two values are normally not equal, but related. In the semiconductor example discussed here, they are nearly the same value. Hall mobility is measured using the Hall bar pictured in Figure 2.29. An electric field is created along the $x$ axis with current $J_x$, and a magnetic field $B_z$ is applied perpendicular to the substrate (i.e., on the $z$ axis). The resulting Lorentz force imparts an acceleration to electrons in the direction of the $y$ axis. Because there is no current along the $y$ axis, a field, called the Hall field $E_y$, is created by the deflection of electrons in that direction. The Hall field represents

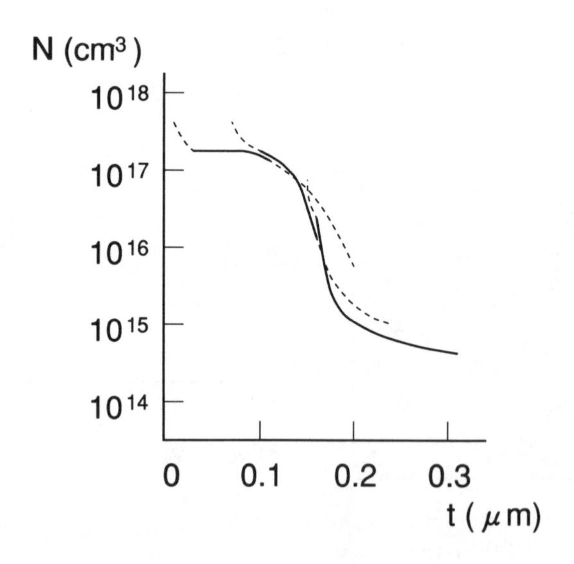


**Figure 2.28** Typical carrier profile of an epitaxial layer obtained by step-etching.

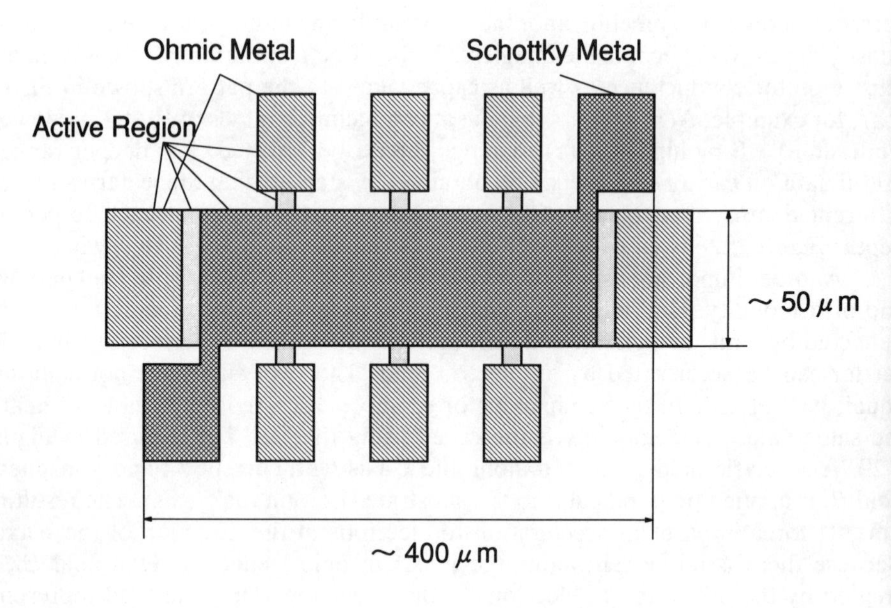


**Figure 2.29** Schematic view of a Hall bar with Schottky junction metal.

e force required to neutralize the Lorentz force on an electron. The relationship tween these quantities is expressed by

$$E_y = R_H J_x B_z \tag{2.100}$$

here $R_H$ is the Hall coefficient. Once the Hall coefficient is found by measuring e field $E_y$, the Hall mobility $\mu_H$ can be determined by

$$\mu_H = |R_H \sigma| \tag{2.101}$$

here $\sigma$ represents conductance. The relation of Hall mobility to layer depth can found by applying a voltage to the Schottky electrode on the surface of the Hall r and relating the result to the measurement of capacitance.

Drift mobility directly affects FET properties. Placing ohmic electrodes separated by a distance $L$ on an active layer of width $W$, we apply a voltage $V$ between e electrodes. Assume that the field created by $V$ is sufficiently weak (−2 kV/n) such that the electron velocity does not reach saturation. The current $I$ under ese conditions can be written

$$I = \frac{q\mu_d N_d WaV}{L} \tag{2.102}$$

here $N$ is the donor density and $a$ the thickness of the donor layer. From this we n determine the average drift mobility $\mu_d$ from the formula

$$\mu_d = \frac{IL}{qN_d WaV} \tag{2.103}$$

## 5.2 Flow of the Wafer Manufacturing Process

he processes used to manufacture a FET are different for each type of FET ructure. They can be broadly divided into those processes that form the Schottky ate after the source and drain electrodes, and those that form the gate first. Figure 30 shows an outline of the process flow in FET manufacturing.

## .5.3 Isolation

nless ion implantation is used, the active layers are actually connected, stratified yers within the wafer. This makes it necessary to eliminate the unwanted layers

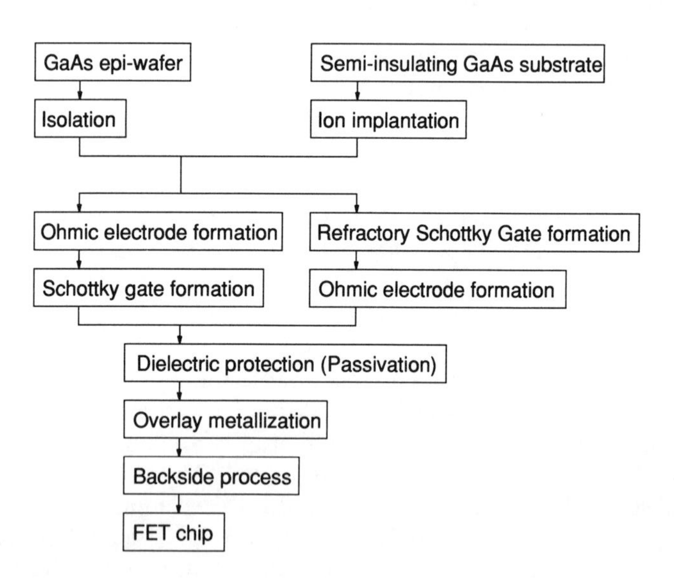


**Figure 2.30** Flow of the FET manufacturing process.

from the active region of the FET. This isolation can be achieved in two ways. The first is by chemically etching away the unwanted active layers using a technique called *mesa isolation*, and the second is by using ion implantation to render the dopant inactive, called *implantation isolation*.

In mesa isolation, the desired active area is protected by a resist layer, and the GaAs is etched by chemicals (e.g., $H_2SO_4/H_2O_2/H_2O$) to remove the remaining active layers. The term *mesa* is derived from the way the remaining active layers are left standing like isolated tablelands. The taper of the mesa edge differs according to the crystal orientation and etchant. This method has advantages in that there is complete physical separation of the elements, and the process is simple. Its disadvantage is that the flatness of the wafer is spoiled, and fine pattern details such as gates are easily cut off by the mesa edge. Careful attention must be given to the orientation of the gates with respect to the crystal, and to the choice and handling of the etchants.

In implantation isolation, the desired portion of the active layer is protected by a resist or insulating film, and then ionized boron, oxygen, or other elements are implanted. The implanted ions break the bonds of the GaAs crystal lattice and act as deep traps, effectively deactivating the dopant. This method has the advantage that the wafer retains its surface flatness, especially important in FETs with fine gate structures. Its disadvantage is that additional high-temperature processes at around 800°C cannot be performed on the wafer. There is no difference in isolation between the two methods.

To verify that complete isolation has occurred, a pattern such as the one shown in Figure 2.31(a) can be used to test for leakage current and breakdown voltage. As explained in the section on drain breakdown current, breakdown voltage in the test pattern must be increased by placing the active layer outside two rows of facing electrodes, so that the electric fields do not concentrate at the electrode edges. Actual FETs such as that shown in Figure 2.31(b), where the source and drain electrodes are close to each other, also locate the active area outside the source and drain electrodes in order to increase the breakdown voltage.

### 2.5.4 Ohmic Electrode Formation

When a contact is formed between a metal and a semiconductor, the result is a Schottky junction incorporating a potential barrier. The height of the barrier $\phi_{bi}$ is determined by the semiconductor and the metal, and for GaAs with several different metals it is in the range of 0.5 to 0.9 eV. Two types of current components can cross the barrier. The first is called *thermionic emission*, in which heat distribution includes electrons with energy in excess of the barrier height, which flows over the Schottky barrier. The other is called *tunneling field emission*, in which a tunnel effect causes electrons to pass through the barrier. There is also an intermediate phenomenon known as *thermionic field emission*. Referring to Figure 2.32, the thinner the barrier, the greater the proportion of the current that crosses it

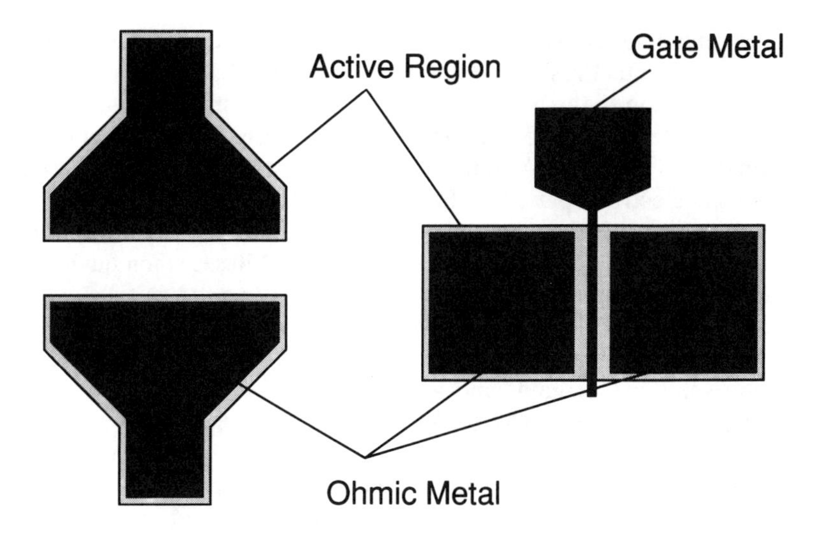


**Figure 2.31** Test pattern for the measurement of leakage current and breakdown voltage of the substrate and typical pattern configuration of an FET.

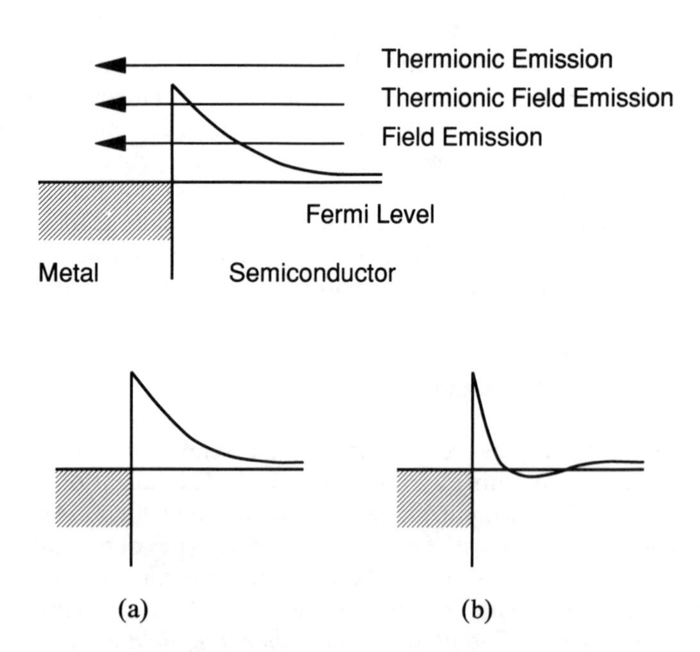


**Figure 2.32** Energy-band diagram of a metal-semiconductor contact: (a) Schottky contact and (b) ohmi contact.

that is due to tunneling, until the Schottky barrier becomes essentially undetectable At this point the contact resistance due to the junction barrier has become insignificant in comparison to the bulk resistance of the GaAs itself, and is called *ohmi resistance*. The material most commonly used to create an ohmic contact is a metalli composition of AuGe/Ni/Au. A source-drain resist pattern is applied to an *n*-GaA surface, and the electrode pattern is formed by AuGe/Ni/Au evaporation and lift off. The ohmic contact is then created by an alloying process in which the materia is heated rapidly to around 450°C for between 30 and 120 sec, then quickly cooled

Qualitatively, the alloying process is thought to operate as follows. The gallium is drawn out of the GaAs layer by the gold with which it has a strong bondin affinity. The gallium is replaced by germanium, which creates an $n^+$ layer. Nicke increases the solubility of gold, and also stands between the GaAs and AuGe increasing the tendency of GaAs to flow toward the AuGe, and promoting th metallurgical reaction. Figure 2.32 represents an energy diagram of the ohmi contact. Despite the use of the name ohmic contact, a physical remnant of th barrier is responsible for causing contact resistance of the order of $10^{-5}$ to $10^{-}$ $\Omega/cm^2$. Recently, another method of lowering contact resistance has been developed, called the *nonalloyed ohmic electrode*, but it is not yet in wide use.

Contact resistance is measured by using ohmic electrode patterns at variou intervals on the active area, as shown in Figure 2.33. If we let the space betwee

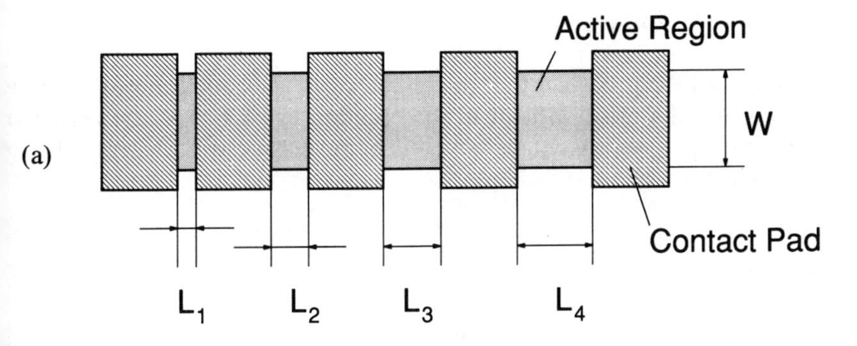


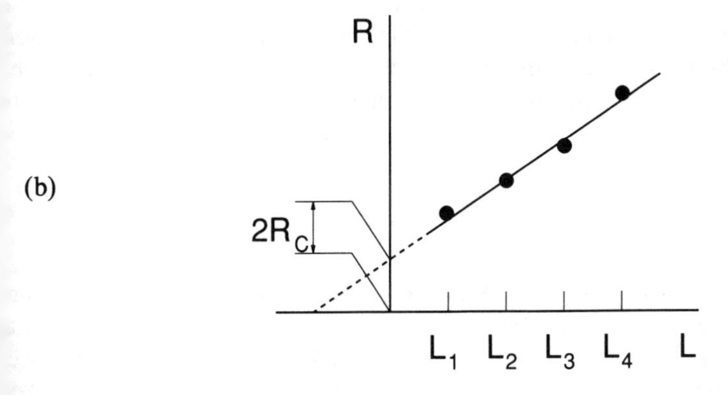


**igure 2.33** (a) Basic test pattern for the measurement of contact resistance and sheet resistance. (b) Plot of the resistance measured as a function of the separation between the electrodes. The gradient of the slope gives the sheet resistance, and the intercept point at $L = 0$ is twice the value of the contact resistance.

lectrodes be $L$, electrode width be $W$, substrate sheet resistance $R_s$, and contact esistance $R_c$, then the resistance between electrodes $R$ can be expressed as

$$R = 2R_c + \frac{R_s L}{W} \tag{2.104}$$

y obtaining data for $R$ over several values of $L$, then extrapolating to the case f $L = 0$, we can calculate a value for the term $R_c$, while $R_s$ is given by the slope f the function. The measurement process must include the correction of the contact esistance of the probe, and accurate values for $L$.

### 2.5.5 Gate Electrode Formation

Gate formation, and the processes that immediately precede and follow it, are the most important and the most difficult in the creation of FET. Specific details of the gate formation process cannot be covered here, but we will present a list of the important factors.

*Gate Orientation*

FET characteristics can vary depending on the direction of the gate fingers. Reces contours, which directly affect breakdown voltage in the high-power FET, ar determined by the orientation of the gate and by the GaAs etchant used. Becaus electron transfer speed varies somewhat according to gate direction, it can als cause differences in transconductance and drain current near pinch-off. Further more, the orientation also effects the piezoelectric polarization charge created b the piezo effect at the edge of the gate finger, which affects saturated drain curren and transconductance.

*Gate Metal Selection*

GaAs Schottky barriers can be made from many different metals, but most ar unsuitable for use as a gate metal because of poor bonding with GaAs, therma instability, or workability. Aluminum gates have long been used in high-powe FETs if the gate length is longer than 0.5 μm. But aluminum has become less we suited to the demands of a technology moving toward shorter gates. Once the met has passed through heat treatment and conductivity testing, recrystallization cause small grains to develop. The grain size is of the order of 0.1 μm, large enough t begin to cause reliability problems in short gate structures. For this reason, Schottk gates are now being made of more highly reliable refractory metals, such as titaniu (Ti), tungsten-silicide (WSi), and titanium-tungsten (TiW). In microwave FET gold is being used in the upper layer of multilayer mushroom-gate or T-shape gate structures because of its resistance to chemical etching agents and the nee for lower gate resistance. A FET using the low-resistance refractory metal LaB was recently announced.

*Gate Patterning*

Gate fingers are the finest part of the FET pattern. Gates less than 0.5 μm lon are too small to be precisely defined by exposure to visible light wavelengths, s hat electron beams, ion beams, and X rays are used instead. For T-shaped gate ormation, a multilayer resist structure is usually employed. Such a structure is a nultilayer coating of resist with different sensitivities. With such finely detailed ;ates, special care must be taken from the time the resist layer is laid down to reate the gate pattern until the metal is deposited. If a positive resist layer is used, t is possible that remnants of the resist material may remain in the grooves of the inger pattern, in which case they are difficult to remove. If the metal is deposited mmediately, good Schottky properties will not be obtained. Instead, the GaAs nust be chemically etched in order to clean the GaAs surface where the metal is o be applied. A recess structure is effective for this purpose, and improves the ET characteristics.

### *reprocessing Before Gate Formation*

The Schottky junction area is the most important part of the FET. Theoretically, he height of the Schottky barrier depends on the gate metal. However, it is lmost always constant, usually about 0.7 eV, regardless of the type of gate metal ised. This appears to be the case because of the presence of layers of either xidized gallium or arsenic between the gate electrode and GaAs junction. In rder to prevent the formation of unwanted inclusions, proper preprocessing is equired.

### *Gate Formation Methods*

Gate metal is normally applied by one of two methods: either the evaporation nethod or the sputtering method. Pattern formation is by either the lift-off nethod or by over-gate patterning, a process in which everything except the gate s etched. The most common combinations are (1) vapor deposition with lift-off and 2) sputter deposition with overgate patterning. Whenever the gates are formed y sputtering, sometimes the metal is apt to get into the recesses, resulting in n increased gate length. Furthermore, when using sputtering with GaAs, it is ecessary to use a heat treatment process in order to eliminate any sputtering amage.

The Schottky junction properties of a gate must be tested and evaluated with espect to both forward and reverse current and voltage. The $C$-$V$ method is useful or measuring junction capacitance and obtaining information about the junction arrier. The gate evaluation process uses a diode similar to the one shown in Figure .27. The current flowing across a Schottky barrier in GaAs is due to thermionic mission. When a forward voltage $V$ is applied to a Schottky junction, the forward urrent density $J$ is

$$J = A^*T^2 \exp\left(-\frac{qV_\phi}{kT}\right)\left[\exp\left(\frac{qV}{kT}\right) - 1\right]$$
$$= J_s \left[\exp\left(\frac{qV}{kT}\right) - 1\right] \qquad (2.105)$$
$$\approx J_s \exp\left(\frac{qV}{kT}\right), \qquad \text{for } V \gg \frac{kT}{q}$$

where $V_\phi$ is the built-in potential of the barrier, $A^*$ is the effective Richardson constant, $k$ is Boltzmann's constant, and $J_s$ is the forward current density extrapolated to $V = 0$. Practical Schottky junctions are described by

$$J = J_s \exp\left(\frac{qV}{nkT}\right) \qquad (2.106)$$

rather than (2.105) where $n$ is called an *ideality factor*, representing the divergence from the ideal Schottky junction ($n = 1$). By differentiating (2.106), we get

$$n = \left(\frac{q}{kT}\right) \frac{\partial v}{\partial (\ln J)} \qquad (2.107)$$

With $J_s$ computed from the measured data, $V_\phi$ can be computed from (2.105) as follows:

$$V_\phi = \left(\frac{kT}{q}\right) \ln\left(\frac{A^*T^2}{J_s}\right) \qquad (2.108)$$

The effective Richardson constant for thermionic emission is [19]

$$A^* = \frac{4\pi q m^* k^3}{h^2} \qquad (2.109)$$

where, for free electrons ($m^* = m_0$), $A = 4\pi qmk^2/h^3 = 120$ A/cm²/K². The effective electron mass in GaAs in a low field is $m^* = 0.068m_0$, so that

$$A^* = \frac{m^*}{m_0} A = 8.2 \text{ A/cm}^2/\text{K}^2 \qquad (2.110)$$

he $C$-$V$ method can be used to calculate $V_\phi$. The junction capacitance $C$ of a chottky diode created in an active area of uniform donor density $N$ can be xpressed as a function of the applied voltage $V$ by

$$C = \sqrt{\frac{q\epsilon N}{2\left(V_\phi - V - \frac{kT}{q}\right)}} \tag{2.111}$$

Vith the assumption that $V \gg kT/q$, the equation becomes

$$\frac{1}{C^2} = \frac{2(V_\phi - V)}{q\epsilon N} \tag{2.112}$$

When this is differentiated with respect to $V$, it becomes (2.98).] This is the same s taking the value at $V = 0$ on a plot of $1/C^2$ with respect to $V$. This has the dvantage of not being dependent on junction area, but it can only be applied vhere the doping density is uniform.

### .5.6 Protective Layers

After the source, drain, and gate electrodes are formed, a protective coating (pas-ivation) is applied to the GaAs channel area. In the high-power FET, this pro-ective layer is important because it maintains reliability by controlling metal migra-ion. The main role of the passivation film is to avoid a reaction between GaAs nd air (oxygen and humidity). The material is normally $SiO_2$ or $Si_3N_4$, and is deposited by sputtering, evaporation, or chemical vapor deposition. Concerns that must be kept in mind at the design stage include the following:

1. *Endurance:* The protective material should be able to withstand high humidity and high temperatures. The quality of the protection film should be controlled by the deposition process.
2. *Stress:* The protective materials, $SiO_2$ and $Si_3N_4$, have different coefficients of thermal expansion from that of GaAs, so that the thermal history of the chip creates stress between the protective film and the GaAs. This stress leads to the creation of a piezoelectric polarization charge (piezo effect), which is a serious problem in digital ICs.
3. *Parasitic capacitance:* When an insulating film is applied, it also has the unintended effect of creating a capacitor between the electrodes on either side. The feedback capacitance between the gate and drain is especially likely to increase when the film is applied.

4. *Surface-state level:* Normally, in the interface between the GaAs and th protective film there is a random energy level called the *surface-state level* considered to be due to remnants at the interface.

### 2.5.7 Overlay Wiring

Because the high-power FET is constructed with several bonding pads, each o which connects a number of FET units, some crossing of electrodes is unavoidable Such crossovers may be constructed above insulating layers, or by building ai bridges in which the upper electrode is literally a bridge over the lower one. I the interdigital FET, the most common type of high-power FET, gate and sourc crossovers are the most frequently used. The source is normally in the uppe position, because of the heavy plating required to carry the drain current. Th crossover is normally constructed from a metal-insulator-metal (MIM) capacito between two metals. The inner insulator layer is selected to provide the capacito with sufficient breakdown voltage. Care must be taken to ensure that there are n pinholes or cracks in the insulating material. In the air bridge, the structure of th upper electrode is critical. A multilayered metal structure of Ti/(Pt)/Au/plated A is normally used for capacitance and strength. To ensure reliability, the bridg should be no more than 50 to 70 $\mu$m long. There should also be sufficient insulatio above the lower electrode to prevent chip failure in the event that the bridge break down. For reliability, the current density in the drain and source electrodes shoul be no more than $10^5$ A/cm$^2$.

### 2.5.8 Backside Processing and Via-Hole Connections

Once the upper side of the wafer is completely formed, the backside processe begin. The wafer must be thinned to the specified thickness, 20 to 80 $\mu$m in th case of GaAs. This is normally done by mechanical processes. The next step is t create the via-hole openings as required by the pattern. Both dry and wet etchin methods can be used for this process. For thin wafers, wet etching alone is use but for thicker wafers dry etching or a combination of the two processes provide better uniformity and control. The shape of the hole is determined by the proces used, a straight, sharp-edged hole from the dry etching method, and a rounde concave depression from the wet method. The etching method to be used, therefor is a consideration at the design stage. The next step is metal deposition on th backside surface. If via-holes have been created by the dry etching method, the the metal may have to be applied by sputtering instead of evaporation. Then, layer of gold plating is applied to reduce thermal resistance in the high-power FE by forming a plated heat sink. Figure 2.34 shows a cross section of a chip with via-hole structure. Because intrusion of solder into the via-hole area during moun

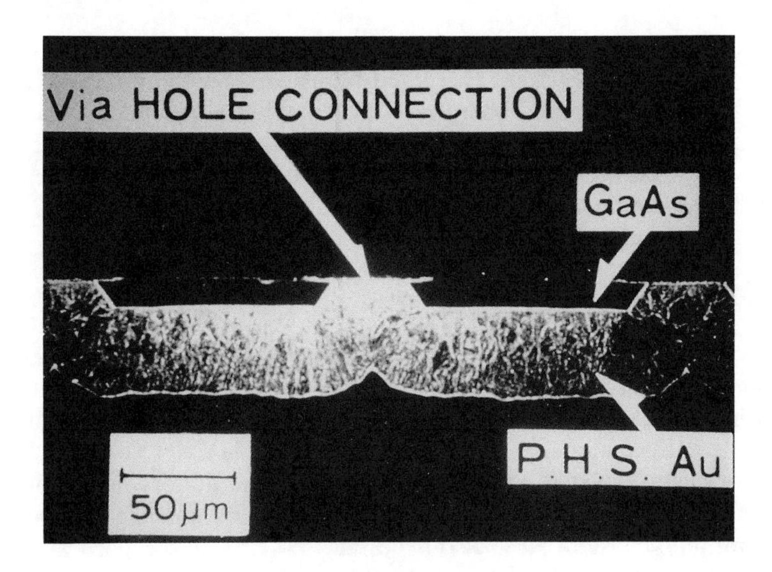


gure 2.34 SEM photograph of an FET chip with a via-hole.

g can cause chips to crack, it is safer to first fill the holes with a soft metal such gold. Thus via-hole and PHS technologies are well-suited for use together.

## 6 EVALUATION

### 6.1 Evaluation of dc Characteristics

easurement of the dc characteristics is fundamental to the evaluation of a FET. he most widely used instrument for this purpose is the curve tracer. There are so computer-controlled measurement systems with power supplies possessing the ırrent and voltage meters for each individual FET terminal. The most important spects to be aware of in high-power FET evaluation are oscillations, heat, and easurement cables.

*scillations*

igh-power FETs are prone to negative resistance areas on the output side where scillation may occur. In some wafers where grounding is difficult as a result of e fabrication process used, the drain current and voltage can be hard to measure roperly. In this case, it is a good idea to create FET monitor cell patterns, which ın be measured as an intermediate check step during processing. The high imped-

ance of small FETs makes oscillation unlikely. For the evaluation of complet
FET chips, oscillation must be prevented by using magnetic cores or capacito
Coaxial cable should be used with the shield completely grounded. Alternativel
the chip unit may be placed into a high-frequency measurement circuit with bi
feed circuits (or bias tees) to suppress oscillation. The high-frequency circuits a
normally terminated with a proper load such as 50$\Omega$.

*Heat*

Because current generates heat, the characteristics of the FET chip may chan
during the process of measurement. To measure under the conditions that the FE
are used in, it is a good idea to build a test fixture with radiation fins, fan cooli
and other heat dissipation features, and to test in a stable state. To test in conditio
that do not cause elevated channel temperatures, use pulse voltage (or curren
sources. Care must be taken with pulse intervals when measuring responses th
have a long time constant. It is also necessary to verify that capacitors and oth
oscillation preventing devices do not interfere with the measurement.

*Cable*

When dealing with FETs that may on occasion carry several amperes of curren
it is necessary to consider cable resistance. Sensing terminals for voltage monitori
should be placed as close to the FET as possible. The best cable is triaxial wi
three wires for power, sensor, and ground. For measurements of FETs in the waf
stage, measure the contact resistance of the probes first, then compensate for th
during actual measurements.

Measurement techniques need not be discussed in detail here, but the fi
measurement item should be the gate breakdown voltage, an indicator of the lev
of continuously sustainable test voltage, because this provides information f
avoiding unintentional damage during testing. Although not considered a tr
operating characteristic, FET light emission, already discussed in a previous sectio
provides a means of confirming uniformity of operation. Another method, call
*electron-beam-induced current* (EBIC), can be used for a more detailed observati
of the depletion layer. A photograph of the EBIC process is shown in Figure 2.3

### 2.6.2 Output Power Measurement

FET input and output power characteristics can be measured using a system su
as the one illustrated in Figure 2.36. Input and output impedance matching shou
be as close as possible to the device under test. Approximate matching is ma

(a)

(b)

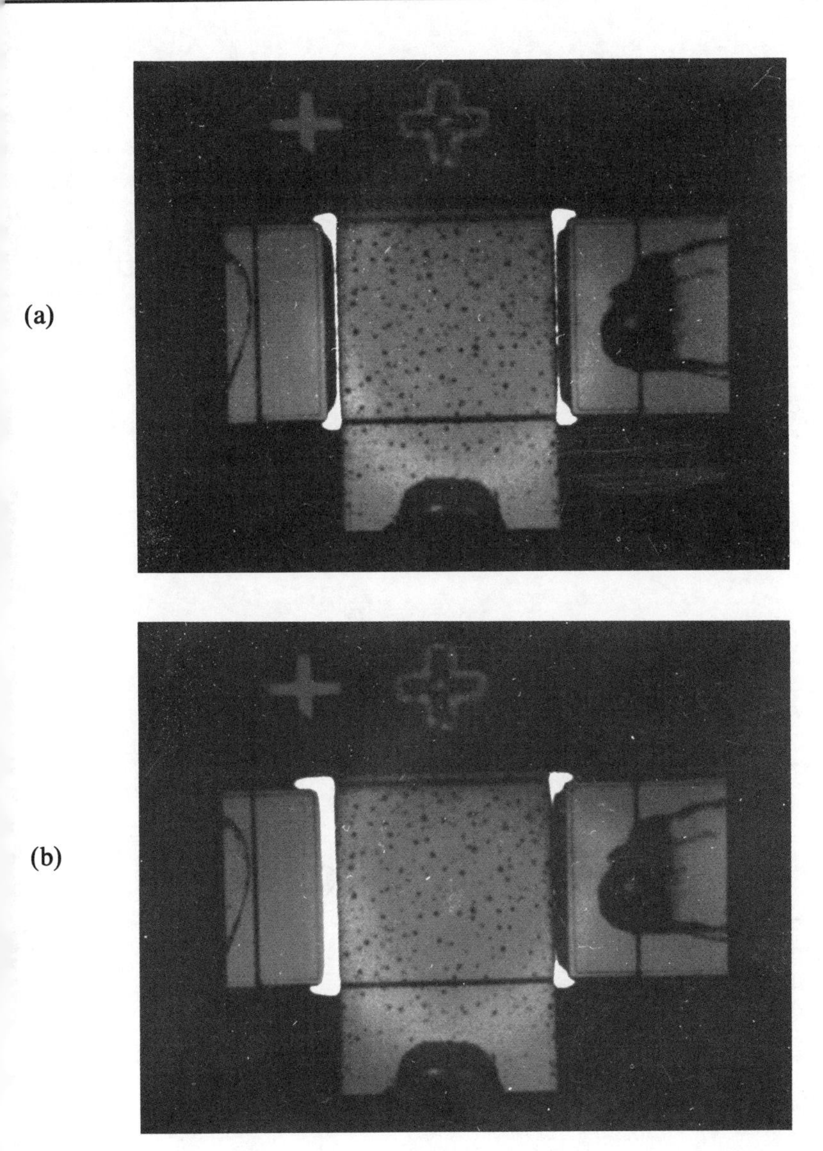

gure 2.35 Operation of a fat FET observed by EBIC: (a) $V_g = -|V_P|$, $V_{ds} = 0\text{V}$ and (b) $V_g = -|V_P|$, $V_{ds} = 5\text{V}$.

ith matching circuits above the microstrip line inside the test fixture and then ne-tuned by an external tuner. It is necessary to measure the high-frequency loss the circuit between the RF source and the device under test, and between the

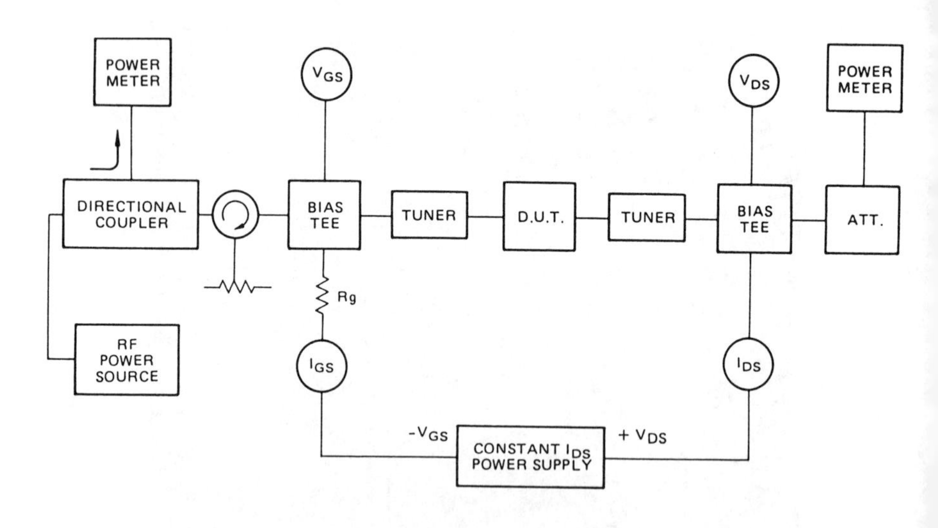


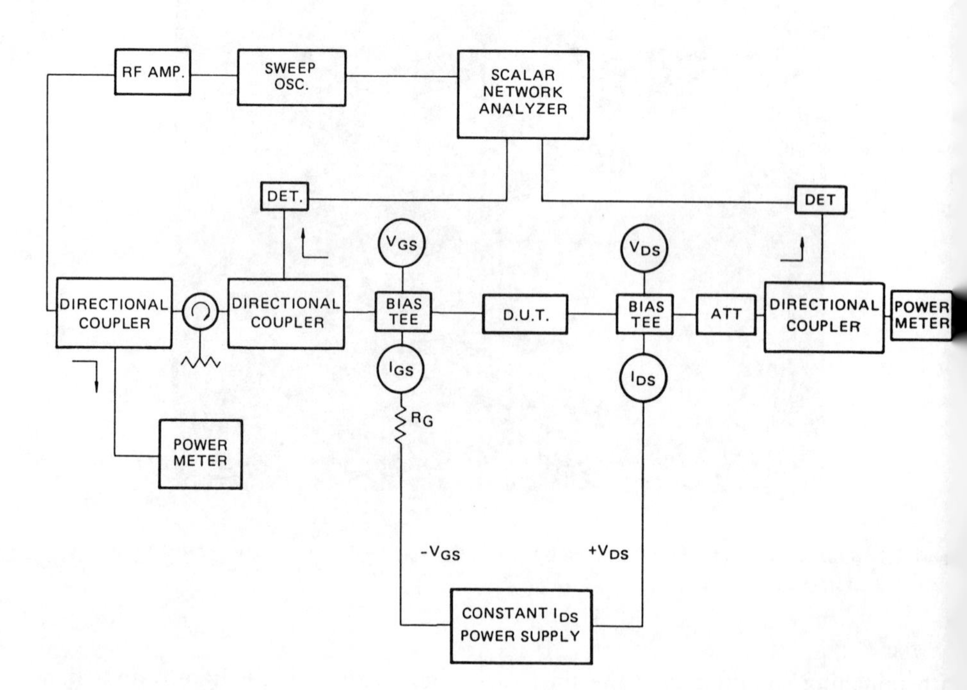


**Figure 2.36** Block diagram of the $P_{in}$ versus $P_{out}$ measurement system for (a) chip FETs or discrete (unmatched) FETS and (b) internally matched FETs or amplifiers.

ower meter and the device under test as part of the calibration process. Also, it advantageous to incorporate a spectrum analyzer on the output side to detect ny parasitic oscillations. The scalar network analyzer is an aid to calibration when is necessary to match specific bandwidths. The bias power supply should naturally e free of surges and be equipped with a current limiter. A protective resistor hould be placed on the gate side. Next, we will discuss items of concern during e measurement process.

As mentioned earlier, the matching impedance varies according to input wer level. Thus matching at reduced input levels is small-signal matching, and ill result in high gain but insufficient power output. Impedance matching should made at input power levels corresponding to the 1-dB gain compression point. s we will see later, the input impedance does not vary much with input power vel, but the output impedance varies considerably.

An important part of the measurement circuit in Figure 2.36 is the attenuator. he attenuator must be chosen with respect to the output power and gain of the evice being measured, measurement system loss, and the range and sensitivity of e power sensor.

Figure 2.36 also does not show any special measures used for higher harmonic equencies. For example, suppose that the transmission lines in this system are t up to accommodate signals of up to 18 GHz, and that it is being used to measure -band characteristics. With output power in a saturated condition, high-frequency armonics are created. The second- and third-order harmonics are not attenuated d enter the power sensor, where all output power, including harmonics, will be roneously measured as fundamental output power. To measure the fundamental mponent alone, an isolator and filter must be placed before the power sensor that the high-frequency harmonics are not measured and are not reflected from e sensor back toward the device. Harmonic components can be monitored by a ectrum analyzer. In addition, as mentioned earlier, reflecting harmonics at certain pedance levels has the effect of dramatically increasing efficiency—or, in other ords, allowing harmonics to enter a measurement system raises the concern that e devices may be overrated.

Gain measured by the system in Figure 2.36 is transducer gain, that is, power easured at the input end is read from a signal source and is not the same as the wer input to the device. To accurately measure power input to the device, flected power must be measured at the input port. This will be discussed in more tail in the section on load-pull. Reflection at the input end should be monitored a scalar network analyzer to verify that it is suitably low. With this exception, in measured by the system in Figure 2.36 is the same as normal power gain. nple measurements will be presented in the following sections, as various devices e introduced.

### 2.6.3 Measurement of Distortion Features

Measurement techniques are presented for the distortion characteristics normal
defined as:

1. AM-AM, AM-PM distortion;
2. Harmonic components;
3. Intermodulation distortion.

*AM-AM, AM-PM Distortion*

AM-AM and AM-PM distortion can be computed from measurement of the inp
and output power characteristics, which are normally determined by using a ne
work analyzer. Figure 2.37 shows a block diagram of the measurement syster
First, the device is removed and the input and output parts of the test fixture a
directly connected. Then the input power level is varied to measure the referen
phase and amplitude. Then the device is inserted, a bias applied, and the pha
and amplitude again measured as a function of the input power level. The deviatio
of gain and phase from their values at the lowest input level characterize AM-A
and AM-PM distortion as a function of the input power level. No element oth
than the device itself is assumed to exhibit a nonlinear response to variations
signal level.

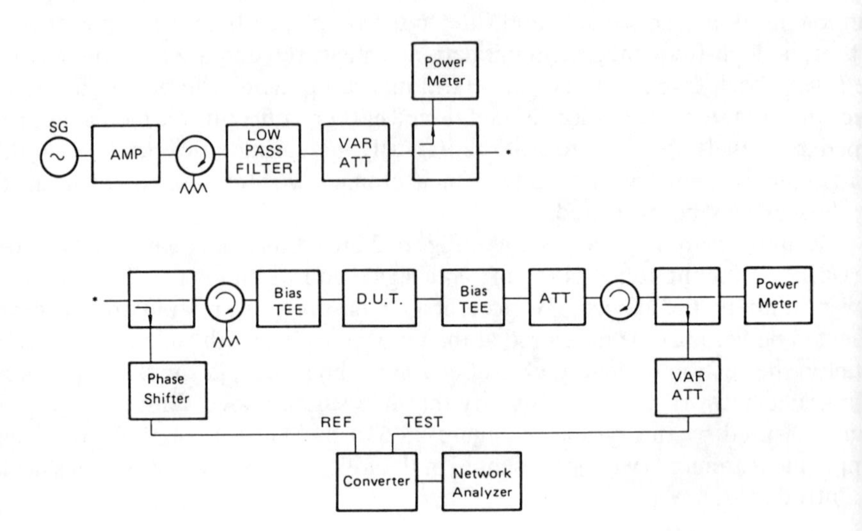


**Figure 2.37** Block diagram of AM-PM measurement system.

## *larmonic Components*

'he measurement system is basically the same as that for input-output power ıeasurements, except that the power sensor at the output end is a spectrum ana-'zer. System components must be carefully selected to ensure that they will carry gnals of at least three times the frequency of the fundamental.

## *ıtermodulation Distortion*

. block diagram of the measurement system is shown in Figure 2.38. Both the 'equency and power of the input signal are monitored. It is better to include an olator or similar element at the output end to ensure that harmonics are not :flected back into the device. If the input amplifier in Figure 2.38 does not produce nough power to drive the device, it will be necessary to add a driver amplifier efore the device under test (DUT). In this case a preliminary measurement is :quired to obtain the $IM_3$ of this amplifier, $IM_3(D)$. ($IM_3$ is defined as the ratio f the power of the third-order intermodulation products in a single sideband to ıe power of the fundamental.) If the driver amplifier has an output power $P$, then ıe output power of the sideband with an $IM_3$ value of $IM_3(D)$ will be $P \cdot IM_3(D)$. /hen these signals are applied to the DUT, then the DUT, which has a power ain $G$, generates a fundamental signal of power $P \cdot G$ and a sideband signal f power $P \cdot G \cdot IM_3(D)$ in addition to generating its own third-order intermodulation istortion signal of $P \cdot G \cdot IM_3(F)$. These two sideband components may have any

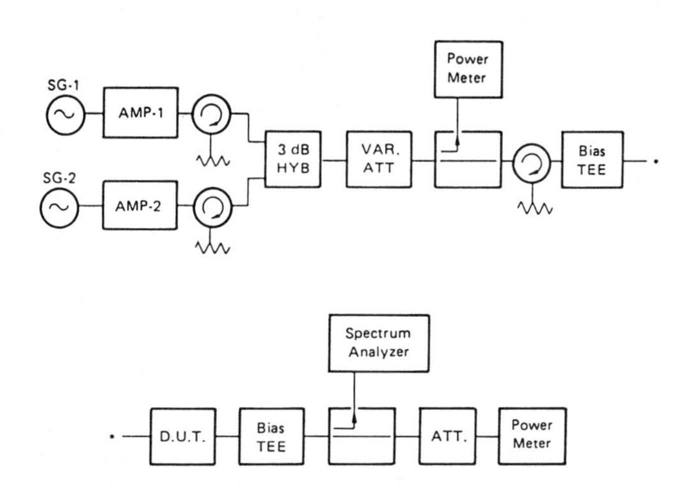


**gure 2.38** Block diagram of third-order IMD ($IM_3$) measurement system.

phase relationship to each other, but the worst case output power of the sideban
frequencies occurs when they are in phase and is given by a simple summation
i.e., the worst case total measured value of $IM_3$, represented by $IM_3(T)$, will be

$$IM_3(T) = IM_3(D) + IM_3(F) \quad (2.113$$

For reasonable measurement accuracy, $IM_3(D)$ must be at least 10 dB lower tha
$IM_3(F)$, and preferably 30 dB lower.

### 2.6.4 Impedance Measurement and Load-Pull Measurement

Impedance data are essential to the FET user. The *S* parameters, which expre
transmission and reflection characteristics, are essential for matching circuit desig
However, in high-power FETs with nonlinear operation, the optimum load impe
ance point varies with the input power as the output power approaches saturatio
Therefore, output power must be measured and charted at the load side for eac
input power level. This type of measurement is called *load-pull measurement.*
this section, we outline both of these types of impedance measurement.

#### *S Parameters*

The vector network analyzer is the instrument used to measure *S* parameter
Methods of calibration and measurement vary with each type of instrument ar
are not discussed here. Several factors require careful attention when measuri
the *S* parameters of high-power FETs, however. In large FETs (including FE
without matching circuits) the input and output impedances at microwave fr
quencies are substantially smaller than $50\Omega$, and $S_{11}$ and $S_{22}$ are both near t
circumference of the Smith chart. High levels of reflection can significantly affe
the precision of measurements. Conversely, impedance measurements of sever
ohms require much higher precision and smaller margins of error than at 1-$\Omega$ leve
In fact, a measurement of chip *S* parameters becomes more difficult when the si
of the chip exceeds the width of the microstrip line. Rather than measuring lar
FET chips directly, it is more practical to measure FET cell units individually, a
then to mathematically deduce the *S* parameters of the complete FET.

When measuring across wide frequency ranges, the setting of attenuation
very important. In high-power FETs, $|S_{21}|$ will be large in low-frequency bands a
small in high-frequency bands. Here it becomes necessary to use either the prop
attenuator or to divide bands in order to achieve the desired precision.

### ,oad-Pull Measurement

,oad-pull is one of the ways to measure the optimum load side impedance at a iven input power level. With the appearance of automatic tuners and computer-ontrolled measurement instruments, it has become possible to make this most omplex of measurements without difficulty. Figure 2.39 shows a block diagram f a load-pull measurement system. The reflection-transmission properties of the nput side tuner, bias network, and directional coupler, as well as the output side uner, bias network, and attenuator, must first be calibrated through the network nalyzer. Record the position of the tuner stubs (screw or slide) and measure $S$ arameters at each position. Once the reflection-transmission properties have been oted, the measurement process itself consists of observing input and output power evels. First, set the tuner at the source side to a given position (such as 50$\Omega$) and ieasure output power levels for various positions of the tuner at the load side. 'he input side power sensors will monitor the power available from the signal ource $P_A$ and the power reflected by the device $P_R$. The power supplied to the evice $P_{in}$ is the difference between the two, $P_A - P_R$. The output power sensor ieasures the output power $P_L$. The transducer power gain $G_T$ is by definition

$$G_T = \frac{P_L}{P_A} \tag{2.114}$$

'hile the power gain $G_P$ is

$$G_P = \frac{P_L}{P_{in}} \tag{2.115}$$

'hen, repeating output power measurements for various load impedance points, e can obtain the load side impedance point at which maximum output power is eached. With the output tuner at this position, vary the setting of the input tuner nd measure the power. This is called *source-pull measurement*. Now set the input de tuner at the position of maximum output power and perform load-pull mea-irements again. By alternately repeating load-pull and source-pull measurements, ie locations of the optimum input and output impedance for output power can e obtained. Finally, by varying the magnitude of the input signal, it is possible to epeat this measurement process to observe the change of the optimum input and utput impedance with respect to the input signal level. A sample of such a mea-irement is shown in Figure 2.40. The input impedance has little dependence on ie input power level, and one or two source-pull measurements are normally ifficient. The output impedance, however, is highly dependent on signal level. In

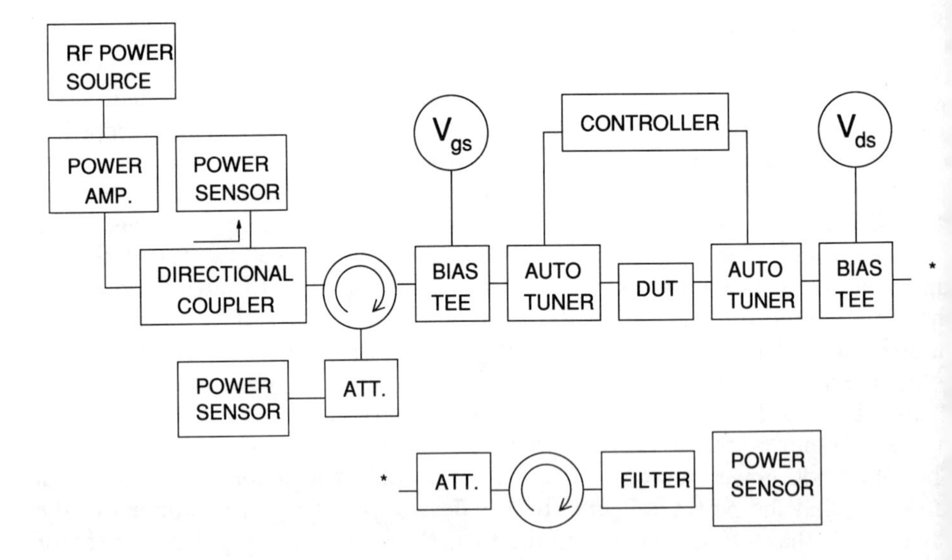


**Figure 2.39** Typical block diagram for load-pull measurement system.

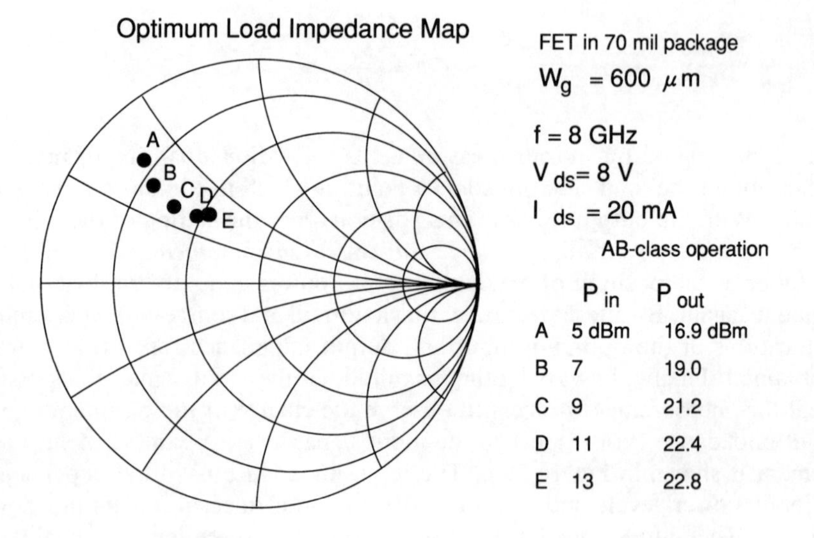


**Figure 2.40** Example of load-pull measurement.

an FET chip, the optimum output impedance at low input power levels will be close to the impedance values in a small-signal system, but as the input power increases, the real part of the optimum impedance will approach the values given in Table 1.1.

## 2.7 CURRENT FETS

In this section we present several FET chips available as of 1992 that incorporate the basic design technologies discussed in the preceding sections.

### 2.7.1 Standard FET Chips

Many high-power GaAs FETs use gate electrodes made of aluminum. The Al-gate FET already has a long history, and can probably be considered as perfected both in the manufacturing process and its design. Here we discuss the technologies in an Al-gate high-power FET chip [32]. An overall view of the chip, whose size is 0.56 by 2.61 mm, is shown in Figure 2.41. It is composed of units with gate finger widths of 125 $\mu$m with a total FET gatewidth of 18.0 mm. Each of the 12 bonding pads holds a single FET cell with a gatewidth of 1.5 mm, so that the 12 single FET cells are connected in parallel. The bonding pads reached by gates at the side of the chip are for connection to other chips when several chips are used together, in order to coordinate operation and suppress oscillation. The chip is 25 $\mu$m thick. Source electrodes are shorted to the back of the chip through via-hole connections and a 35-$\mu$m gold-plated heatsink reduces thermal resistance. Air-bridge construction is adopted at the gate-source intersection. This FET is designed to be used in the C-band to X-band, and Ku-band range. One chip has a power output ($P_{1dB}$) of approximately 7W. Figure 2.42 shows an SEM photograph and schematic diagram of the channel cross section. The pattern has a gate length of 0.6 $\mu$m, defined

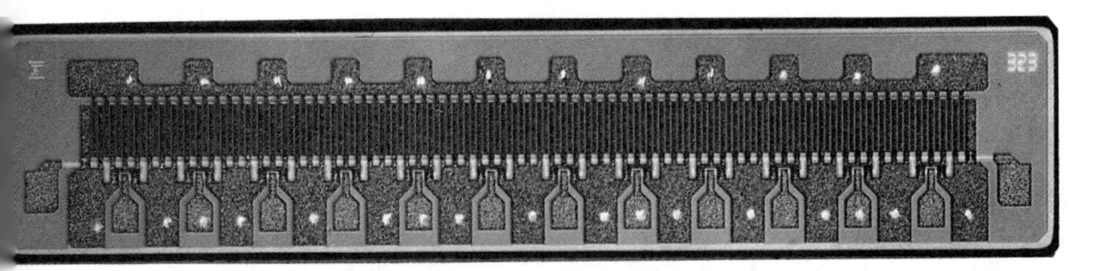

**Figure 2.41** Overview of C- and X-band high-power GaAs MESFET chip. Chip size is 0.56 by 2.61 mm.

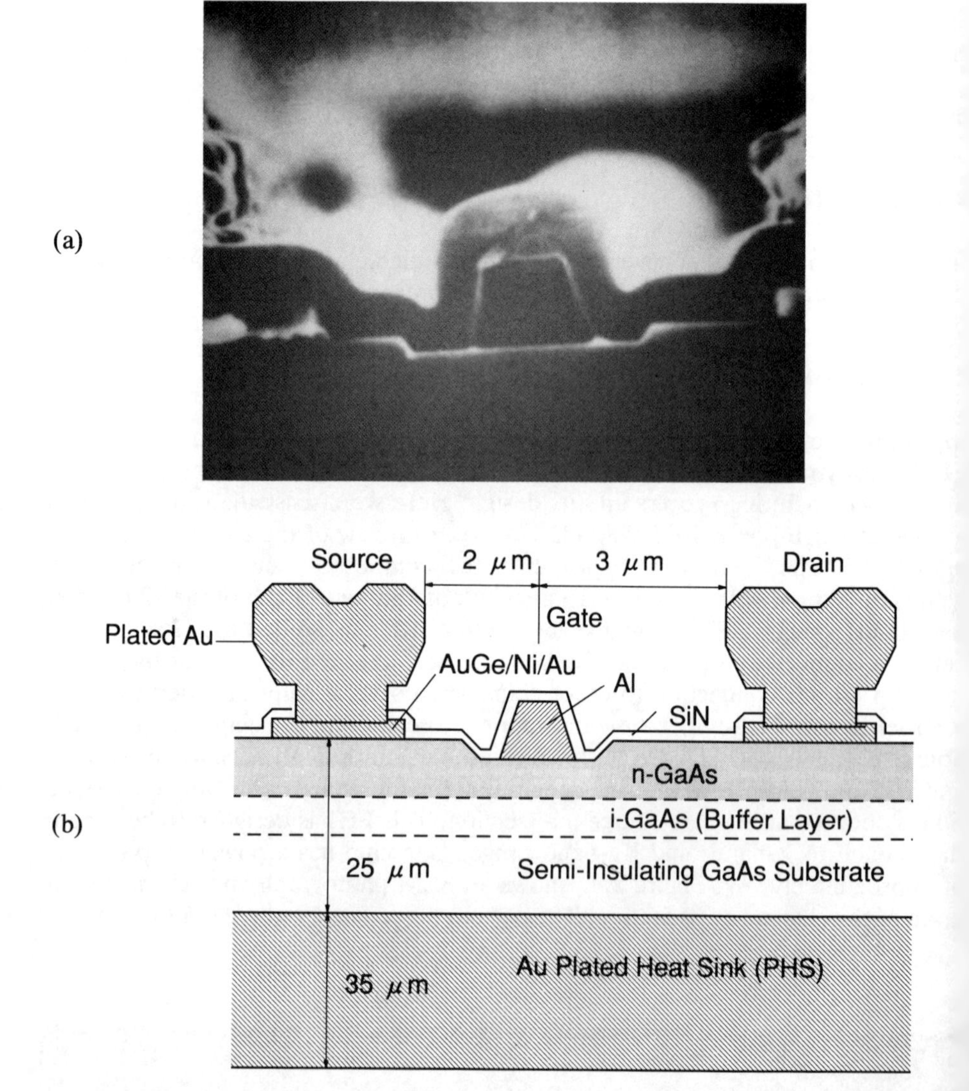


**Figure 2.42** (a) SEM photograph and (b) cross-sectional view of the channel of the FET in Figure 2.41

by electron-beam lithography, and formed by aluminum evaporation and lift-off. Gradual recesses have been formed at a location offset from the drain toward the source to achieve a high gate and drain breakdown voltage. The GaAs recess has an aspect ratio (gate length/channel thickness) in excess of 5. The GaAs surface s coated with a SiN film to reduce variance of characteristics and increase reliability. Source and drain electrodes have a multilayer construction of AuGe/Ni/Au/Ti/Pt/Au/Au plating to provide sufficient current handling capacity. The GaAs activated layer has a doping density of $1.7 \times 10^{17}$ $cm^{-3}$, and is grown by the VPE method. The buffer layer is semi-insulating GaAs with a Cr doping. The features and reliability evaluation of this chip are discussed later, in Section 2.7.2 on internally matched FETs.

The development of higher frequency high-power FETs has been accompanied by the appearance of high-gain FETs. We now discuss a high-power GaAs FET developed for K-band use (Figure 2.43). Since higher gain requires increased transconductance and lower gate capacitance, the doping density of the active region has been increased, and the gate length shortened to 0.3 $\mu$m. The total gatewidth is 1200 $\mu$m, with two individual FET cells connected in parallel. Each cell has 12 50-$\mu$m-wide gates for a gatewidth of 600 $\mu$m. These specifications were the result of the design methods discussed earlier. Tournament-type gate signal feed lines eliminate the phase differential in signals reaching the gate fingers. The gate spacing is 13 $\mu$m, determined by the required thermal resistance and source-drain capacitance. An air-bridge construction is used for the gate-source crossovers. Via-hole connections and a PHS are also used. An SEM photograph and an analytical diagram of the channel cross section are shown in Figure 2.44. The Schottky gate length is 0.3 $\mu$m with a T-type (mushroom-type) multilayer WSi/Ti/Au/Au

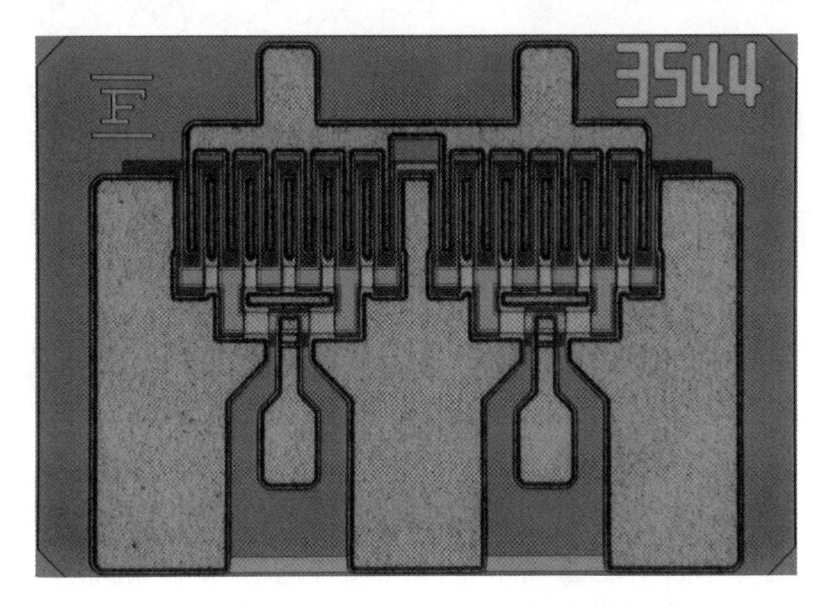


**Figure 2.43** Overview of Ku- and K-band high-power GaAs MESFET chip (gatewidth = 1200 $\mu$m).

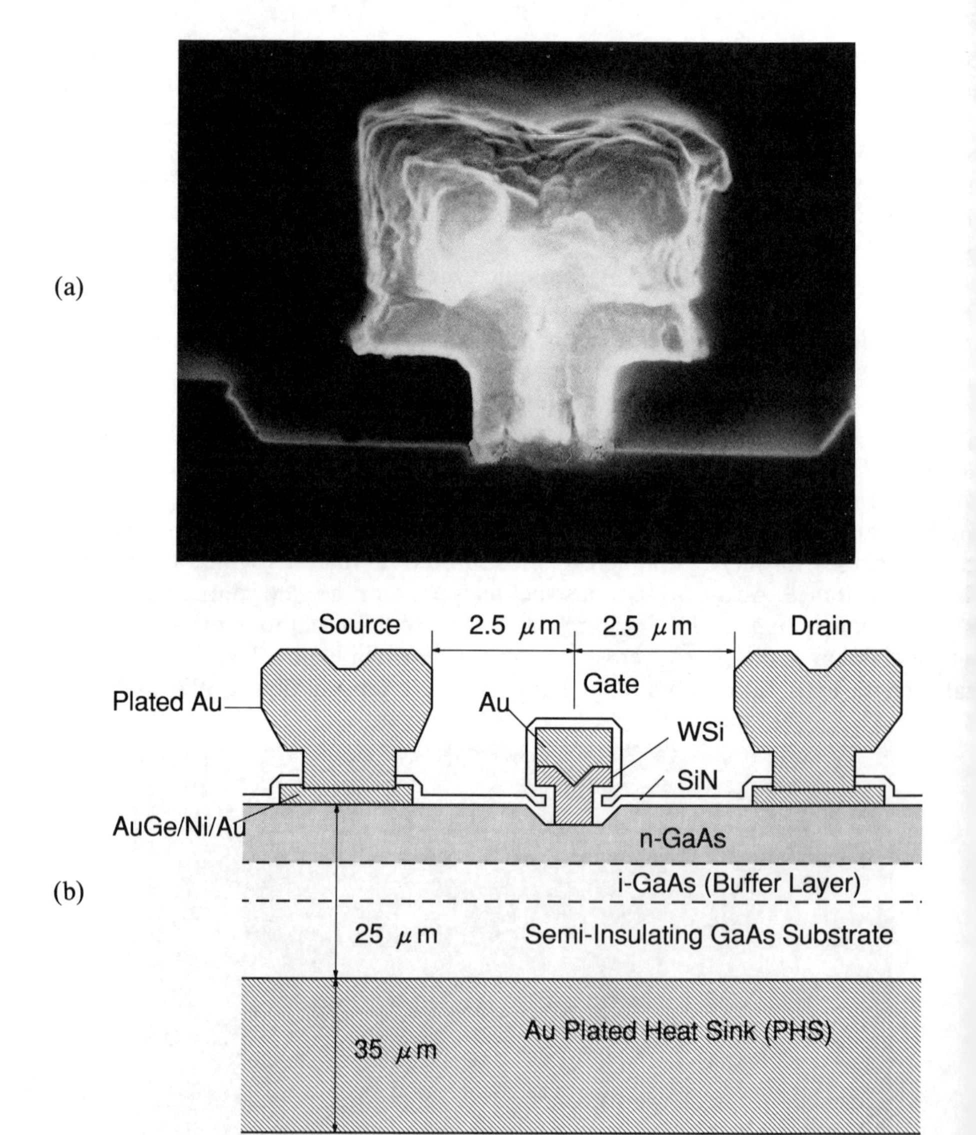


**Figure 2.44** (a) SEM photograph and (b) cross-sectional view of the channel of the FET in Figure 2.4

plated construction. The use of WSi, a refractory metal, in the Schottky junctio area provides high reliability and the Au overlay reduces the gate resistance. Th source and drain electrodes are of multilayer AuGe/Ni/Au/Ti/Pt/Au/Au plate construction. The doping density of the active layer is determined by applying the scaling rule to the previous FET and is $3.5 \times 10^{17}$ cm$^{-3}$. The GaAs surface from the source to drain and the T-type gate are covered with a SiN protective film coating. The short gate length does not allow the use of mesa etching for isolation; hence, element isolation is achieved by ion implantation instead. The saturated drain current $I_{dss}$ is 200 mA/mm, the knee voltage is 1.2V, and the transconductance ($g_m$) is 100 mS/mm. The drain breakdown voltage at gate pinch-off is 20 to 22V. Input and output power characteristics and the power-added efficiency are shown in Figure 2.45. The FET is available as a single-cell FET ($w_g$ = 600 $\mu$m), or with two cells ($w_g$ = 1200 $\mu$m) or four cells ($w_g$ = 2400 $\mu$m). The 600-$\mu$m FET at 18 GHz and $V_{ds}$ = 8V achieves an output power of $P_{1dB}$ = 23.8 dBm (400 mW/mm), a linear gain of $G_L$ = 9.0 dB, and a power-added efficiency of $\eta_{add}$ = 35%. Using S parameter scaling, the gain is $G_{amax}$ = 6.5 dB at 30 GHz. The 1200-$\mu$m FET correspondingly has $P_{1dB}$ = 26.3 dBm (355 mW/mm), $G_L$ = 8.5 dB, and $\eta_{add}$ = 35%, and the 2400-$\mu$m FET has $P_{1dB}$ = 29.0 dBm (330 mW/mm), $G_L$ = 8.0 dB, and $\eta_{add}$ = 35%.

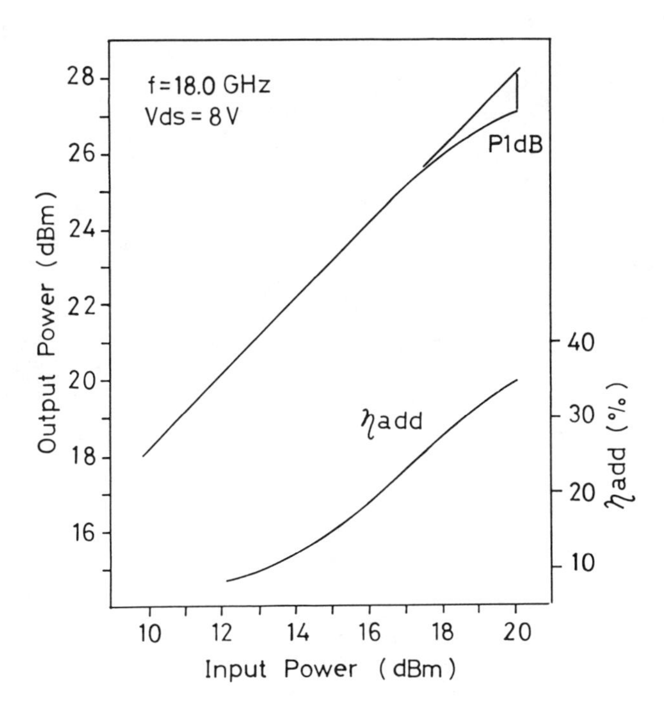


Figure 2.45 Output power and power-added efficiency versus input power of the FET in Figure 2.43. (Conditions: $f$ = 18 GHz; $V_{ds}$ = 8V; $I_{ds}$ = 0.6 $I_{dss}$)

### 2.7.2 Internally Matched FETs

A high-power FET chip generally has low input and output impedances and canno achieve the desired gain and output power without the use of an impedance match ing circuit to 50$\Omega$. Because impedance matching must be carried out as close t the chip as possible, an FET with matching circuits in the desired frequency ban placed inside the package—the internally matched FET—has been developed an is now in widespread use. As examples, we will discuss the Fujitsu FLM 3742-1 and FLM 4450-25D, the former from the standpoint of design for high reliability and the latter from the standpoint of matching circuit design.

The Fujitsu FLM 3742-10 is an internally matched FET for use in the 3.7- t 4.2-GHz frequency band. The inside of the package is pictured in Figure 2.46 There are two Al-gate FET chips (previously discussed) connected in parallel. Th input side of the matching circuit consists of a Wilkinson-type power divider/com biner, which transforms the 50-$\Omega$ input terminal impedance to two 25-$\Omega$ branches followed by a lumped (*L-C-L-C-L*) low-pass circuit, which matches to the FE

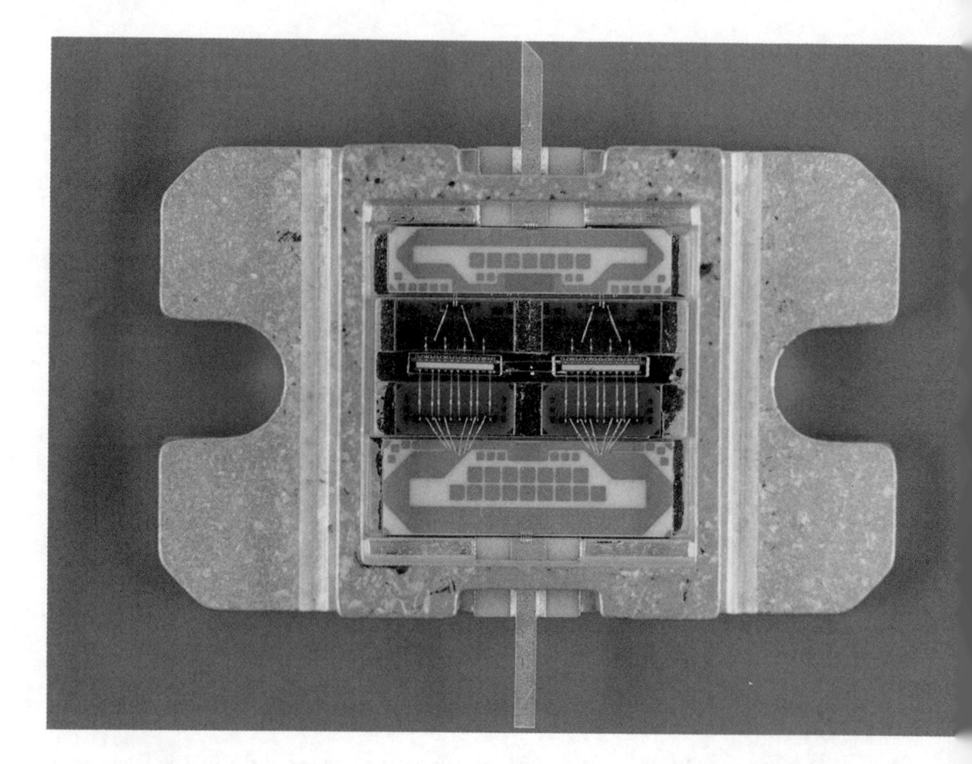

**Figure 2.46** Internal view of Fujitsu's FLM 3742-10.

nput impedance (approximately 1Ω). On the output side, the structure is similar, xcept that the low-pass lumped element circuit is (*L-C-L*). The power divider/ ombiner is fabricated on an alumina substrate. The lumped low-pass circuits are abricated with capacitor patterns and gold wires on a high dielectric constant ubstrate.

Power input and output characteristics of the FET are shown in Figure 2.47. t operates as a Class AB device, with more than 50% efficiency. This FET was lesigned for use in high-reliability applications, including satellite communications, nd it easily meets the reliability goals demanded. Accelerated life testing (see Chap. 5) was carried out at channel temperatures between 100° and 280°C. Figure .48 shows the mean time to failure (MTTF, see Sect. 5.6.2) in the life test as a unction of channel temperature. The test was carried out at temperatures of 270°, 50°, and 230°C. The failure mode was a gate short due to burnout. Figure 2.49 hows an SEM photograph of a failed FET chip after removal of the protective nsulating layer, with voids visible in the aluminum gate metal. This phenomenon s due to migration of Al into the GaAs, a failure mode that is unavoidable as long s Al is used as a gate metal. A new FET with refractory metal gates has been leveloped to overcome the problem. The value of activation energy (see Sect. .6.1), calculated from Figure 2.48 is 1.17 eV, which at room temperatures (channel

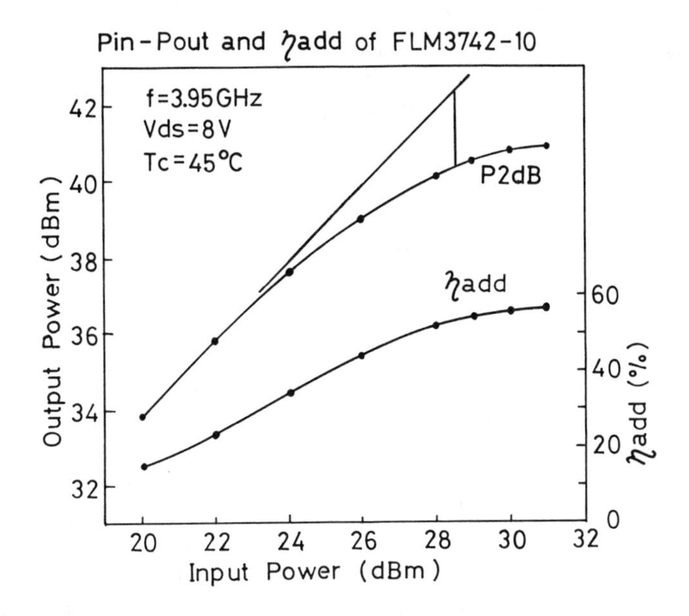


**igure 2.47** Output power and power-added efficiency of the FLM 3742-10.

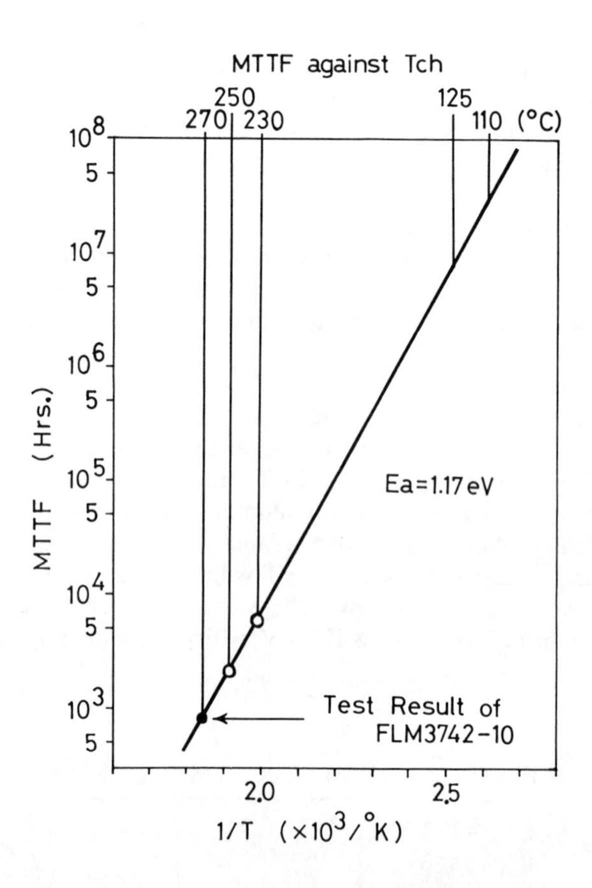


**Figure 2.48** MTTF versus channel temperature for the FLM 3472-10.

temperature of 110°C) gives an MTTF of 2.9 × $10^7$ hr, well in excess of the lifetim required for satellite applications.

The Fujitsu FLM 4450-25D [33] is an internally matched FET for the 4.4- t 5.0-GHz range, developed for use in multiple-channel digital communications. Th application naturally demands high output power, high gain, and high reliabilit as well as low intermodulation distortion to suppress interference from adjace channels. Gain, output power, and $IM_3$ had to be considered at the design stag Because the target output power was 25W, four 7-W FET chips are connected i parallel. Because each of the FET chips had a gatewidth of 18 mm, the extreme low input and output impedance made it difficult to obtain the necessary data fo circuit design. The data were obtained from the unit FET cell that made up th FET. The FET cell had a gatewidth of 1.5 mm, and 12 cells were linked to for the FET chip. The FET cells had a gatewidth of 125 $\mu$m and a gate length of 0

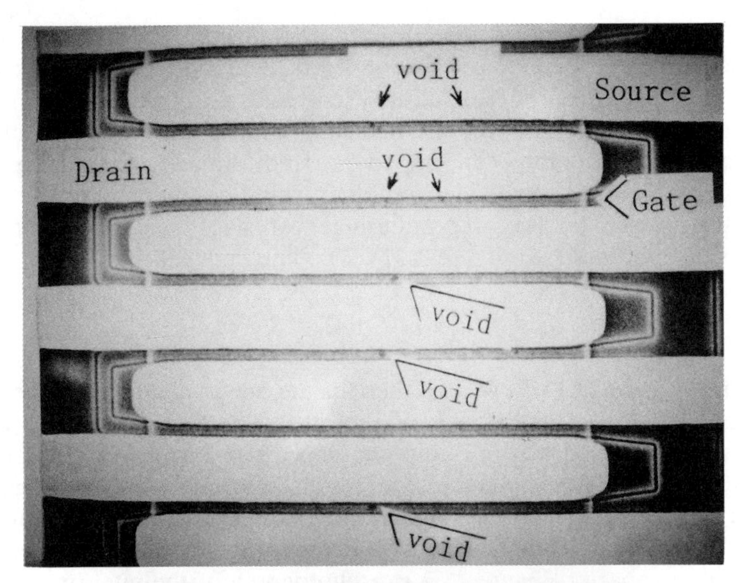


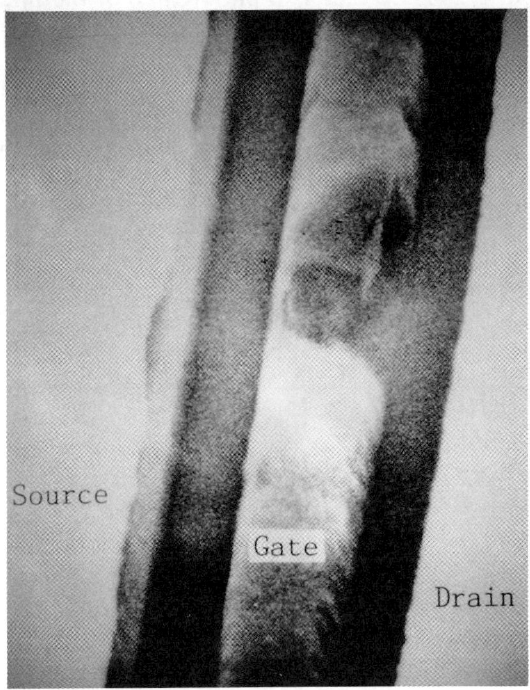


**Figure 2.49** SEM photograph of (a) a failed FET chip and (b) a gate finger.

μm. The load-side derived $P_{1dB}$ contours of output power, gain, and $IM_3$ calculate from measurement of the FET cell $S$ parameters, load-pull and $IM_3$ measurement are shown in Figure 2.50. The hatched area represents the optimum load impedanc region for the features desired. The matching circuits are composed of a four branch power divider/combiner and impedance transformer. In C-band frequencies a quarter-wavelength power divider/combiner would be too large to use. We sur mounted the problem by designing and using a two-stage divider/combiner of one eighth-wavelength lines (see Figure 2.51). The line impedances are

$$Z_1 = \sqrt{Z_S Z_L}, \qquad Z_2 = 2\sqrt{Z_S Z_L}$$

According to Figure 2.51, $Z_L$ is 50Ω, thus for the power divider/combiner VSWF to be within the design limit of 1.4, $|Z_S|$ must be between 20 and 30Ω. Betwee the divider/combiner and the FET chip, the impedance is transformed by an (L-C circuit. According to the impedance conversion ratio, there are two steps on th input side and one on the output side. On the output side careful attention is give to the circuit linewidth, plating thickness, and the number of bonding wires require for the current capacity. Figure 2.52 is a photograph of the interior of the FLM

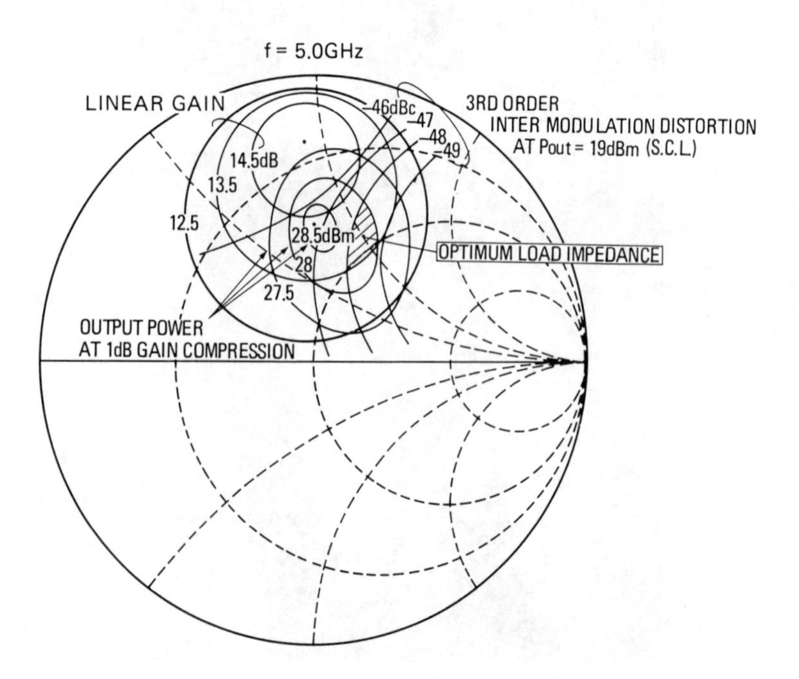


**Figure 2.50** Load impedance contours for constant $P_{1dB}$, $G_L$, and $IM_3$ of the FLM 3742-10 unit FE cell.

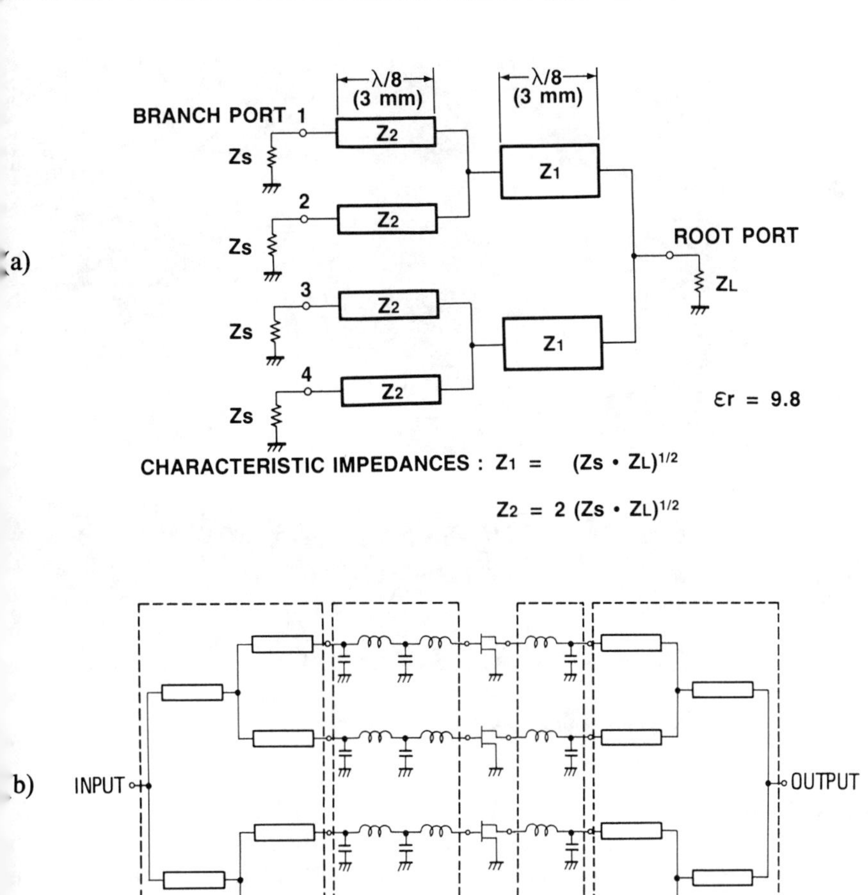


**Figure 2.51** Diagram of the matching circuit of the FLM 3472-10: (a) four-way power divider/combiner and (b) overall features of the matching circuit.

4450-25D. The external dimensions of the package are 17.4 by 24.0 mm. The power divider/combiner pattern is formed on a 0.38-mm-thick alumina substrate, with a dielectric constant of 9.8. The lumped capacitor is made from a 0.15-mm-thick high dielectric constant material with a dielectric constant of 140. High-frequency characteristics are shown for power output versus power input [Figure 2.53(a)], third-order intermodulation distortion [Figure 2.53(b)], and power output and gain versus

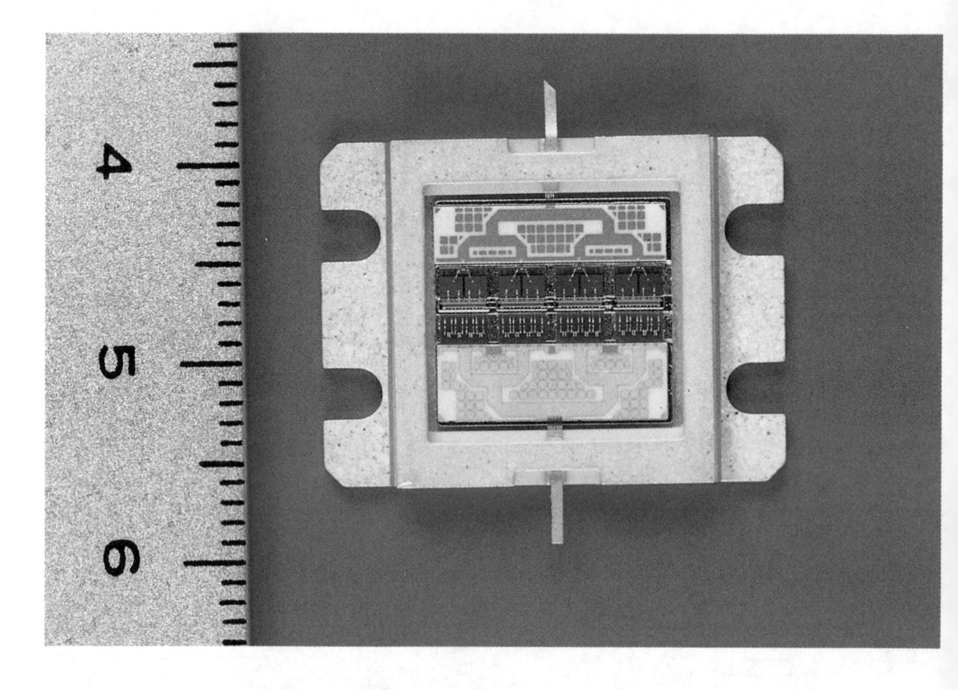


**Figure 2.52** Internal view of Fujitsu's FLM 4450-25D.

frequency [Figure 2.53(c)]. At 4.7 GHz, $P_{1dB}$ = 44.3 dBm (26.9W), $G_L$ = 12. dB, $\eta_{add}$ = 40%, and $IM_3$ at an output power of 35 dBm (single-carrier level) i −46 dBc. These features are more than sufficient for use in digital communications Over the frequency range of 4.4 to 5.0 GHz, $P_{1dB}$ flatness is better than 0.3 dB and $G_L$ flatness is better than 0.5 dB.

As an example of an application of internally matched FETs, Figure 2.5 shows an internal photograph of a 14-GHz amplifier for the VSAT applicatio Fifty-ohm internally matched FETs and modular chip amplifiers, with an outpu power and gain scaled to prevent power saturation in each interstage, are use Half of the space in the amplifier is used for bias feed.

### 2.7.3 MMIC Power Amplifiers

At K-band and higher frequencies, the impedance varies with the length of th wires leading from the FET chip, making matching difficult. This has led to th development of the *monolithic microwave integrated circuit* (MMIC), which has i matching circuit on the chip. The Fujitsu 38-GHz MMIC pictured in Figure 2.5 is an example. This chip is constructed from four parallel FETs, each cell havin

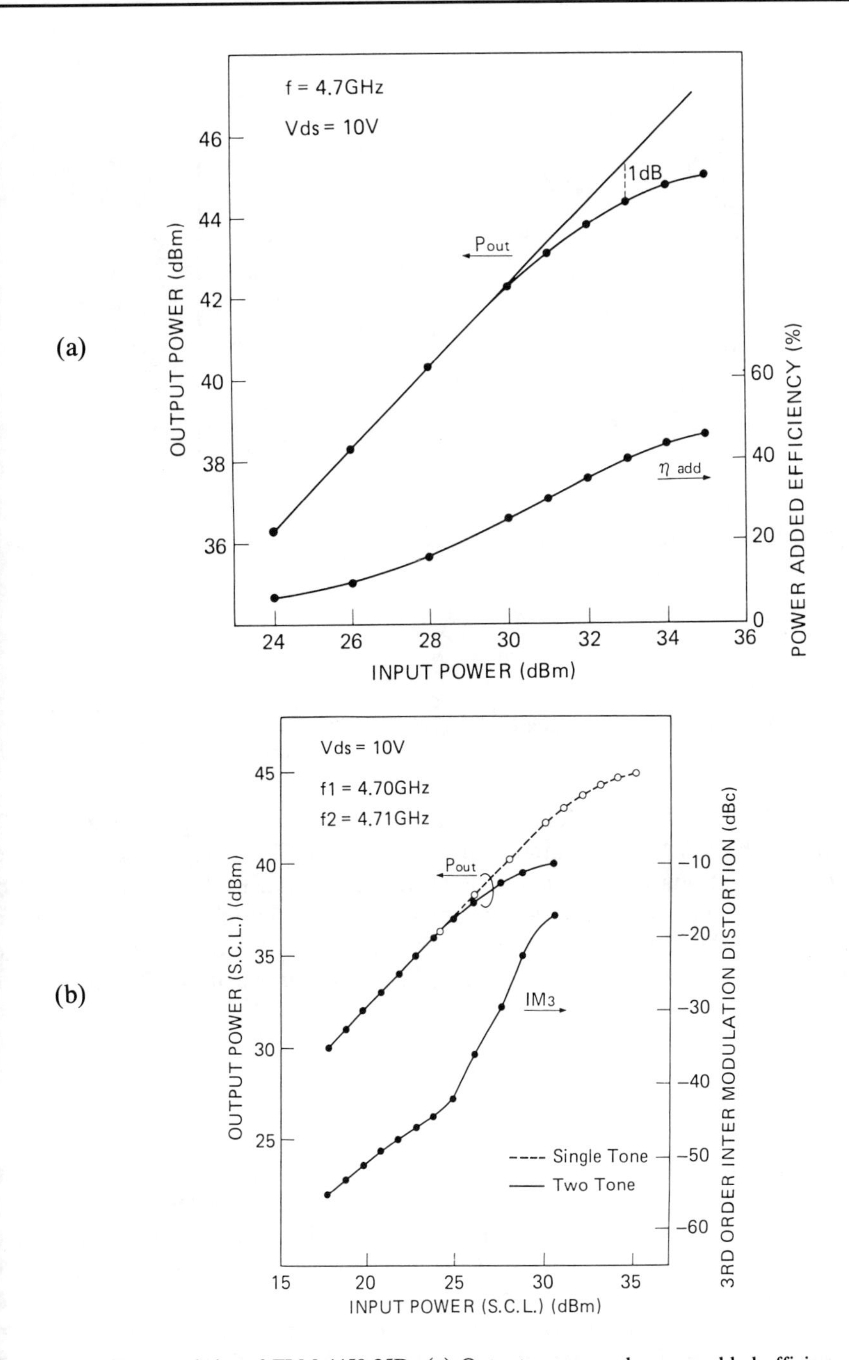


**Figure 2.53** Characteristics of FLM 4450-25D: (a) Output power and power-added efficiency versus input power, (b) output power (single-carrier level) and third-order intermodulation distortion versus input power (single-carrier level), and (c) frequency response of $P_{1dB}$ and $G_L$ ((c) appears on p. 136).

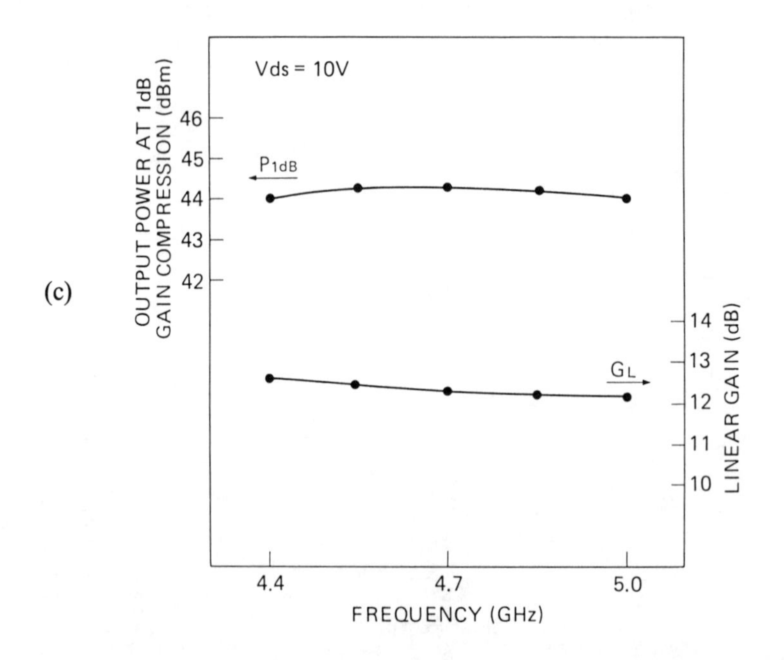


**Figure 2.53** (Cont.) (c) Frequency response of $P_{1dB}$ and $G_L$.

a gatewidth of 400 $\mu$m (gate finger width of 50 $\mu$m $\times$ 8). The gate length is 0.2 $\mu$m, with a WSi/Ti/Au/Au plated construction. All FET source electrodes ar grounded through via-hole connections beside them. All FETs are connected b high-impedance lines for phase synchronization. The dimensions of the chip ar 1.6 by 2.0 mm. At 38 GHz, $P_{1dB}$ is 24.7 dBm and $G_L$ is 5.1 dB. The FET marke at K-band and higher frequencies is expected to see wider use of MMIC constructio in the future.

## 2.8 TRENDS IN TECHNOLOGY

Advances in materials technology (particularly crystal manufacturing), proces technology, and circuit design technology have created new devices aimed at replac ing the GaAs FET. Examples are the indium phosphide (InP) MISFET and HEMT which achieve higher frequencies and higher gain by virtue of a higher electro saturation velocity than GaAs. Another example is the GaAs/AlGaAs heterc junction bipolar transistor, which replaces the FET with a bipolar transistor. Alsc there are devices with new structures, entirely different from the structures c previous devices, such as the permeable base transistor (PBT) and hot electro

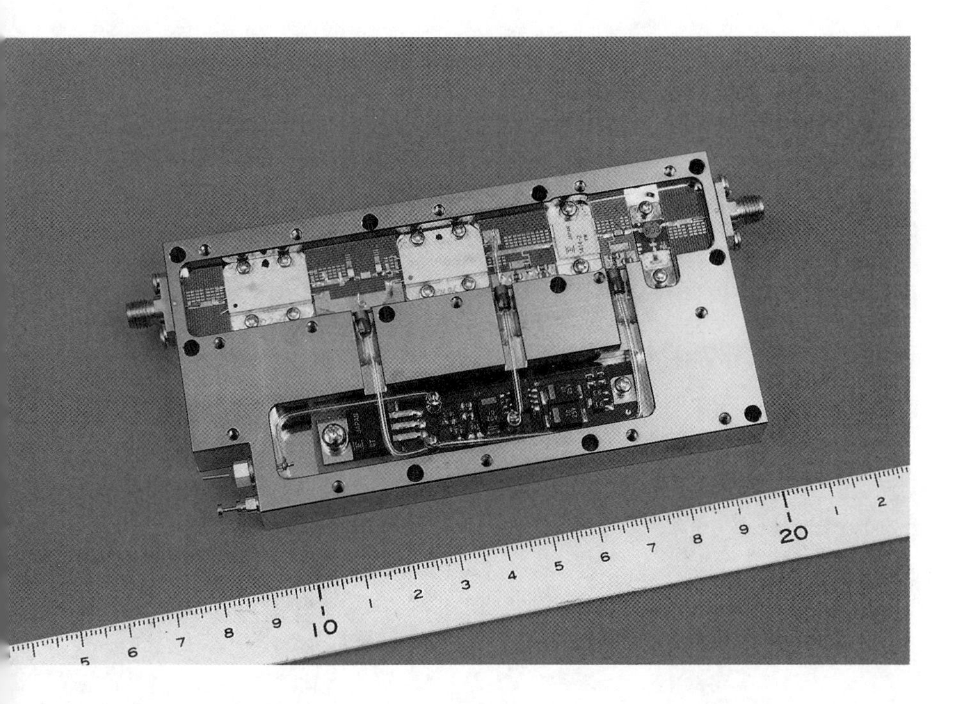

igure 2.54 Internal view of 14-GHz amplifier for VSAT system using internally matched FETs and FET module amplifiers.

ansistor (HET). In this section we introduce several new high-power microwave evices. The practicality of these devices has yet to be demonstrated, but each of em provides an indication of the direction of future technology.

### .8.1 Material Technology: The InP MISFET

ompared to GaAs, InP has two significant characteristics:

1. Thermal conductivity is higher (GaAs: 0.455 W $cm^{-1}$ $°C^{-1}$ versus InP: 0.68 W $cm^{-1}$ $°C^{-1}$).
2. Saturated electron velocity in high fields is higher, and reaches saturation at a higher electric field strength.

These qualities make InP an outstanding material for high-power devices. lthough its bandgap at 300K is slightly lower than that of GaAs (GaAs: 1.42 eV rsus InP: 1.27 eV), InP has outstanding insulating properties. The most promising gh-power InP device is the MISFET. Of course the MISFET is not necessarily e ideal device for high-power applications. GaAs MISFETs have been studied, ıt with GaAs a high-density surface-state layer is created in the interface between

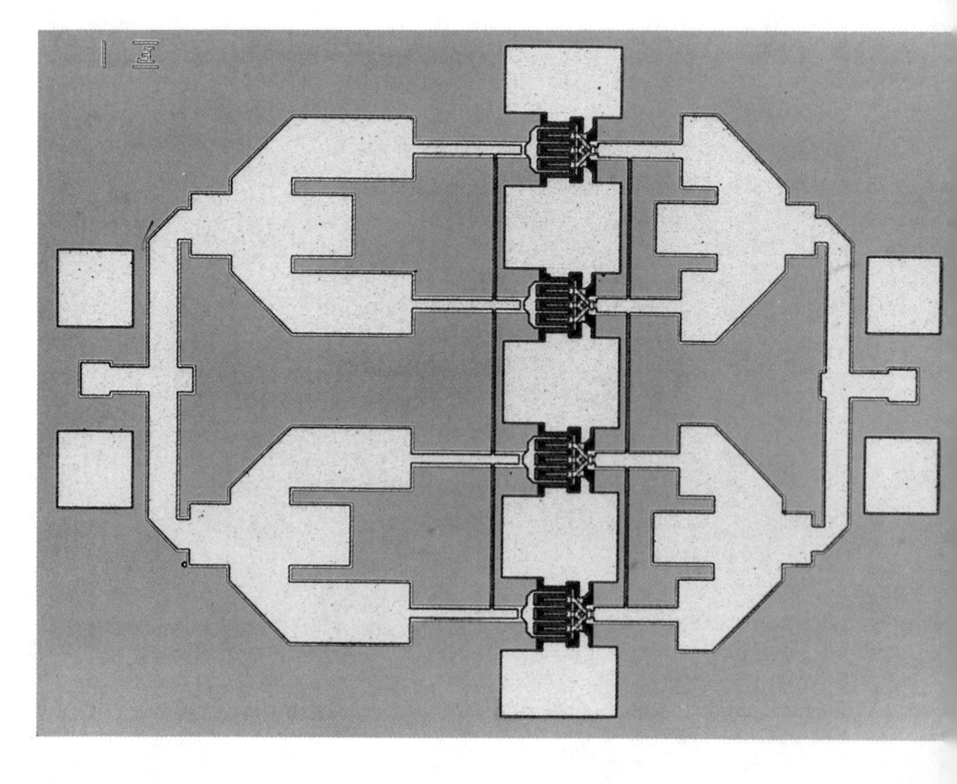

**Figure 2.55** A 38-GHz MMIC power amplifier.

the insulating layer and the semiconductor making it difficult to create *n*-typ channels. The surface-state layer density of InP is one order of magnitude less tha that of GaAs, and this makes it possible to create the MISFET. On the other han it is difficult to create a good Schottky gate on InP, which makes it hard to bui a MESFET. Advances are being made in developing the InP MISFET as a bas element for ultrahigh-speed composite semiconductor integrated circuits and as th base element for integrated optoelectronic integrated circuits. Itoh and Ohata [3 developed a recess-type self-aligned MISFET, with a *chemical vapor depositic* (CVD) $SiO_2$ layer and a channel length of 0.85 $\mu$m, that achieved a power gain 7.2 dBm at a frequency of 11.5 GHz with a noise figure of 4.4 dB. The maximu output power was 1 W/mm. The channel cross section is shown in Figure 2.5 Tokuda et al. [35] used an indirect plasma CVD method to deposit a PSG lay on an inverse recess structure element. With a gatewidth of 960 $\mu$m, the devi achieved $P_{1dB}$ of 0.51W at 30 GHz and $G_a$ of 3 dB. These developments offer ho that advances in insulating film technology and process technology will lead to t development of superior high-frequency electrical devices. Also, Ohata [36] h

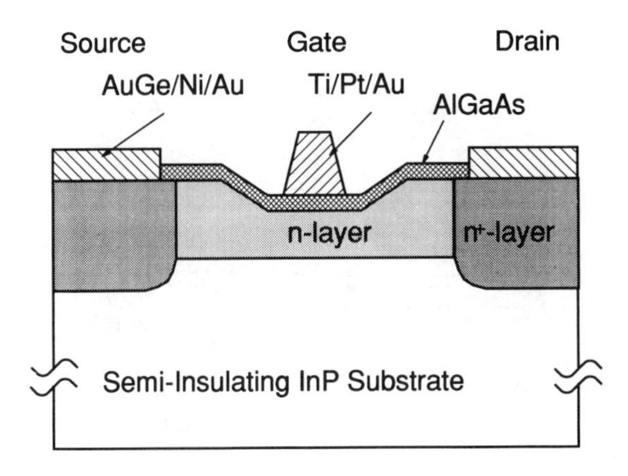


igure 2.56 Cross-sectional structure of typical InP MISFET.

eveloped a MISFET using an insulating layer of AlGaAs, in which the low surface-ate density of the *n*-InP/*i*-AlGaAs interface produced outstanding results.

### .8.2 The Heterojunction FET: HEMT and Heterostructure MISFET

Vith advances in MBE and MOCVD methods has come a number of new devices sing GaAs/AlGaAs heterojunctions. Even in the GaAs MESFET, new technol-gies are being developed using AlGaAs or GaAs/AlGaAs superlattices. More ggressive uses of the strengths of the heterojunction are the HEMT and het-rostructure MISFET.

Aluminum and gallium ions in tetrahedral covalent crystal configuration both ave a diameter of 1.26Å. Therefore, both GaAs and AlAs have identical lattice onstants, and AlGaAs likewise has the same lattice constant regardless of its onstitution. This fact means that the GaAs/AlGaAs heterojunction should produce potential difference for electrons with no stress in the crystal lattice. HEMTs are widespread use as low-noise elements. Their high gain affords outstanding char-cteristics in high-frequency applications, and they are being developed for high-ower applications at millimeter wavelengths and higher frequencies. Because the IEMT normally has an electron donor layer density of around $1 \times 10^{18}$ cm$^{-3}$, it difficult to develop high gate breakdown voltages. Therefore, development is roceeding in the direction of increased output power by means of higher current. ikosaka et al. [37] have developed a multichannel HEMT in which the electron ansit layer has a two-dimensional, multilayer structure, with the wafer epilayer ructure shown in Figure 2.57. This device uses high current levels of $I_{dss}$ = 330 A/mm and $I_F$ = 530 mA/mm. The gate-drain breakdown voltage is between 7

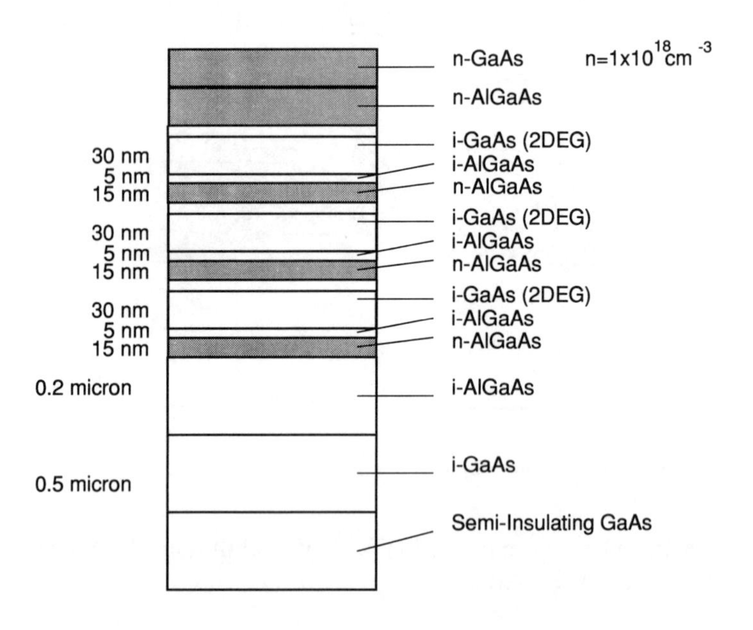


**Figure 2.57** Epistructure of multichannel power HEMT [37].

and 8V. The device has a gatewidth of 2.4 mm, and at 30 GHz it achieves an outp power of 1W and a gain of 3.1 dB with an efficiency of 15.6%. These features compare favorably to the MESFET.

One attempt at a GaAS MISFET, a heterostructure MISFET with an ins lating layer of undoped AlGaAs, merits attention. Its strengths are:

1. GaAs/AlGaAs produces an extremely good heterojunction interface, wi no surface-state problems.
2. Because the AlGaAs is undoped, there is no "DX center" of deep electr traps.
3. AlGaAs has a higher Schottky barrier than GaAs by approximately 0.2 e allowing a greater input power margin.

Kim et al. [38] achieved an output power of 630 mW with a gain of 7 dB and efficiency of 37% at 10 GHZ in a heterostructure MISFET with a gatewidth 750 $\mu$m. The MISFET cross-section is shown in Figure 2.58. Further optimizati of structural properties led to the announcement of elements with even bett characteristics. Hida et al. [39] created a structure combining a heterostructu MISFET and MESFET, called a doped-channel MIS-like FET (DMT), with gatewidth of 280 $\mu$m, which at 28.5 GHz achieved an output power of 18 dBm 15% efficiency with $G_L$ = 6.4 dB. This device was not necessarily designed f

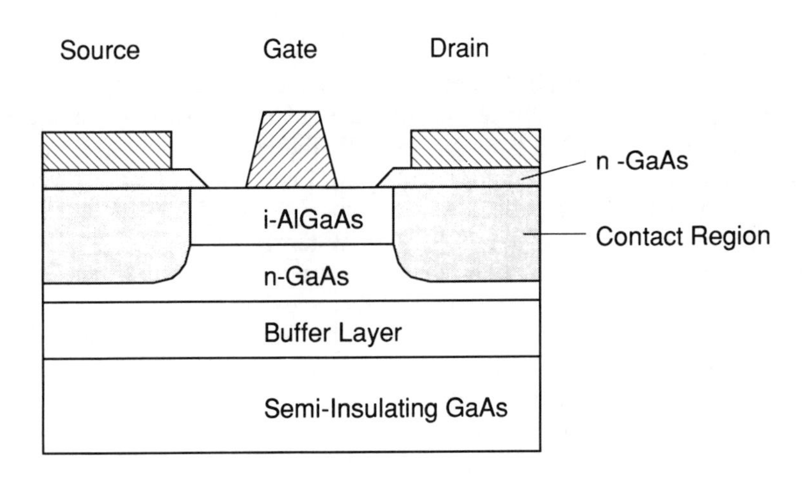


**gure 2.58** Cross-sectional structure of typical GaAs heterojunction MISFET.

gh-output applications, so that its output by itself is not particularly noteworthy, ıt the gain level is significant.

These applications have been serious attempts to use heterojunctions to con- ol the operation of elements, but there have been other attempts made to use ·terojunctions to improve performance characteristics. One example is a GaAs ESFET buffer layer constructed from an AlGaAs/GaAs heterojunction and a ·terojunction superlattice, created to reduce the effect of the GaAs substrate. nother example is the insertion of thin AlGaAs or GaAs layers to improve process ntrol by using the difference in etching rates between GaAs and AlGaAs. Both these techniques suggest future directions for the use of heterojunctions in ·velopment work.

### 8.3 The Heterojunction Bipolar Transistor

nlike the FET, in which current flows across the semiconductor surface, the polar transistor has a current that flows vertically up and down from the surface rough the chip. This gives such devices considerable current driving capability d makes them well suited for high output power applications. The silicon bipolar ınsistor has always been a high-power device and is widely used at frequencies ) to L-band, but the device that has attracted attention for high-frequency, high- wer applications is the *heterojunction bipolar transistor* (HBT) with the superior gh-frequency properties of GaAs. By using a GaAs/AlGaAs heterojunction as emitter base, it is possible to lower the emitter doping density with respect to e base doping density with no decrease in emitter injection efficiency. This makes it possible to reduce the base resistance and emitter capacitance. Figure 2.59 show a cross section of a typical HBT. The HBT is being intensively developed for us in ultrahigh-speed digital applications, but there have been several announcement of microwave high-power HBT devices as well. Tsuda et al. [40] have reported o an HBT with a total emitter length of 320 $\mu$m, a maximum output power of 1W a power gain of 6 dB, and a collector efficiency of 49% at 5 GHz. Wang et al. [41 achieved an output power of 0.358W (3.58 W/mm), a gain of 11.4 dB, and a efficiency of 43% at 18 GHz in a common base configuration. These announcemen demonstrate that the HBT shows millimeterwave band characteristics that ar superior to the MESFET. The complex manufacturing process required has ke it from reaching the mass production stage; however, advances in process tec nology are sure to bring such devices onto the market.

In addition to the HBT, the PBT, and HET have shown promise for th development of future high-frequency, high-power devices.

## 2.9 CONCLUSION

In this chapter, we have discussed the design and evaluation of, primarily, the hig power GaAS MESFET. By now, this technology can be called mature, but advanc in the technologies of crystal growth, manufacturing processes, and circuit desig combined with the introduction of new concepts from heterojunctions and ne materials, continue to open up new directions for development. Such developme continues to be motivated by market demands. The FET continues to advance ar

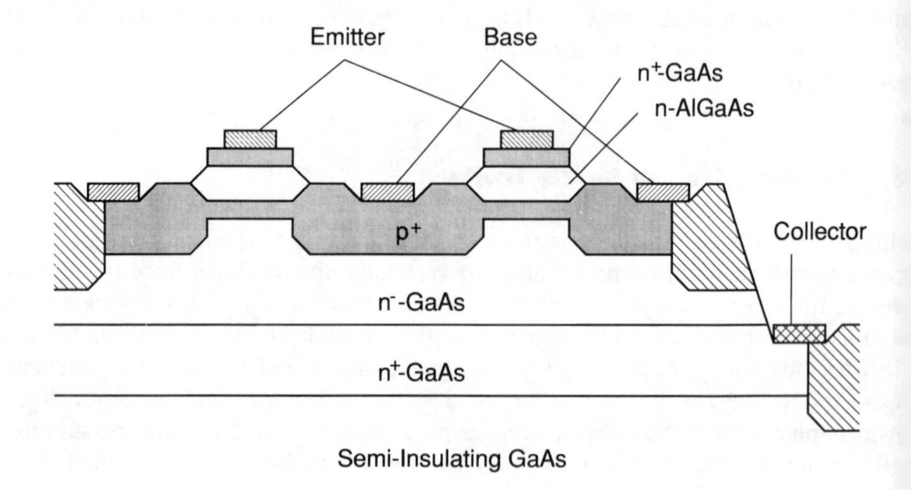


**Figure 2.59** Cross-sectional structure of a microwave HBT.

hange in response to wider and more demanding applications of digital multiplexer ommunications, ever-higher frequencies from UHF to microwave and millimeerwave bands, and the high reliability demands of satellite applications. We coninue to look forward to the new horizons that emerge as communication systems nd electronic devices surpass old limits and encourage each other to open new narkets and technologies.

## REFERENCES


1. Mead, C. A., "Schottky Barrier Gate Field Effect Transistor," *Proc. IEEE*, Vol. 54, February 1966, pp. 307–308.
2. Baechtold, W., W. Walter, and P. Wolf, "X and Ku band GaAs MESFET," *Electron. Lett.*, Vol. 8, January 1972, pp. 35–37.
3. Liechti, C., E. Gowen, and J. Cohen, "GaAs Microwave Schottky-Gate FET," *IEEE Int. Solid-State Circuit Conf. Tech. Digest*, 1972, pp. 158–159.
4. Fukuta, M., T. Mimura, I. Tsujimura, and A. Furumoto, "Mesh Source Type Microwave Power FET," *IEEE Int. Solid-State Circuit Conf. Tech. Digest*, 1973, pp. 84–85.
5. Napoli, L. S., R. E. DeBrecht, J. J. Hughes, W. F. Reichert, A. Dreeben, and A. Triano, "High-Power GaAs FET Amplifier—A Multigate Structure," *IEEE Int. Solid-State Circuit Conf. Tech. Digest*, 1973, pp. 82–83.
6. Fukuta, M., K. Suyama, H. Suzuki, K. Nakayama, and H. Ishikawa, "Power GaAs MESFET with a High Drain-Source Breakdown Voltage," *IEEE Trans. on Microwave Theory and Techniques*, Vol. MTT-24, June 1976, pp. 312–317.
7. Higashisaka, A., Y. Takayama, and F. Hasegawa, "A High-Power GaAs MESFET with an Experimentally Optimized Pattern," *IEEE Trans. on Electron Devices*, Vol. ED-27, June 1980, pp. 1025–1029.
8. Fukaya, J., M. Ishii, M. Matsumoto, and Y. Hirano, "A C-Band 10 Watt GaAs Power FET," *IEEE MTT-S Int. Microwave Symp. Digest*, San Francisco, CA, May 30–June 1, 1984, pp. 439–440.
9. Mimura, T., S. Hiyamizu, T. Fujii, and K. Nanbu, "A New Field-Effect Transistor With Selectively Doped GaAs/n-$Al_xGa_{1-x}As$ Heterojunctions," *Jpn. J. Appl. Phys. Lett.*, Vol. 19, May 1980, pp. L225–L227.
10. Williams, R. E., and D. W. Shaw, "Graded Channel FET's: Improved Linearity and Noise Figure," *IEEE Trans. on Electron Devices*, Vol. ED-25, June 1978, pp. 600–605.
11. Smith, P. M., M. Y. Kao, P. Ho, P. Chao, K. H. G. Duh, A. A. Jabra, R. P. Smith, and J. M. Ballinball, "A 0.15μm Gate-Length Pseudomorphic HEMT," *IEEE MTT-S Int. Microwave Symp. Digest*, Long Beach, CA, June 13–15, 1989, pp. 983–986.
12. Bodway, G. E., "Two Port Power Flow Analysis Using Generalised Scattering Parameters," *Microwave J.*, Vol. 10, May 1967, pp. 61–66.
13. Mimura, T., H. Suzuki, and M. Fukuta, "Visible Light Emission from GaAs Field-Effect Transistor," *Proc. IEEE.*, Vol. 65, September 1977, pp. 1407–1408.
14. Yamamoto, R., A. Higashisaka, and F. Hasegawa, "Light Emission and Burnout Characteristics of GaAs Power MESFET's," *IEEE Trans. on Electron Devices*, Vol. ED-25, June 1978, pp. 567–573.
15. Higashisaka, A., "Research on Schottky-Barrier Gate Field Effect Transistors," Ph.D. thesis, Tokyo Institute of Technology, 1981 (in Japanese).
16. Fukuta, M., "Research on High-Frequency, High-Power Field Effect Transistor," Ph.D. thesis, Nagoya University, 1977 (in Japanese).

17. Wemple, S. H., W. C. Niehaus, H. M. Cox, J. V. Dilorenzo, and W. O. Schlosser, "Control o Gate-Drain Avalanche in GaAs MESFET's," *IEEE Trans. on Electron Devices*, Vol. ED-27, Jun 1980, pp. 1013–1018.
18. Hikosaka, K., Y. Hirachi, and M. Abe, "Microwave Power Double-Heterojunction HEMT's," *IEEE Trans. on Electron Devices*, Vol. ED-33, May 1986, pp. 583–589.
19. Sze, S. M., *Physics of Semiconductor Devices*, 2nd ed., New York: John Wiley & Sons, 1981.
20. Fukui, H., "Determination of the Basic Device Parameters of GaAs MESFET," *Bell Syst. Tech J.*, Vol. 58, March 1979, pp. 771–797.
21. Hasegawa, F., "Power GaAs FETs," Chapt. 3 in *GaAs FET Principles and Technology*, J. V Dilorenzo, and D. Khandelwal, Eds., Norwood, MA: Artech House, 1982.
22. Aono, Y., A. Higashisaka, T. Ogawa, and F. Hasegawa, "X- and Ku-Band Performance of Sub micron Gate GaAs FETs," *Jpn. J. Appl. Phys.*, Vol. 17, Suppl. 17-1, 1978, pp. 147–152.
23. Hirachi, Y., Y. Takeuchi, M. Igarashi, K. Kosemura, and S. Yamamoto, "A Packaged 20-GH 1-W MESFET with a Novel Via-Hole Plated Heat Sink Structure," *IEEE Trans. on Microwav Theory and Techniques*, Vol. MTT-32, March 1984, pp. 309–316.
24. Takahashi, H., T. Igarashi, S. Inaba, and Y. Hasegawa, "A 0.25μm Gold Gate GaAs Power FET," *Proc. Int. Workshop on Solid State Power Amplifiers for Space Application*, ESA WPP-013 (ESA ESTEC), November 1989, pp. 217–222.
25. D'Asaro, L. A., J. V. Dilorenzo, and H. Fukui, "Improved Performance of GaAs Microwav Field-Effect Transistors with Via-Connections through the Substrate," *Int. Electron Devices Meetin Tech. Digest*, 1977, p. 371.
26. Pengelly, R. S., *Microwave Field-Effect Transistors—Theory, Design, and Applications*, New York John Wiley & Sons, 1982.
27. Kohn, E., "Dimensionierung und Technologie Integrationsfähiger GHz—MESFETs aus galli umarsenid," Ph.D. dissertation, Technische Hochschule Aachen, 1975, pp. 64–67.
28. D'Asaro, L. A., J. V. Dilorenzo, and H. Fukui, "Improved Performance of GaAs Microwav Field-Effect Transistors with Low Inductance Via-Connections Through The Substrate," *IEE Trans. on Electron Devices*, Vol. Ed-25, October 1978, pp. 1218–1221.
29. Hirachi, Y., and M. Fukuta, "High Power GaAs FET—Design and Characterization," Chap. 2 i *Handbook on Compound Semiconductor Devices*, R. Itoh, Ed., Science Forum, 1986 (in Japanese)
30. Fukuta, M., and Y. Hirachi, *Introduction to GaAs Field-Effect Transistor*, Tokyo: IEICE, 1991 (i Japanese).
31. Sumitani, K., M. Komura, M. Kobiki, Y. Higaki, Y. Mitsui, H. Takano, and K. Nishitani, "A High Aspect Ratio Via Hole Dry Etching Technology for High Power GaAs MESFET," *Tech Digest GaAs IC Symp.*, 1989, pp. 207–210.
32. Takase, S., Y. Hirano, Y. Tanguchi, T. Ohno, and K. Ohta, "High Gain, High Power Adde Efficiency GaAs Power FET for Satellite Use," *Workshop Proc. on High Power Solid State Ampli fiers*, Wembley, UK, Sep. 8, 1989.
33. Taniguchi, Y., Y. Hasegawa, Y. Aoki, and J. Fukaya, "A C-Band 25 Watt Linear Power FET," *IEEE MTT-S Int. Microwave Symp. Digest*, Dallas, TX, May 8–10, 1990, pp. 981–984.
34. Itoh, T., and K. Ohata, "X-Band Self-Aligned Gate Enhancement-Mode InP MISFET's," *IEE Trans. on Electron Devices*, Vol. ED-30, July 1983, pp. 811–815.
35. Tokuda, H., H. Ishimura, H. Sasaki, F. Sasaki, Y. Oda, S. Yanagawa, and M. Higashiura, "Deple tion-Mode Power InP MISFETs Operated at 30GHz," *Proc. 1988 Int. Symp. GaAs & Relate Compounds*, 1989, pp. 475–480.
36. Ohata, K., "Microwave Heterojunction Devices," *European Microwave Conf. Digest*, 1990, pp 136–146.
37. Hikosaka, K., N. Hidaka, Y. Hirachi, and M. Abe, "A 30-GHz 1-W Power HEMT," *IEEE Electro Device Lett.*, Vol. EDL-8, 1987, pp. 521–523.

8. Kim, B., H. Q. Tserng, and H. D. Shih, "Microwave Power GaAs MISFETs with Undoped AlGaAs as an Insulator," *IEEE Electronic Device Lett.*, Vol. EDL-5, 1984, pp. 494–495.
9. Hida, H., A. Okamoto, H. Toyoshima, and K. Ohata, "An Investigation of i-AlGaAs/*n*-GaAs Doped Channel MIS-Like FET's (DMT's)—Properties and Performance Potentialities," *IEEE Trans. on Electron Devices*, Vol. ED-34, July 1987, pp. 1448–1455.
10. Tsuda, K., et al. "AlGaAs/GaAs Heterojunction Bipolar Transistors for Microwave Power Amplifier Application," *Extended Abstracts of the 19th Conf. on Solid-State Devices and Materials*, Tokyo, 1987, pp. 271–274.
11. Wang, N. L., N. H. Sheng, W. J. Ho, M. F. Chang, G. J. Sullivan, J. A. Higgins, and P. M. Asbeck, "18 GHz High Gain, High Efficiency Power Operation of AlGaAs/GaAs HBT," *IEEE MTT-S Int. Microwave Symp. Digest*, Dallas, TX, May 8–10, 1990, pp. 997–1000.

# Chapter 3
# Computer-Aided Design of GaAs FET Power Amplifiers

**O. Pitzalis**
*EEsof*

## 3.1 INTRODUCTION

Large-signal microwave *computer-aided design* (CAD) became widely available for microwave applications following an adaptation of the SPICE time-domain program. In 1986, by incorporating single- and coupled-microstrip line models plus RF power sources and measurements, mwSPICE™ became the first CAD tool specifically suited for the large-signal design/analysis of microwave amplifiers, mixers, and oscillators. Then in 1988, the first of several commercial harmonic balance simulation tools appeared with the introduction of Libra™ [1]. Since that time harmonic balance has almost entirely replaced SPICE-based time domain for large-signal, steady-state, microwave design and analysis. The obvious advantages of harmonic balance are greater ease of use and, for most circuits, greater simulation speed. In addition, the ability of harmonic balance to use conventional linear models and $S$ parameter characterization data for any linear circuit components in the network greatly expanded the versatility and accuracy of large-signal analysis. Today, harmonic balance tools provide an efficient means of large-signal design and optimization of many circuits, including amplifiers, mixers, and oscillators.

Prior to large-signal CAD, limited linear design analysis based on $S$ parameter characterization of nonlinear semiconductor devices was used for impedance matching and stability analysis of power amplifiers. Hand-calculated load lines and, later, load-pull impedance characterization measurements provided the only means for linear design optimization of load matching networks. No effective way existed to analyze gain, saturation characteristics, dc-to-RF efficiency, or $Q$-point bias shifting under overdrive, or to determine the sensitivity of these characteristics to network tolerances. Breadboard prototypes were essential. Prototype amplifiers were th only means for design evaluation and improvement of large-signal circuits.

State-of-the-art MICs and MMICs, however, are fabricated using such sophis ticated process technology that the implementation of prototype designs is no practical. For example, a breadboard MMIC would require a few months fo completion of the MMIC process cycle together with an unacceptably high devel opmental cost for most projects. In addition, hardware prototypes have som limitations. For instance, it is usually impossible to access points internal to th circuit to make actual measurements of current and voltage waveforms that coul be critical to the circuit's operational life. However, using large-signal analysis i is possible to confidently simulate current and voltage waveforms at all points i a circuit.

The evolutionary CAD technology advances that have been made in recen years have been greatly supported through the parallel development of more vali models for both active devices and passive circuit elements. In particular, accurat large-signal performance predictions for microwave circuits depend absolutely o accurate large-signal models for MESFETS. Since these MESFET models ar nonideal, the designer can benefit from an understanding of the features an limitations of the widely used large-signal MESFET models so that he or she ca prudently apply the models in design and analysis. Discussion of the difference among the popular models plus insight into what can be expected in simulation is a goal of this chapter.

Following the MESFET large-signal modeling discussion, the capabilities c such a model are demonstrated in a series of general electrical performance sim ulations of a C-band MESFET power amplifier stage. The simulation results ar compared with all the available measurement data. In addition, a series of simulate signal waveforms provides valuable insight into large-signal MESFET operationa behavior.

## 3.2 GaAs FET NONLINEAR MODELS

The large-signal MESFET models popularly used for circuit design and analysi are empirical models that incorporate analytical functions to represent MESFE nonlinearities dependent on $V_{gs}$ and $V_{ds}$. The analytical functions usually do nc have any relevance to the behavioral physics of the MESFET.

These large-signal MESFET models can be considered to consist of two part The first is a static dc $I$–$V$ portion that relates to the device's resistive element and the transconductance. The second part is the dynamic portion that account for the reactive capacitive elements. If extrinsic interconnections or a package ar to be included in the device model, then the extracted associated inductance apacitances, or distributed transmission elements would form additional elements n the dynamic portion.

## .2.1 The MESFET Large-Signal RF Equivalent Circuit

he MESFET large-signal model is best introduced by starting from the MESFET nall-signal equivalent circuit. The linear, lumped element, equivalent circuit of igure 3.1 has long served as an accurate small-signal model for virtually all GaAs IESFETs and HEMTs with gate lengths ranging from 0.1 $\mu$m to several microns nd with gatewidths ranging from a few microns to several centimeters. This equivent model is capable of accurately reproducing measured $S$ parameters to frequencies of at least 60 GHz, and probably to beyond 100 GHz. It should be pointed ut that for accurate modeling at these frequencies the equivalent circuit is simply tended by defining bond pad and other electrode and interelectrode capacitances additional extrinsic elements.

Some obvious points regarding the small-signal model that should be stated e:

1. Since $g_m$, $\tau$, $C_{gs}$, $C_{gd}$, $R_i$, and $R_{ds}$ vary with $I_{ds}$ and $V_{ds}$, the small-signal model is only valid at the specific quiescent bias point $I_{dsQ}$, $V_{dsQ}$ for which it has been extracted. The domain capacitance $C_{dc}$ accounts for the charge that would be a Gunn domain should one exist. Gunn domain formation, once common in power MESFETs, is now rarely seen in state-of-the-art devices.

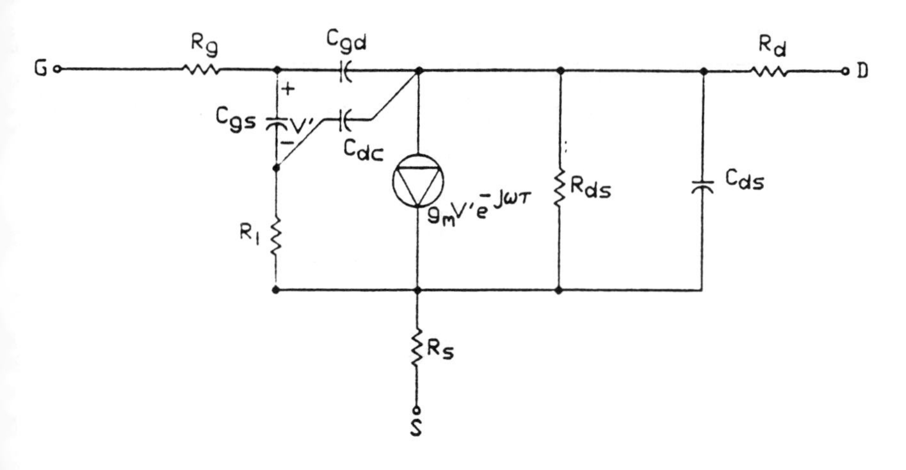


**ure 3.1** The MESFET small-signal equivalent circuit model.

2. The small-signal MESFET equivalent circuit model is a subset of the large signal MESFET equivalent circuit model. That is, an ideal large-signal model biased at any specified quiescent point within the active *I–V* range of operation but operated in the small-signal range, produces an accurate small-signal model.

The SPICE MESFET large-signal equivalent circuit model that is widely used is shown in Figure 3.2. The diodes represent the gate-to-source and the gate-to drain Schottky barrier junctions. In this model $C_{gs}$ and $C_{gd}$ are modeled as Schottky barrier diode capacitances, each varying as a function of the respective junction voltage according to the equations:

$$C_{gs} = G_{gs0}/(1 + V_{gs}/V_{\phi})^{0.5} \quad (3.1)$$

$$C_{gd} = C_{gd0}/(1 + V_{gd}/V_{\phi})^{0.5} \quad (3.2)$$

where $C_{gs0}$ and $C_{gd0}$ are the zero-voltage junction capacitances and $V_{\phi}$ is the built in junction potential (typically, 0.8V for GaAs MESFETS). This capacitance model is of limited accuracy, particularly for $C_{gs}$, which varies considerably with both $V$ and $V_{gs}$.

The junction conductances $G_{gs}$ and $G_{gd}$ are negligibly small in normal Class A bias operation where both junctions are in reverse bias. However, under extremes of quiescent bias or large-signal operation, the conductances apply if either or both of the junctions should become forward biased.

The output resistance $R_{ds}$ is dispersive with frequency [2] as shown in Figure 3.3. The value of $R_{ds}$ at dc is much larger than ac values, being typically 3 to

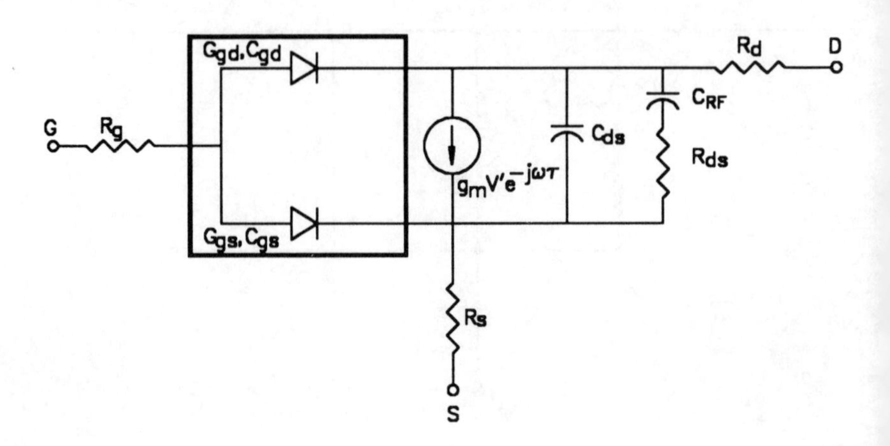


**Figure 3.2** The SPICE MESFET large-signal equivalent circuit model.

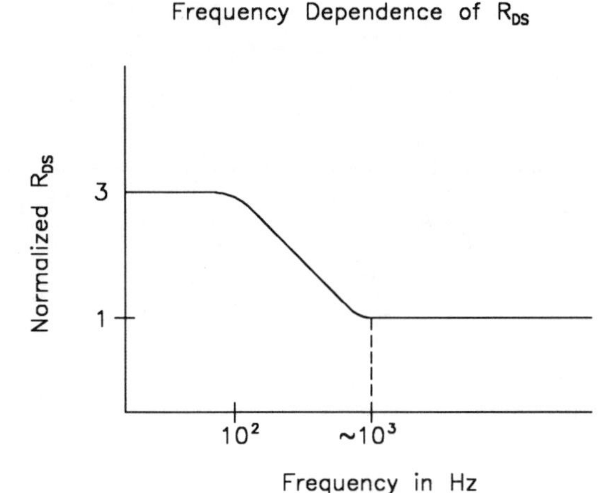


**igure 3.3** The frequency dispersion of the drain-to-source resistance $R_{ds}$.

imes larger. The transition frequency, which depends on the semiconductor mate-ial characteristics, can typically be anywhere between 1 and 100 kHz. The output esistance $R_{ds}$ indicated in Figure 3.2, is the RF value. The capacitance $C_{RF}$ sets he transition frequency for $R_{ds}$.

A similar dispersion exists for transconductance with the ac value of $g_m$ settling t a plateau that is typically 10% to 15% below the dc value of $g_m$. The popular arge-signal models to be discussed are unable to account for the $g_m$ frequency ispersion. Instead, the models portray the dc $g_m$ that is derived from dc $I$–$V$ neasurements. Recently, a feasible method [3, 4] for simply incorporating the RF $_m$ (as well as $R_{ds}$) in the models has been demonstrated. Unfortunately, the pro-osed approach requires specialized pulsed $I$–$V$ measurements that are not simply nplemented.

In Figure 3.4 the large-signal model is redrawn in a form that more closely esembles the small-signal equivalent circuit topology from Figure 3.1. The symbols lentify $g_m$, $\tau$, $C_{gs}$, $C_{gd}$, $R_i$, and $R_{ds}$ as nonlinear elements. Two elements not found 1 the small-signal equivalent circuit are the nonlinear gate-to-source conductance $_{gs}$ and the gate-to-drain conductance $G_{gd}$, which model forward bias junction onduction.

The $V_{gs}$, $V_{ds}$ dependency of these equivalent circuit elements is illustrated in e curves in Figures 3.5 to 3.11. The parameterized $V_{gs}$, $V_{ds}$ data came from xtracting the small-signal equivalent circuit values from $S$ parameter measurements ken over a broad range of $V_{gs}$, $V_{ds}$ quiescent bias points. The MESFET chosen a TriQuint 0.5$\mu$m gate length by 300-$\mu$m gatewidth device processed using ion

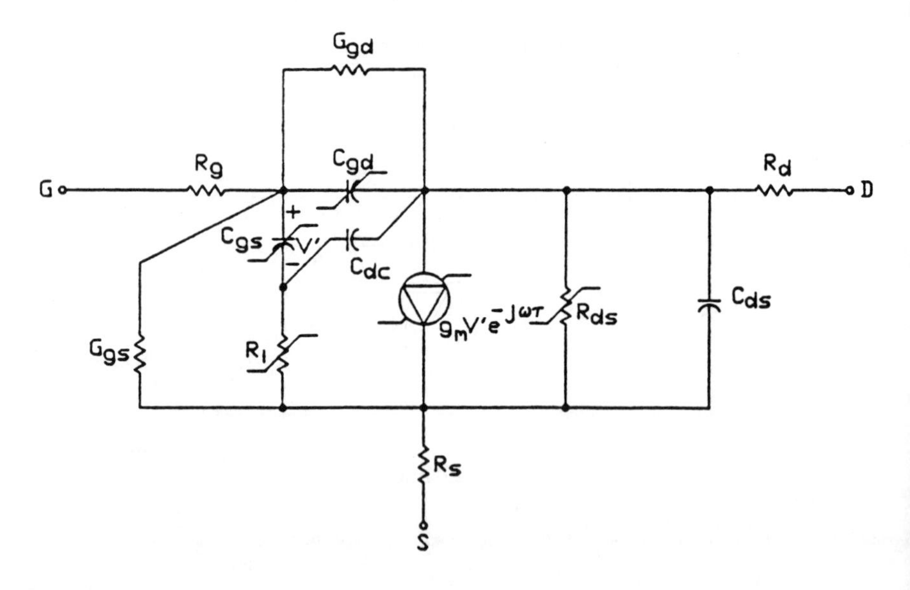


**Figure 3.4** The MESFET large-signal equivalent circuit model redrawn in the form of the small-signal equivalent circuit model.

implantation. The bias dependencies illustrated for this MESFET will generally apply to most MESFETs, even those of differing gate lengths and gatewidths, or for MESFETs fabricated using a different material process such as an epitaxial construction. To better illustrate the bias dependencies, plots are presented for each element versus both $V_{ds}$ and $V_{gs}$.

Transconductance $g_m$ is plotted versus $V_{gs}$ for several $V_{ds}$ values in Figure 3.5(a). The plot indicates that $g_m$ grows monotonically with $V_{gs}$, with a minor dependence on $V_{ds}$ in saturation. The greatest $g_m$ nonlinearity occurs in the $V_{gs}$ range near pinch-off. In Figure 3.5(b) the behavior of the family of $g_m$ versus $V_{ds}$ curves with $V_{gs}$ as a parameter closely resembles a family of dc $I$–$V$ curves of $I_{ds}$ versus $V_{ds}$. Of all the equivalent circuit elements, $g_m$ is the most nonlinear. In large-signal Class A operation, where waveforms approximate unclipped sinusoids, nonlinearity in $g_m$ dominates the generation of harmonic and *intermodulation distortion* (IMD) products. An even greater nonlinearity, which occurs in large-signal operation, involves clipping of the $I_{ds}$ and $V_{ds}$ waveforms. Such clipping will occur when a class A stage input is overdriven forcing the stage into Class AB operation.

The transconductance time delay $\tau$ corresponds to the transit time of the majority carriers from the source to the drain passing under the bias-dependent region defining the operating gate length. The accompanying bias dependence of $\tau$ is illustrated in Figure 3.6. As $V_{ds}$ increases, the drain-to-gate depletion region

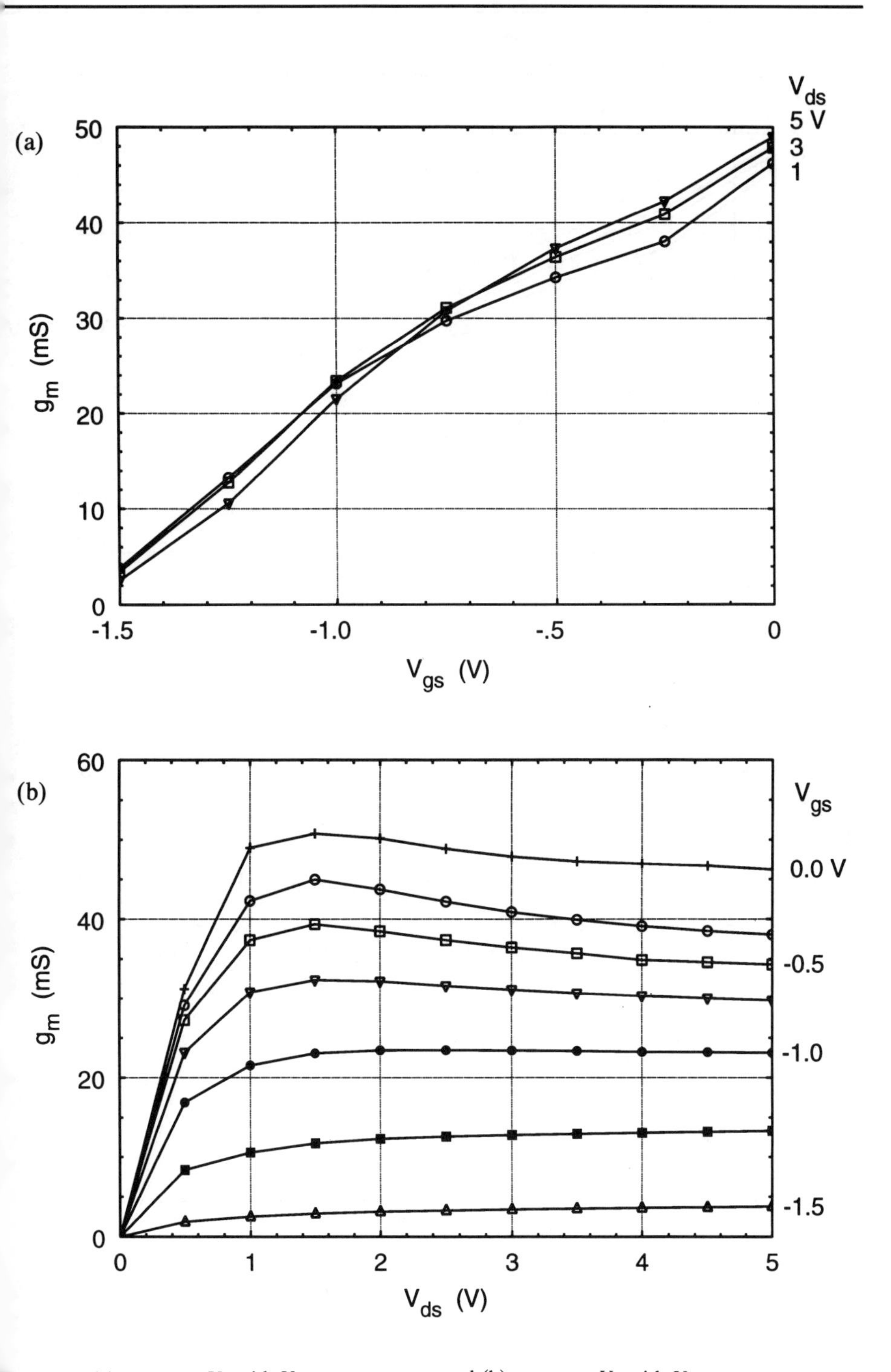


**Figure 3.5** (a) $g_m$ versus $V_{gs}$ with $V_{ds}$ as a parameter and (b) $g_m$ versus $V_{ds}$ with $V_{gs}$ as a parameter.

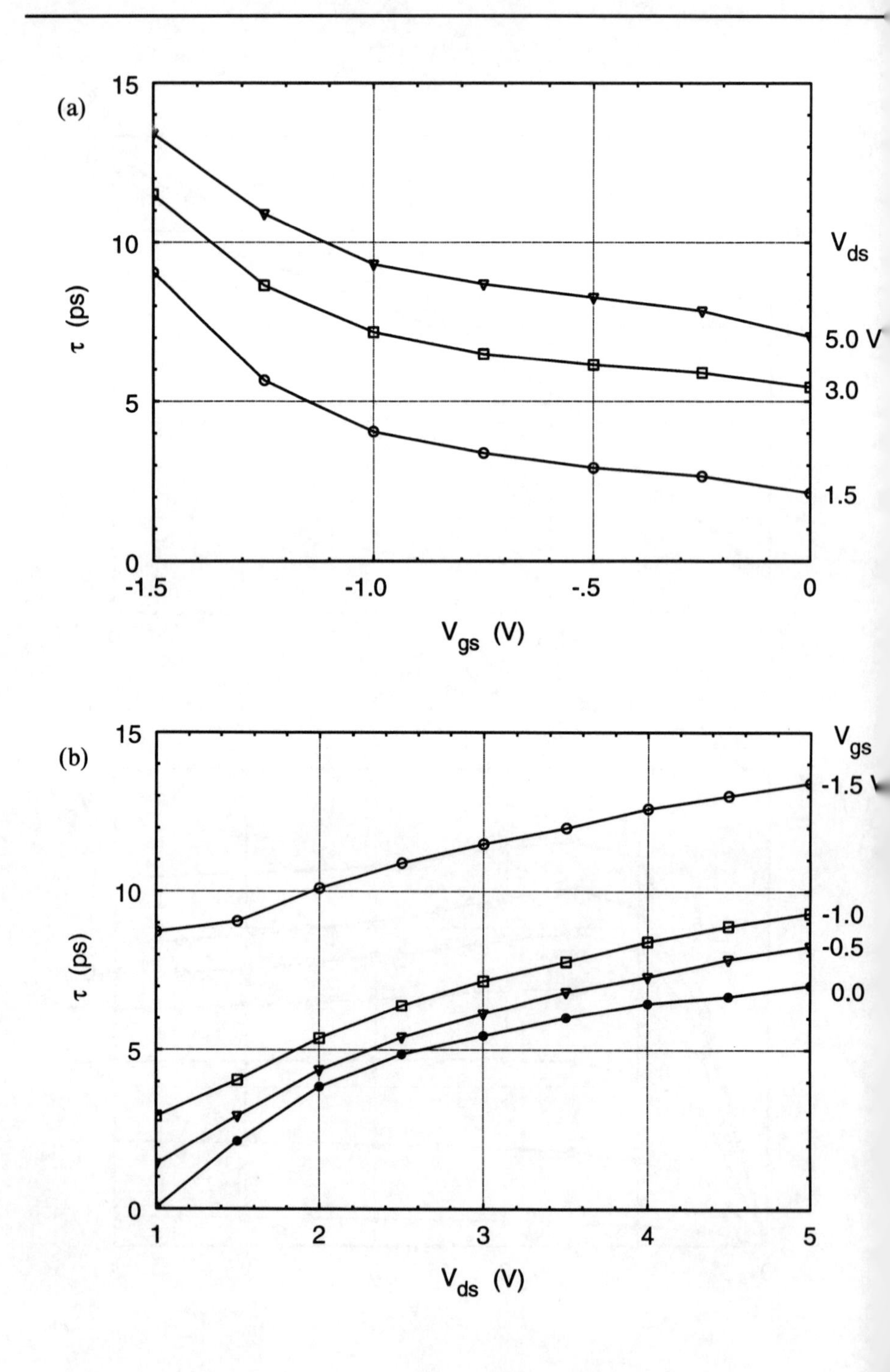


**Figure 3.6** (a) $\tau_{gm}$ versus $V_{gs}$ with $V_{ds}$ as a parameter and (b) $\tau_{gm}$ versus $V_{ds}$ with $V_{gs}$ as a parameter

idens defining a longer effective gate length leading to an increase in $\tau$ with $V_{ds}$. s $V_{gs}$ increases, $\tau$ is seen to decrease in correspondence to the increases that are multaneously occurring for $g_m$.

The family of $C_{gs}$ versus $V_{gs}$ curves of Figure 3.7(a) displays trends similar to ie $g_m$ versus $V_{gs}$ curves. The $V_{ds}$ dependence of $C_{gs}$ is also significant. The $C_{gs}$ oubles as $V_{ds}$ changes from 5V to 0V for $V_{gs}$ greater than $-1$V. Not all MESFETs iow such a large $V_{ds}$ dependency. This $V_{ds}$ dependency for $C_{gs}$ is only modeled in ome of the more advanced MESFET models [5–7]. The $V_{gs}$, $V_{ds}$ variations in $C_{gs}$ ccount for the next major nonlinearity after $g_m$. The $C_{gs}$ nonlinearity produces armonic powers that are typically between 6 and 10 dB below those of $g_m$, making ie $C_{gs}$ harmonic contributions relatively insignificant. Nevertheless, the accurate iodeling of the $C_{gs}$ nonlinearity is important for other reasons. For example, the actance variation from $C_{gs}$ with $V_{gs}$ causes amplitude-dependent transmission hase variations or AM-PM conversion in amplifiers, and FM noise in oscillators.

The power gain $G$ of a tuned MESFET power amplifier stage at an operating equency $f$ can be closely approximated using:

$$G = \left[\frac{f_t}{f}\frac{R_{\text{out}}}{(R_{\text{out}} + R_{\text{LL}})}\right]^2 \frac{R_{\text{LL}}}{R_{\text{in}}} \tag{3.3}$$

here $R_{\text{out}}$ is the transistor output resistance, i.e., $R_{\text{out}} \approx R_{ds}$; $R_{\text{LL}}$ is the dynamic ad line resistance in parallel with $R_{\text{out}}$; and $R_{\text{in}}$ is the transistor input resistance, ., $R_{\text{in}} \approx (R_g + R_i + R_s + 2\pi f_t \cdot L_s)$. From (3.3) it is apparent that the MESFET ower gain nonlinearity is primarily a function of the short-circuit current gain andwidth product $f_t$ where

$$f_t = g_m/2\pi(C_{gs} + C_{gd}) \approx g_m/2\pi C_{gs} \tag{3.4}$$

rom (3.4) it is obvious that the bias nonlinearity of $f_t$ is governed by the bias onlinearities of $g_m$ and $C_{gs}$. As is apparent from Figures 3.5 and 3.7, both $g_m$ the $_{gs}$ have similar variations with $V_{gs}$ and $V_{ds}$. Therefore, in the ratio of $g_m/C_{gs}$, the , and $C_{gs}$ variations partially compensate one another so that $f_t$ variations with $_{gs}$ and $V_{ds}$ can be expected to be considerably smaller than those for $g_m$. This ompensation is evident in the curves of $f_t$ versus $V_{gs}$ and $V_{ds}$ presented in Figures 8(a) and (b), respectively. In these figures, $f_t$ is seen to be nearly constant for $_{gs}$ varying from $-1.0$ to 0V. The $f_t$ variation with $V_{ds}$ is seen to be somewhat eater. The peak $f_t$ occurs at a $V_{ds}$ of about 1.5V. Thereafter, $f_t$ decreases by as uch as 30% as $V_{ds}$ reaches 5V.

Figures 3.9(a) and (b) show that the $V_{gs}$ and $V_{ds}$ dependence of $C_{gd}$ is slight cept when $V_{ds}$ is lower than 1V. Since there is only a small bias variation of $C_{gd}$ the active region of transistor operation, $C_{gd}$ can be considered to act as a fixed

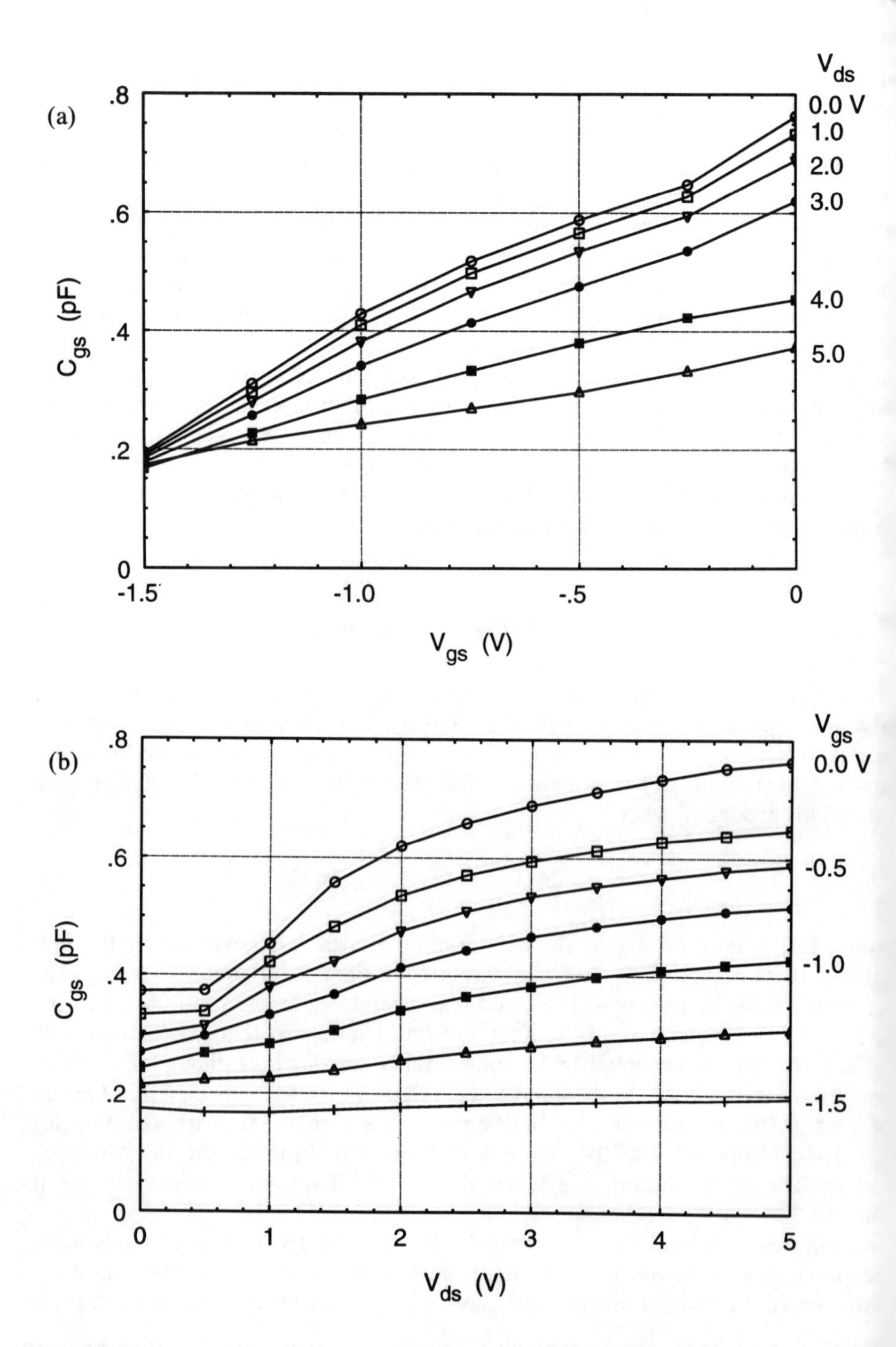


**Figure 3.7** (a) $C_{gs}$ versus $V_{gs}$ with $V_{ds}$ as a parameter and (b) $C_{gs}$ versus $V_{ds}$ with $V_{gs}$ as a parameter.

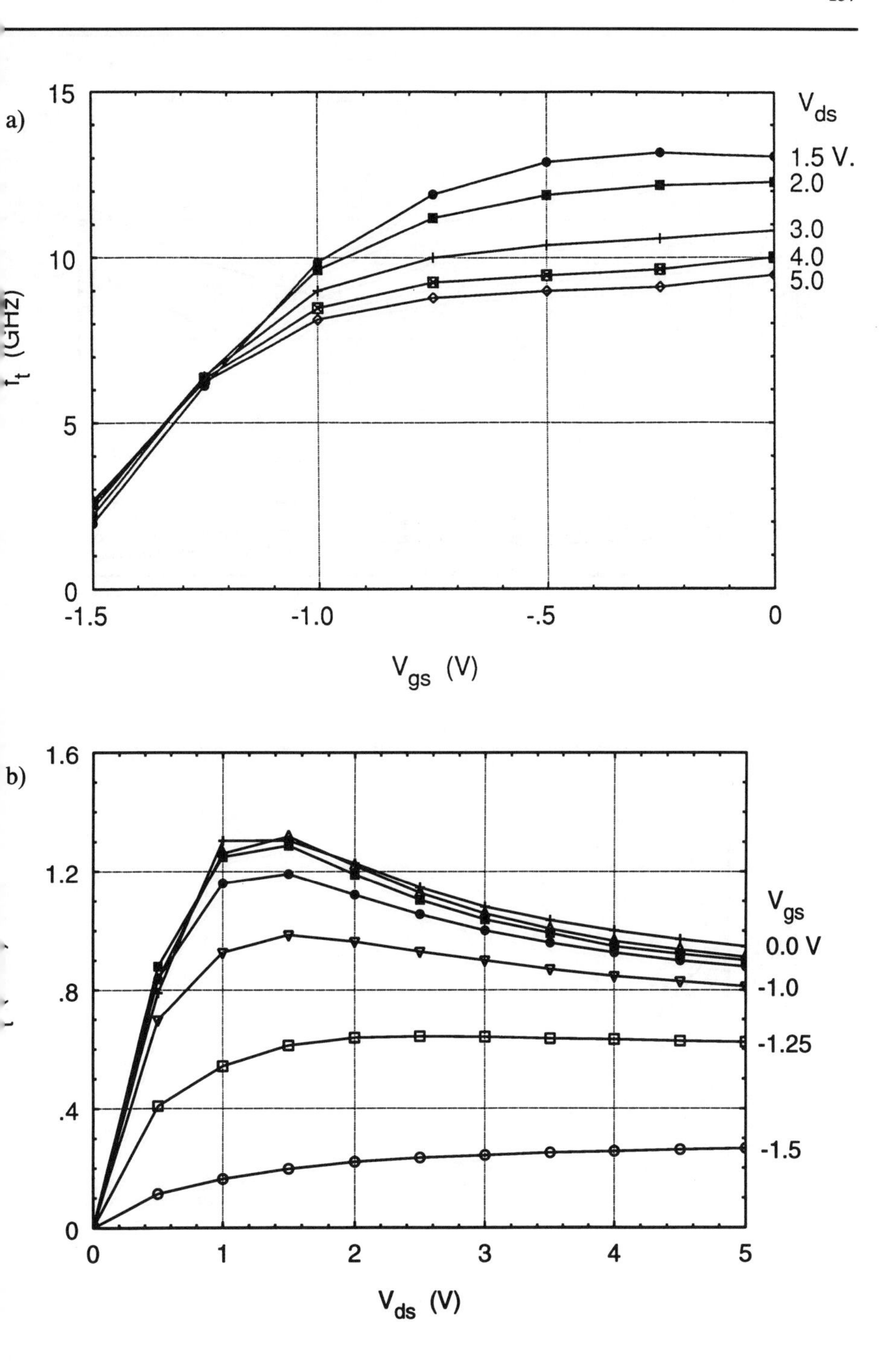


**Figure 3.8** (a) $f_t$ versus $V_{gs}$ with $V_{ds}$ as a parameter and (b) $f_t$ versus $V_{ds}$ with $V_{gs}$ as a parameter.

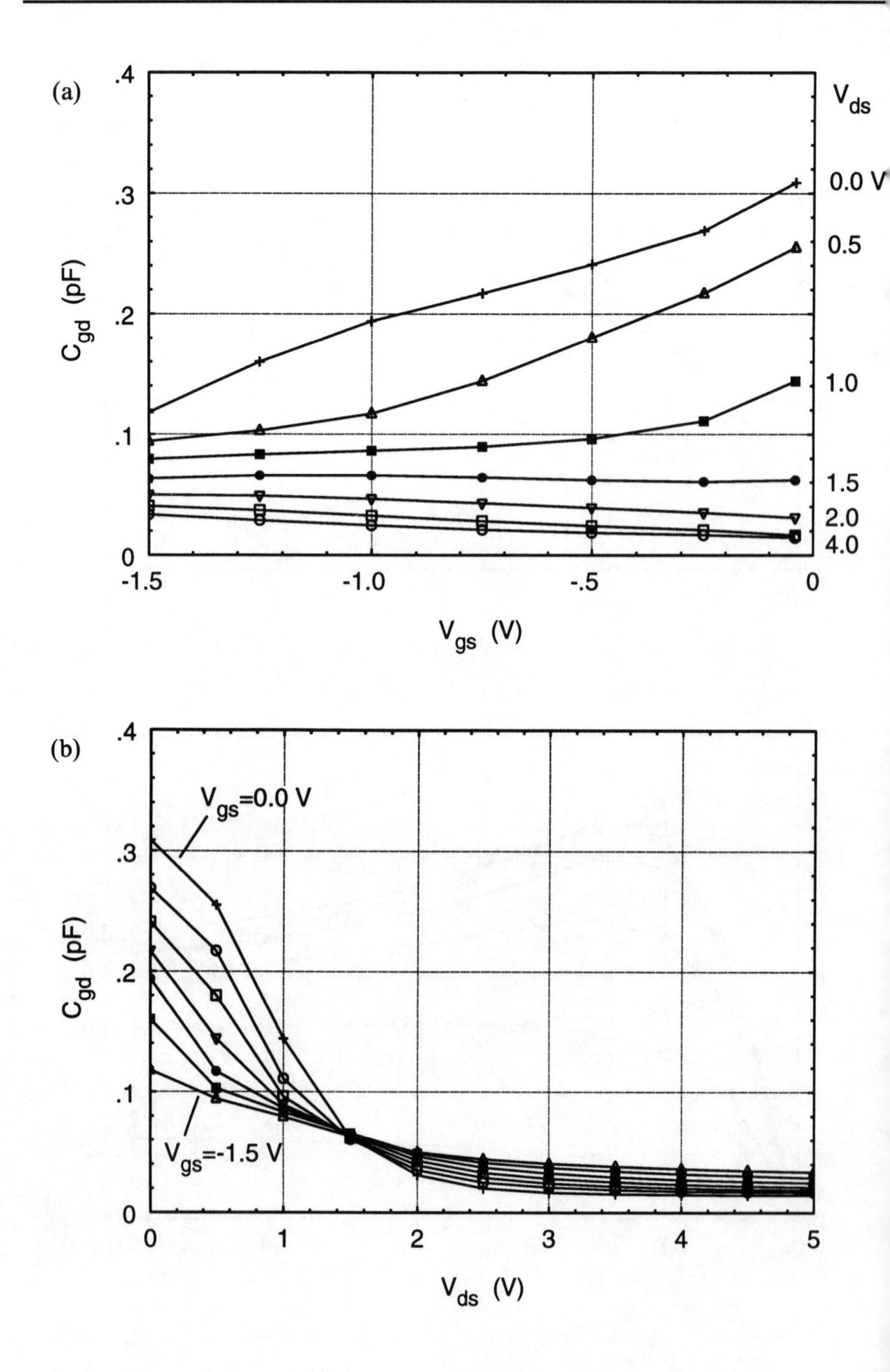


**Figure 3.9** (a) $C_{gd}$ versus $V_{gs}$ with $V_{ds}$ as a parameter and (b) $C_{gd}$ versus $V_{ds}$ with $V_{gs}$ as a parameter

apacitance. For this reason, $C_{gd}$ is indicated as fixed in the large-signal equivalent ircuit of Figure 3.3.

The charging resistance for $C_{gs}$ is $R_i$. As shown in Figures 3.10(a) and (b), ne value of $R_i$ ranges between 0 and 7.5$\Omega$ with a strong dependence on both $V_{gs}$ nd $V_{ds}$. However, the $R_i$ contributions to distortion are insignificant in the presence f the nonlinearities from the other elements. The variation of $R_i$ with $V_{gs}$ could ave a minor effect on AM-PM conversion at the higher frequencies of device peration. The bias dependence of $R_i$ is a complicated nonlinearity to implement n the model, but omitting the bias dependency of $R_i$ eliminates the bias dependence f the MESFET gate-to-source input resistance.

The drain-to-source channel resistance $R_{ds}$ has a $V_{gs}$ and $V_{ds}$ dependency as nown in Figure 3.11. In Figure 3.11(b), $R_{ds}$ is seen to vary almost linearly with $_{ds}$, but as $V_{gs}$ drops below $-1$V and approaches pinch-off, $R_{ds}$ increases rapidly s the conductive drain-to-source channel constricts. The most popular MESFET ynamic model of Figure 3.2 uses a fixed resistance that represents either an average alue for $R_{ds}$ or else the $R_{ds}$ value extracted at the quiescent bias point. A fixed alue for $R_{ds}$ is an acceptable approximation for large-signal Class A operation. lowever, under Class B or C operating conditions, where the device should be onconducting for at least half the cycle, $R_{ds}$ should be very large. A fixed low alue for $R_{ds}$ results in significant current flowing through $R_{ds}$ under Class B and conditions. This results in power dissipation, which causes simulations of dc-to-.F conversion efficiency to be lower than actual. In fact, Class B efficiency simlations will be much more accurate if $R_{ds}$ is eliminated by setting $R_{ds}$ to be large that it approximates an open circuit.

Effects associated with the value of $R_{ds}$ would be most validly modeled through ne static *I–V* analytical function fitted to pulsed *I–V* characterization data in place f the series $R_{ds}$, $C_{RF}$ elements in Figure 3.2. In most cases, the nonlinearity from $_{ds}$ variation with $V_{gs}$ and $V_{ds}$ contributes insignificantly to either harmonic or phase istortion. However, one important distortion contribution from $R_{ds}$ relates to the mulation of the final stage of a transmitter. In this specialized application, mixing ccurs between the signal being transmitted and any signal from an alternative ource that enters the transmitter antenna and is incident on the drain of the IESFET. The nonlinear $R_{ds}$ element produces mixing between these signals, and ne mixing products that are produced at frequencies in the channel bandwidth onstitute intermodulation distortion products which are termed *back IMD*. Simlation of back IMD requires effective modeling of the $R_{ds}$ nonlinearity. The simlation of unbiased MESFET mixers [8] provides another example in which an ccurate representation of $R_{ds}$ nonlinearity is important.

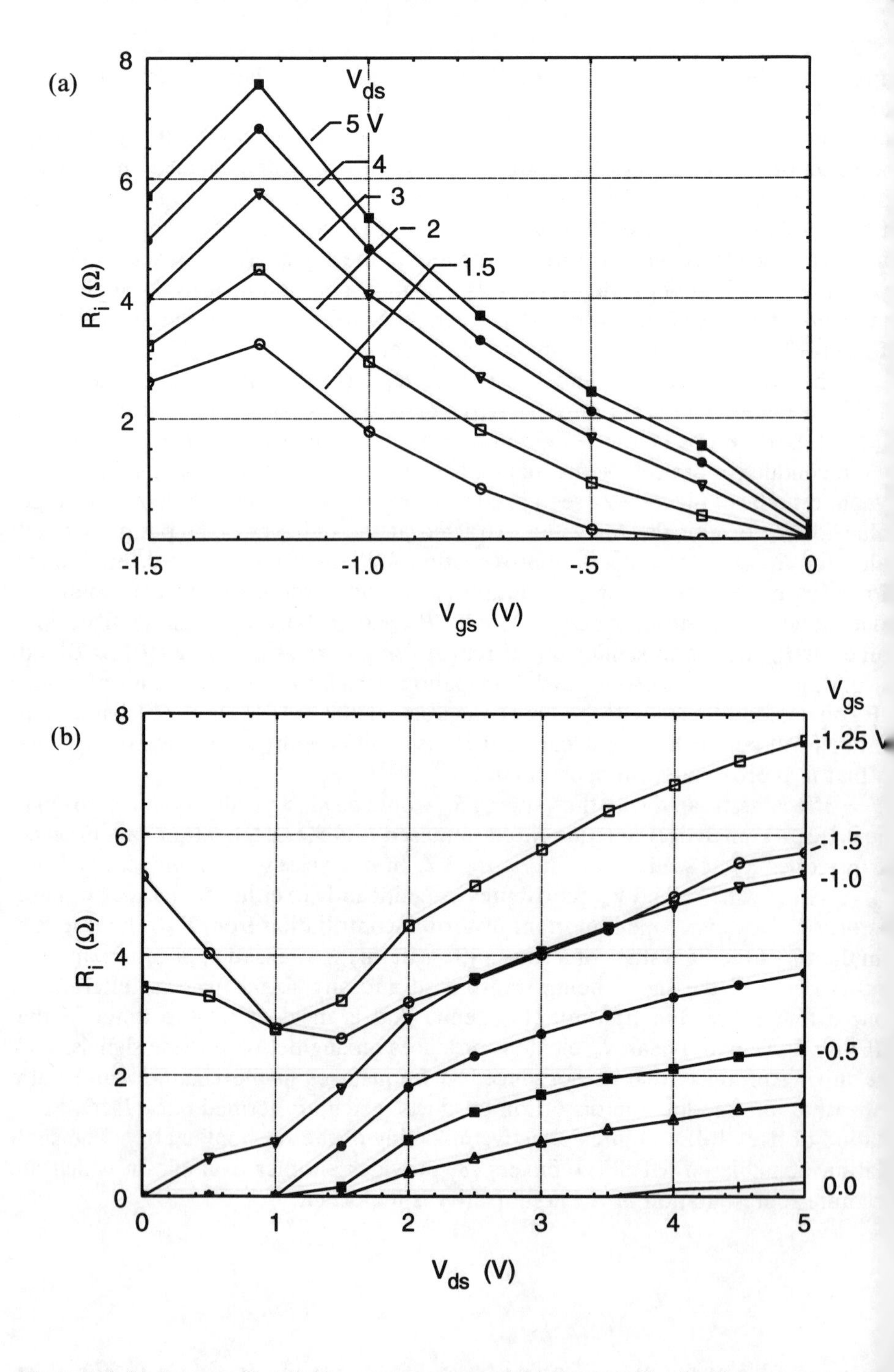


**Figure 3.10** (a) $R_i$ versus $V_{gs}$ with $V_{ds}$ as a parameter and (b) $R_i$ versus $V_{ds}$ with $V_{gs}$ as a parameter.

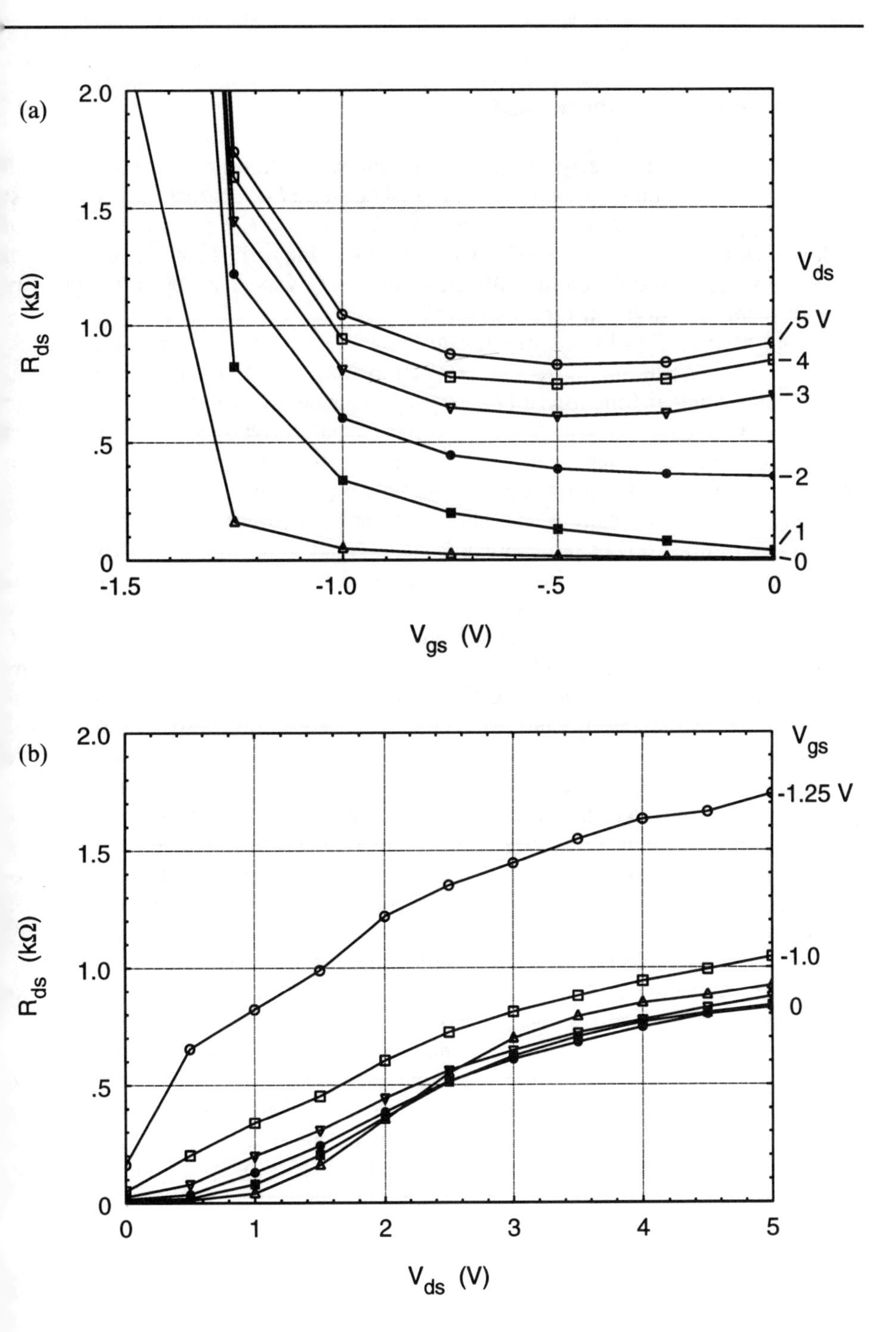


**Figure 3.11** (a) $R_{ds}$ versus $V_{gs}$ with $V_{ds}$ as a parameter and (b) $R_{ds}$ versus $V_{ds}$ with $V_{gs}$ as a parameter.

### 3.2.2 The MESFET Static dc Model

The static $I_{ds}$–$V_{ds}$ characteristics must define the simulation limits for current and voltage in both dc and microwave operation. The static $I$–$V$ characteristics are most commonly presented as curves of $I_{ds}$ versus $V_{ds}$ with $V_{gs}$ as a parameter. A set of such curves for a typical MESFET is illustrated in Figure 3.12(a). This family of curves obscures some fundamental aspects of the $I_{ds}$ versus $V_{gs}$ behavior that are evident from the curves in Figure 3.12(b).

The MESFET follows a nearly square-law behavior near pinch-off, that is, where $(V_{gs} + |V_P|)$ is small. As $(V_{gs} + |V_P|)$ grows, the behavior changes with $I_{ds}$ approaching a linear function of $(V_{gs} + |V_P|)$. This behavior is clearly apparent in the curves of Figure 3.12(b). Since $g_m$ is the partial derivative of $I_{ds}$ with respect to $V_{gs}$ along these curves, a close analytical fit to these curves is essential for accurately modeling the nonlinearity of $g_m$ with $V_{gs}$ and $V_{ds}$.

A few key static voltage and current parameters together with the dynamic load line determine power saturation and, to a large extent, the dc-to-RF efficiency of the amplifier stage. The inherent transistor limits to $I_{ds}$ are controlled by the gate-to-source voltage varying from pinch-off up to a forward bias that limits at approximately $V_{gs} = +0.7\text{V}$. The value of $I_{ds}$ at $V_{gs} = +0.7\text{V}$ corresponds to $I$ for the transistor. The limits of drain-to-source voltage vary from some minimum $V_K$, defined by the linear region of the $I_{ds}$ versus $V_{ds}$ dc characteristics, up to the drain-to-gate avalanche breakdown limit $V_{dgB}$. Whether these device voltage extremes are reached depends on the input signal drive level together with the limits imposed by the circuit. When the MESFET is overdriven, the quiescent bias point, the dynamic load line impedance, and the characteristics of the gate and drain bias circuitry can sometimes limit $I_{ds}$ and $V_{ds}$ prior to reaching the $I$–$V$ limits just described. Regardless of the cause, limiting or clipping of the current or voltage waveforms will constitute the dominant nonlinearity to be encountered in large signal operation.

The large-signal models allow specification of the reverse breakdown voltage of the gate-to-source and gate-to-drain junctions. Unfortunately, static dc measurements of the reverse avalanche breakdown overestimate RF voltage breakdown since the avalanche breakdown mechanism is dependent on the time-duration of the applied voltage as well as the active channel temperature. The voltage peaks of an RF signal are of such short duration that they can typically exceed the dc breakdown threshold without inducing avalanche breakdown. Consequently, the conventionally characterized dc breakdown is not useful for specifying breakdown under RF operation. Recently, Platzker et al. [3] described a pulsed measurement of breakdown that corresponds better with RF operational conditions.

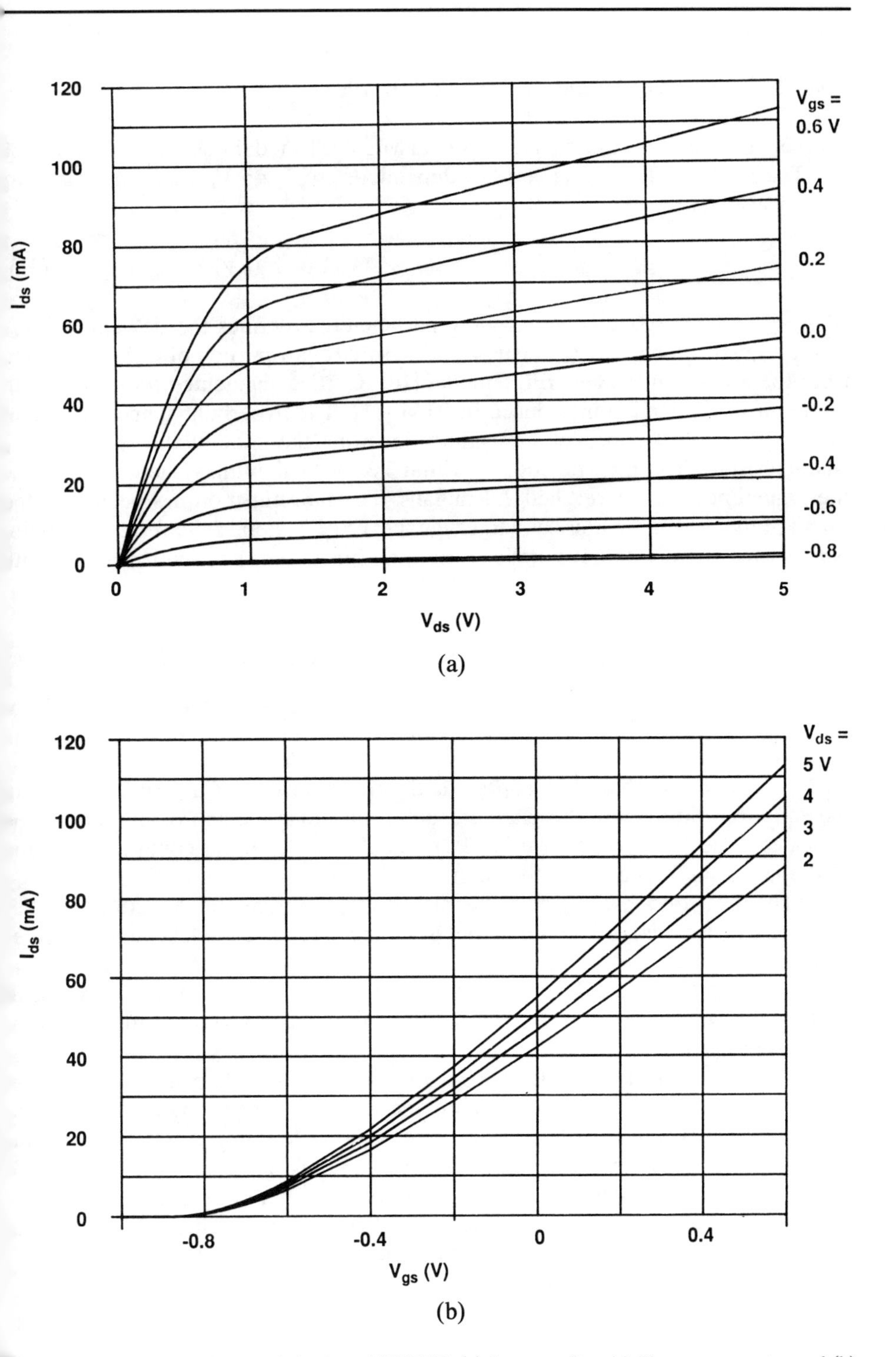


**Figure 3.12** Static $I$–$V$ characteristics for a MESFET: (a) $I_{ds}$ versus $V_{ds}$ with $V_{gs}$ as a parameter and (b) $I_{ds}$ versus $V_{gs}$ with $V_{ds}$ as a parameter.

### *The Curtice Quadratic Function MESFET Model*

The Curtice quadratic function large-signal MESFET model was introduced in 1980 [9]. The model assumes a square-law dependence of $I_{ds}$ on $V_{gs}$ using the following analytical function:

$$I_{ds} = \beta(V_{gs} + |V_P|)^2 (1 + \lambda V_{ds}) \tanh(\alpha V_{ds}) \tag{3.5}$$

where $\beta$ is a transconductance scaling parameter, $\lambda$ defines the dc value of $R_{ds}$ and $\alpha$ adjusts the knee of $I_{ds}$ versus $V_{ds}$. The square-law portion of (3.5) is adopted from the Shichman-Hodges JFET model [10]. Curtice augmented this model with the $\tanh(\alpha V_{ds})$ function introduced by Taki [11]. The hyperbolic tangent is a continuous function that provides versatility in fitting the knee of the $I_{ds}$ versus $V_{ds}$ characteristics. The tanh function is within 1% of its limiting value of unity once the argument $\alpha V_{ds}$ has reached 2.5 radians. The tanh function has been adopted for all of the popular, empirical, large-signal MESFET models. Statz et al. [6] point out that for computational simplicity the tanh function can be approximated with good accuracy by

$$\tanh(\alpha V_{ds}) \approx \begin{cases} 1 - (1 - \alpha V_{ds}/3)^3 & \text{for } 0 < V_{ds} < 3/\alpha \quad (3.6a) \\ 1 & \text{for } V_{ds} \geq 3/\alpha \quad (3.6b) \end{cases}$$

At $V_{ds} = 0$ both tanh and its approximation have a slope of $\alpha$. For simplicity in the equations throughout this chapter, tanh rather than the approximation will be indicated. The $(1 + \lambda V_{ds})$ term in (3.5) models the dc output conductance slope of the $I_{ds}$ versus $V_{ds}$ curves in the saturated region.

The measured $I_{ds} - V_{gs}$ characteristic of a typical transistor is plotted together with the Curtice quadratic fitted model in Figure 3.13. The model equation agrees with the measured data at only two $V_{gs}$ points, namely, at pinch-off, where $V_{gs} = -|V_P|$, and at $I_{dss}$, where $V_{gs} = 0$V. An alternative fitting would be to favor a fit at the quiescent bias current $I_{dsQ}$. In that case the model would sacrifice the fit at $|V_P|$ and at $I_{dss}$.

Since few MESFETs exhibit a square-law characteristic over a broad range of $V_{gs}$, the quadratic model can only provide a compromise fitting of $|V_P|$, $g_m$, and $I_{dss}$ for typical MESFETS. Also, the square-law model for the $g_m$ nonlinearity means only second harmonics can be generated by $g_m$; thus, $g_m$ does not contribute to third-order harmonic and intermodulation distortion. Therefore, in simulating Class A amplifiers this model can be expected to underestimate third harmonic and third-order IMD output power. The merit of the quadratic model is its simplicity, which leads to quick computations and good convergence qualities.

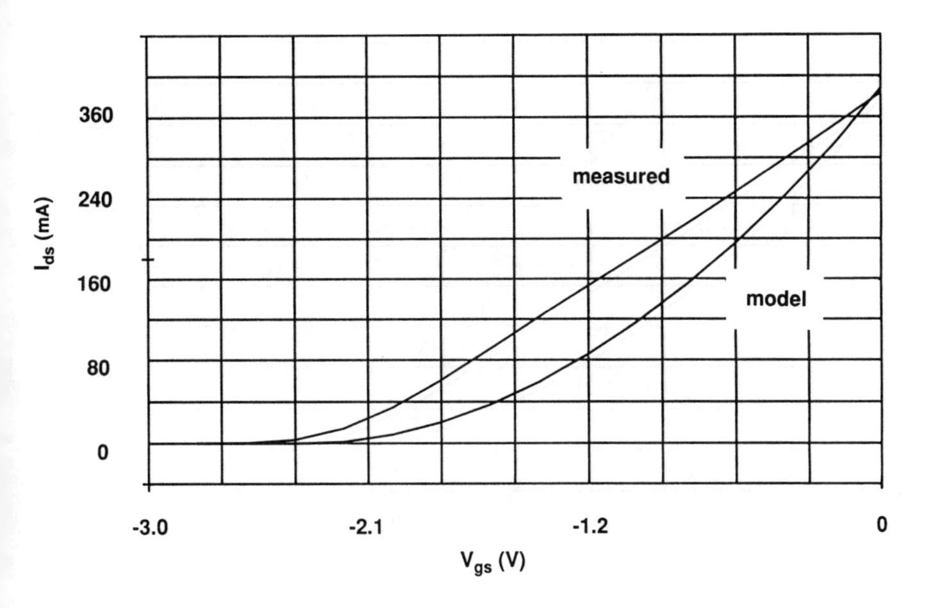


Figure 3.13 Curtice quadratic model fitting to the $I_{ds} - V_{gs}$ characteristic of a typical MESFET.

### The Materka Model

The dc *I–V* characteristic of many transistors is indicated in Figure 3.14, where the value of $V_{gs}$ at which the drain current is pinched off is seen to vary with $V_{ds}$. This characteristic is referred to as a soft pinch-off. The Curtice quadratic model is unable to fit this soft pinch-off behavior.

The Materka model [12] from 1983 has generated interest largely because it can represent the soft pinch-off characteristic. Other than this feature, the analytical function is a quadratic suited to accurate modeling of MESFETs having a square-law *I–V* characteristic and is given by

$$I_{ds} = (1 + V_{gs}/V'_P)^2 \tanh[\alpha V_{ds}/(V_{gs} + V'_P)] \tag{3.7a}$$

where

$$V'_P = |V_P| + \gamma V_{ds} \tag{3.7b}$$

The coefficient $\gamma$ provides the adjustment of pinch-off voltage $V'_P$ as a function of $V_{ds}$ from a nominal $|V_P|$.

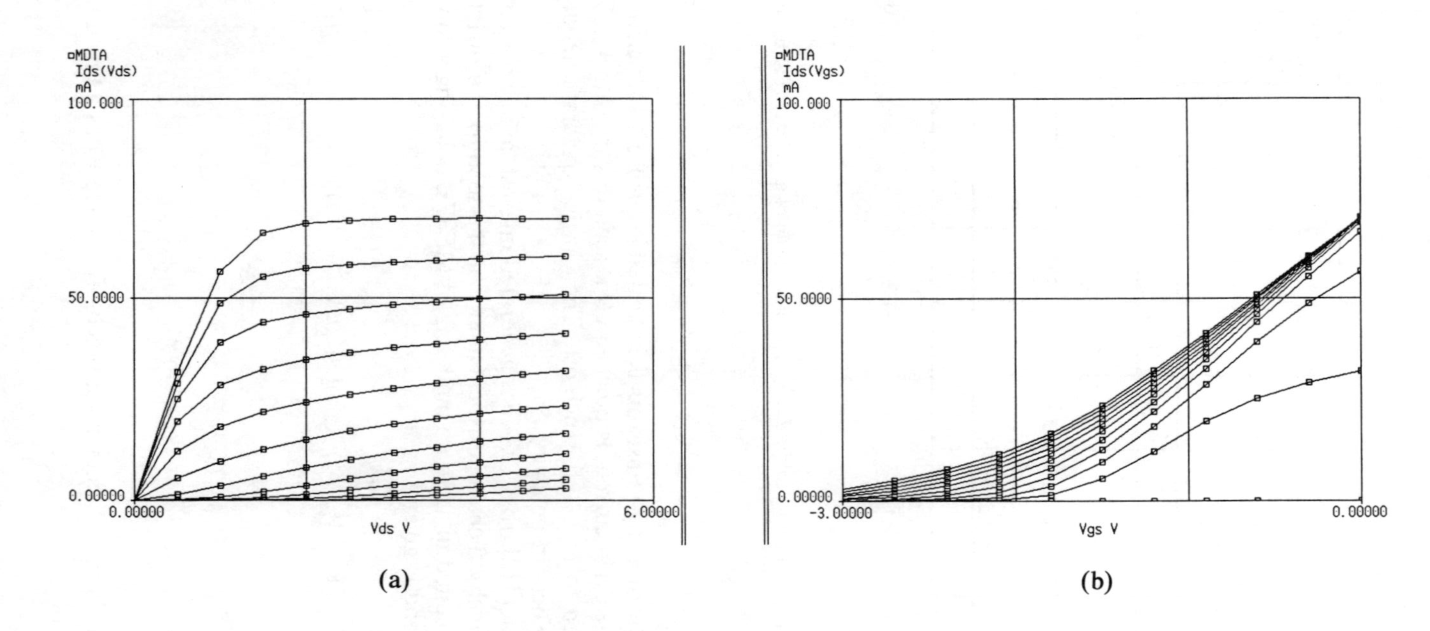


**Figure 3.14** Typical soft pinch-off $I$–$V$ characteristics for a MESFET: (a) $I_{ds}$ versus $V_{ds}$ with $V_{gs}$ as a parameter and (b) $I_{ds}$ versus $V_{gs}$ with $V_{ds}$ as a parameter.

## The Curtice-Ettenberg Model

The Curtice-Ettenberg model, introduced in 1985 [13], uses a third-order polynomial to fit the $I_{ds}$ versus $V_{gs}$ characteristic and is often referred to as the cubic model. It is given by

$$I_{ds} = (A_0 + A_1V_1 + A_2V_1^2 + A_3V_1^3) \tanh(\alpha V_{ds}) \quad (3.8a)$$

where

$$V_1 = V_{gs}[1 + \beta(V_{ds0} - V_{ds})] \quad (3.8b)$$

In (3.8b), $\beta$ controls the change in pinch-off voltage with $V_{ds}$, and $V_{ds0}$ is the drain-source voltage at which the $A_i$ coefficients are evaluated. To some extent, this analytic model function can fit $I$–$V$ characteristics with a soft pinch-off.

Figure 3.15 shows a typical $I_{ds} - V_{gs}$ fit using the cubic model. Favoring the best fit along the major portion of the $I_{ds} - V_{gs}$ curve frequently results in a poor fit to the region of curvature close to pinch-off. The function is capable of replicating

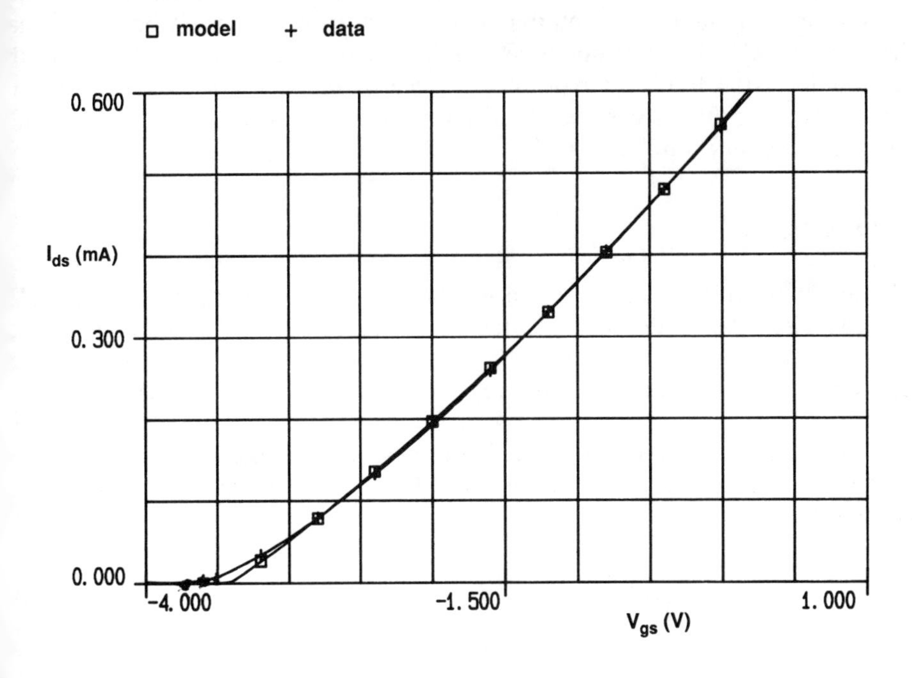


Figure 3.15 Curtice-Ettenberg cubic model fitting to the $I_{ds} - V_{gs}$ characteristic of a typical MESFET.

the curvature more accurately by sacrificing the fit accuracy along other region of the $I$–$V$ curve. For large-signal Class A amplifiers, the model fit near pinch-of is usually less important than fitting the quasilinear $I_{ds} - V_{gs}$ region surroundin the quiescent bias point.

However, for a Class B amplifier stage the accurate fitting of the pinch-of region is important for producing realistic simulations of the output power as function of the input drive signal. For example, whereas the actual $I_{ds} - V_{gs}$ char acteristic in Figure 3.15 has a zero slope at $V_{gs} = -|V_P|$, the slope of the cubi function curve at $I_{ds} = 0$ is positive. The positive slope of the function correspond to a finite $g_m$ value at pinch-off rather than the actual value of zero. The exaggerate $g_m$ will result in predictions of higher than actual gain characteristics for Class operation.

Similarly, for the MESFET operating as a mixer, the dominant $g_m$ nonli earities occur near $V_{gs} = -|V_P|$. The cubic function will typically have reduced th nonlinearity significantly so that simulations can be expected to produce lower tha actual conversion gain. Higher order polynomials could allow additional flexibili in fitting the $I$–$V$ characteristics, but higher order functions have not been wide used because the computation time is increased and also because the functions a more likely to cause simulation convergence problems.

The cubic model can be expected to be capable of more accurate simulatio of third-order harmonic and intermodulation distortion in Class A operation tha the other popular dc $I$–$V$ fitting functions. The coefficient $A_3$ in (3.8a) controls t magnitude of the third-order distortion. Ideally, the extraction of $A_3$ would b made through a fitting between third partial derivatives, $\delta^3 I_{ds}/\delta V_{gs}^3$, for the da and the cubic function.

### The Statz Model

The Statz model [6] introduced two important advances over previous MESFE models. First, the model incorporated an improved analytic dc $I$–$V$ formulati (originally proposed in 1975 as a MOSFET model [14]), and, second, the mod offered an improved charge model representation of $C_{gs}$ and $C_{gd}$ as functions both $V_{gs}$ and $V_{ds}$.

The simple Statz dc $I$–$V$ modeling function, which has proven to be high versatile in fitting a broad variety of MESFETs, contains only two coefficients. is given by

$$I_{ds} = \left[\frac{\beta\,(V_{gs} + |V_P|)^2}{1 + b\,(V_{gs} + |V_P|)}\right](1 + \lambda V_{ds}) \tanh(\alpha V_{ds}) \qquad (3.$$

This equation provides unusually good control over the contour in the transiti of $I_{ds} = f(V_{gs})$ as it goes from square-law to linear behavior. As long as the quantity $b(V_{gs} + |V_P|)$ is $<< 1$, as it is when $V_{gs}$ is close to pinch-off, then behavior is square law since $I_{ds} \approx \beta(V_{gs} + |V_P|)^2$. However, when $b(V_{gs} + |V_P|) >> 1$, then the expression is nearly linear since $I_{ds}$ approaches $I_{ds} = \beta(V_{gs} + |V_P|)/b$. The coefficient $\beta$ is easily evaluated for any choice of $b$ since

$$I_{dss} = \beta|V_P|^2/(1 + b|V_P|) \tag{3.10a}$$

and

$$\beta = \frac{I_{dss}(1 + b|V_P|)}{|V_P|^2} \tag{3.10b}$$

The broad latitude in $I_{ds} - V_{gs}$ fitting using the Statz function is illustrated in the family of curves in Figure 3.16. A hypothetical device with $I_{dss} = 1$A and $|V_P| =$ 3V was assumed throughout. By adjusting $\beta$ and $b$ over a broad range in (3.9), we can accommodate $I_{ds}$ versus $V_{gs}$ behavior that varies from totally square law throughout ($b = 0$) to totally linear throughout (very large $b$). Despite the impor-

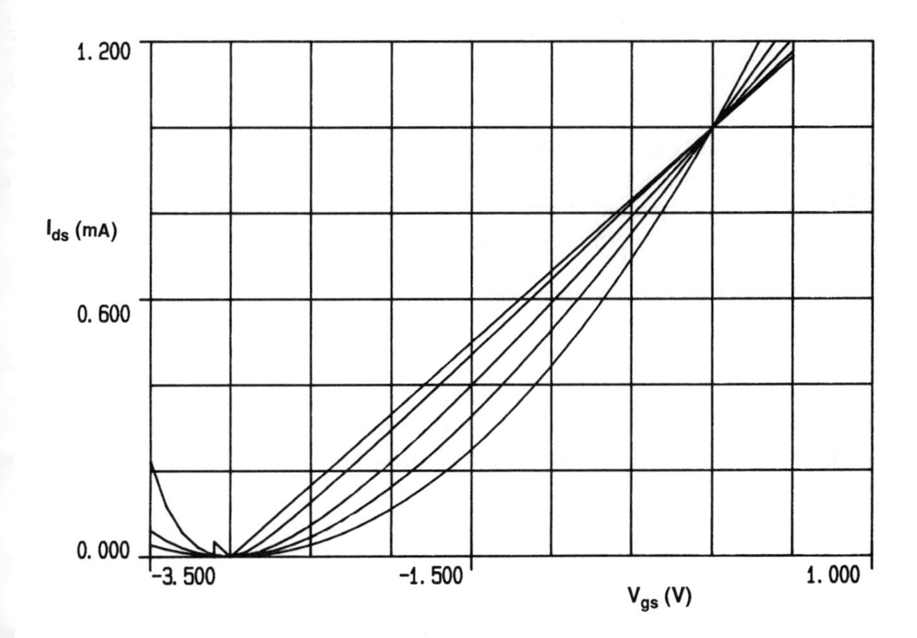


**Figure 3.16** Curves demonstrating the range of Statz model fitting to the $I_{ds} - V_{gs}$ characteristics of MESFETs.

tant advance provided by the Statz formulation of (3.9), it lacks any means of representing the soft pinch-off characteristic common to many MESFETs.

The Statz dc $I$–$V$ function can be expected to produce very good fitting of the $g_m$ nonlinearities near pinch-off, making the model a good choice for simulating MESFET mixers and Class AB and Class B amplifiers. Although $\delta^3 I_{ds}/\delta V_{gs}^3$ for the function exists there is no degree of freedom to adjust the fit at the third derivative as was possible with the Curtice-Ettenberg cubic polynomial function.

The Statz model allows each of the Schottky barrier junction capacitances $C_{gs}$ and $C_{gd}$, to vary as functions of both $V_{gs}$ and $V_{ds}$. Most conventional models represent $C_{gs}$ and $C_{gd}$ as simple diode junction capacitances, each of which varies as a function of a single voltage. The simple diode capacitance representation causes contradictions in defining that $C_{gs} = C_{gd}$ when $V_{ds} = 0$. Similarly, the simple diode junction capacitance dependency improperly models $C_{gs}$ and $C_{gd}$ when $V_{ds}$ is negative, which occurs in applications where the operation of the MESFET is inverted. Statz et al. [6] introduced an improved way of representing the charge distribution in the MESFET to assure continuity of charge. By first defining a total charge under the gate electrode and then apportioning this into a $C_{gs}$ charge and a $C_{gd}$ charge, continuity was assured over all conditions of transistor operation. The improved capacitance equations are of the form

$$C_{gs} = C_{gs0} \cdot f_1 (V_{gs}, V_{gd}) + C_{gd0} \cdot f_2 (V_{gs}, V_{gd}) \quad (3.11)$$

$$C_{gd} = C_{gs0} \cdot f_3 (V_{gs}, V_{gd}) + C_{gd0} \cdot f_4 (V_{gs}, V_{gd}) \quad (3.12)$$

where $f_1$, $f_2$, $f_3$, and $f_4$ are functions of both $V_{gs}$ and $V_{gd}$. The full expressions for $f_1, f_2, f_3$, and $f_4$, are complex and can be found in [6]. The Statz improved capacitance formulations can be expected to simulate AM-PM effects in amplifiers more accurately than does the simple diode capacitance function.

*The TOM Model*

The TOM model (TriQuint's own model) [15] merges some key features of the models just described to create a model that consequently fits a broad range of device dc $I$–$V$ characteristics. The model incorporates the soft pinch-off feature of the Materka model into a significantly modified Statz model. The TOM model is unusual in that it can produce MESFET dc $I$–$V$ families of curves such as the family shown in Figure 3.17. These curves cover $V_{gs}$ ranging from $-2$ to 0.5V in steps of 0.5V. Some of the TOM features demonstrated in these characteristics are:

1. A simple and widely controllable means for fitting $g_m$ as a function of $V_{gs}$ as the transistor changes from square-law dependence on $V_{gs}$ toward a linear dependence;

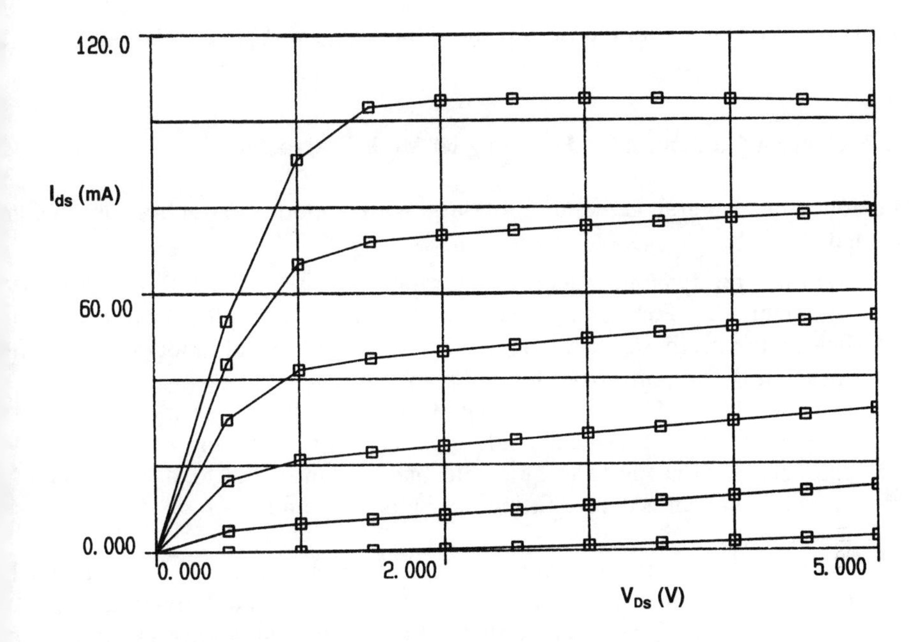


igure 3.17 A family of dc $I$–$V$ curves generated with the TOM model.

2. A soft or gradual pinch-off characteristic, that is, a pinch-off voltage $V'_P$ whose magnitude increases with increasing $V_{ds}$;
3. A simple means for modeling the dependence of $R_{ds}$ on $V_{gs}$, $V_{ds}$ and channel temperature.

'he pinch-off voltage is made a function of $V_{ds}$ as in the Materka model using

$$V'_P = |V_P| + \gamma V_{ds} \tag{3.13}$$

'he flattening of the slope of $I_{ds}$ versus $V_{ds}$ as $V_{gs}$ increases, as illustrated in Figure .17, is fitted using

$$I_{ds} = \frac{I_{ds0}}{1 + \delta V_{ds} I_{ds0}} \tag{3.14a}$$

'here

$$I_{ds0} = \beta(V_{gs} + |V_P| + \gamma V_{ds})^Q \cdot K \cdot \tanh(\alpha V_{ds}) \tag{3.14b}$$

The TOM model has adopted the advanced Statz $C_{gs}$ and $C_{gd}$ formulations in (3.11) and (3.12), respectively.

### 3.2.3 General Guidelines for Large-Signal Model Extraction

The following general guidelines for large-signal model extraction are recommended:

1. Identify the dynamic region of operation for the MESFET on the dc $I$–characteristics. For an amplifier this would mean specifying the dynamic load line region on the $I_{ds}$ versus $V_{ds}$ curves. The large-signal model fitting should focus on the region surrounding this load line.
2. For quasilinear Class A large-signal operation where harmonic and IMD simulations are of principal interest, the accurate fitting of the $I_{ds} - V$ characteristics around the quiescent bias region of the load line should be emphasized. An accurate fitting near pinch-off and for $I_{ds} > I_{dss}$ is of lesser importance.
3. Judgments as to the goodness of fit to $I_{ds} - V_{ds}$ or $I_{ds} - V_{gs}$ data can be misleading in some situations. Fitting to the first partial derivative of $I_{ds}$ data with respect to $V_{gs}$ addresses only the first-order nonlinearity associated with $g_m$. A model that would accurately simulate the third-order harmonic and intermodulation products would result from fitting the third derivatives of the analytic function to the third derivatives of the data. For this approach to be successful, pulse measurements of the static $I$–$V$ data that accurately extract the third-order nonlinearity would be essential.
4. The Curtice-Ettenberg model should be capable of the most accuracy in representing the third-order nonlinearities because, of all the models considered, only the third-order polynomial function offers direct control of third-order nonlinearity through the coefficient $A_3$ of the third-order term.
5. For simulating the electrical performance characteristics of Class B and amplifiers, the focus should be on fitting of the $I_{ds} - V_{gs}$ characteristics near pinch-off and at $I_{dss}$ and above. It is also important to adjust $\tanh(\alpha V_{ds})$ to achieve the best fit to the $I_{ds} - V_{ds}$ knee at $I_{dss}$ and above. These modeling steps address the nonlinear turn-on and the power saturation of the Class and C stages.
6. The fixed value of $R_{ds}$ in the SPICE model of Figure 3.2 should be set to a large value when simulating the operating efficiency of Class AB, B, and stages.

## .3 A LARGE-SIGNAL AMPLIFIER SIMULATION

ieller and Goettle [16] described in detail the design and performance of a 3.7-) 4.2-GHz, 1W, high-efficiency, Class AB, quasimonolithic power amplifier stage. 'his example was previously considered from an analytic standpoint in Chapter 1. 'he amplifier used a commercially available MESFET, the Fujitsu FLK202XV, aking it a practical candidate for simulation. In addition, the paper provided a umber of single-frequency and stepped-frequency measurements of power and fficiency suitable for making the simulation comparisons to be presented here.

### .3.1 The FLK202XV Large-Signal Model

rior to simulation a large-signal model for the FLK202XV chip must be produced. Statz *I–V* model for the Fujitsu FLK202XV MESFET chip was extracted using nly the manufacturer's published [17] *S* parameter and dc *I–V* curves. The dc ·*V* curves for the FLK202XV are shown in Figure 3.18. The soft pinch-off char-:teristic of the transistor is apparent from these curves. Superimposed on the ·*V* curves is an ideal resistive load line approximating the optimum dynamic load ıe. The fitting was focused on $I_{ds}$, $V_{gs}$ points along the load line. Figure 3.19 shows ırves of $I_{ds}$ versus $V_{gs}$ along the load line together with the final Statz fitting, which extremely close. The model function curve can be seen to indicate positive $I_{ds}$

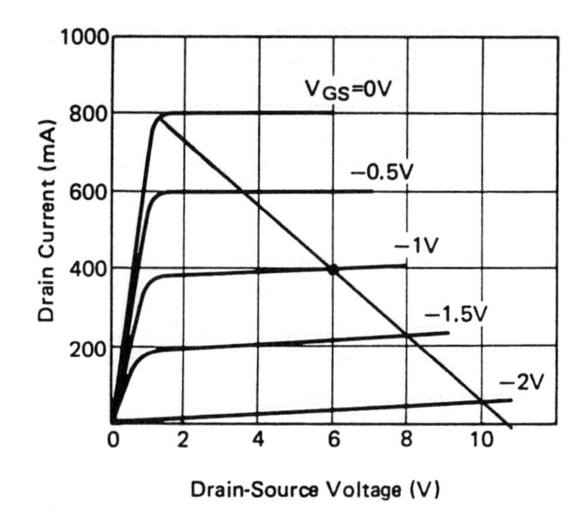


ure 3.18 The dc *I–V* curves for the FLK202XV showing the idealized dynamic load line (taken from [17]).

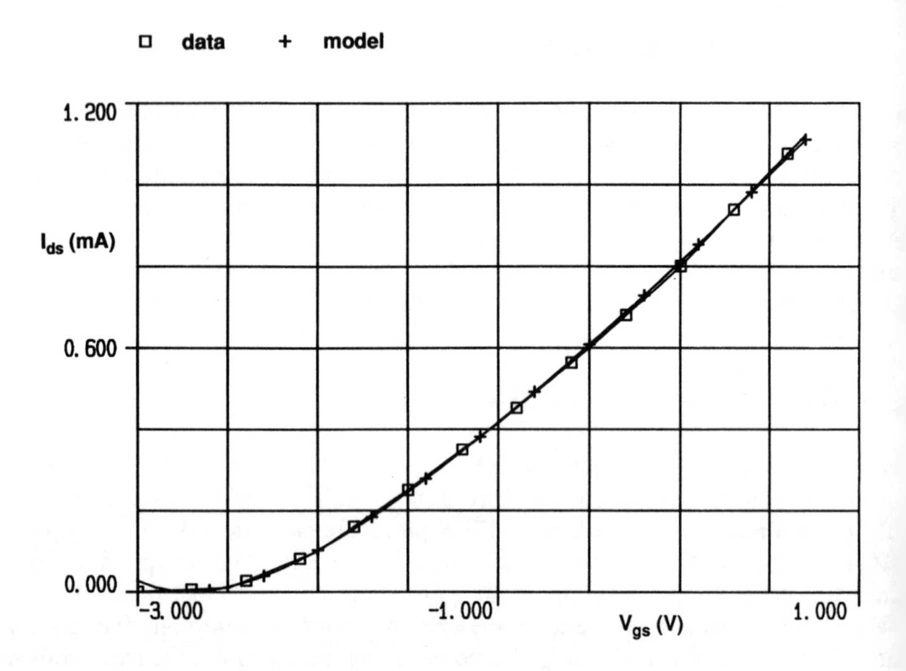


**Figure 3.19** Data and model curves for $I_{ds}$ versus $V_{gs}$ along the load line.

below pinch-off. For simulation the function is clamped at $I_{ds} = 0$ for $V_{gs} < -|V$ Note that by fitting $I_{ds}$ versus $V_{gs}$ along the load line the soft pinch-off behavior included for transistor operation along this load line.

The ac model of Figure 3.2 was fitted to the manufacturer's catalog $S$ parar eter data measured at a quiescent bias of $I_{ds} = 480$ mA and $V_{ds} = 10$V. The Smi chart of Figure 3.20 includes curves comparing $S_{11}$ and $S_{22}$ taken from the larg signal model, biased at $I_{ds} = 480$ mA and $V_{ds} = 10$V, with those from the ma ufacturer's measured data from 1 to 6 GHz. The fits to $S_{21}$ and $S_{12}$ in decibels vers frequency are shown in Figures 3.21 and 3.22, respectively. The model fitting all four $S$ parameters is quite good. The extracted Statz $I$–$V$ model parameters a the dynamic model parameters are shown in Table 3.1. The FLK202XV ch incorporates via-hole source grounding, which constitutes an extrinsic mod parameter. The extracted value of the via inductance of 11 pH is listed Table 3.1.

### 3.3.2 Simulation of the 3.7- to 4.2-GHz, 1-W Class AB Amplifier

The overall microstrip schematic of the stage is shown in Figure 3.23 includi circuit capacitors, bond wire and via inductances, and dc bias voltages. Stubs a

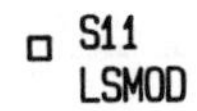

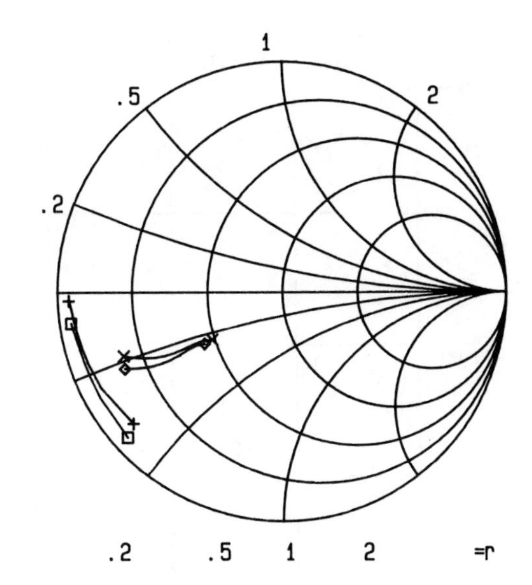


**Figure 3.20** Curves of $S_{11}$ and $S_{22}$ for data and the large-signal model from 1 to 6 GHz where $I_{dsQ}$ = 480 mA; $V_{dsQ}$ = 10V.

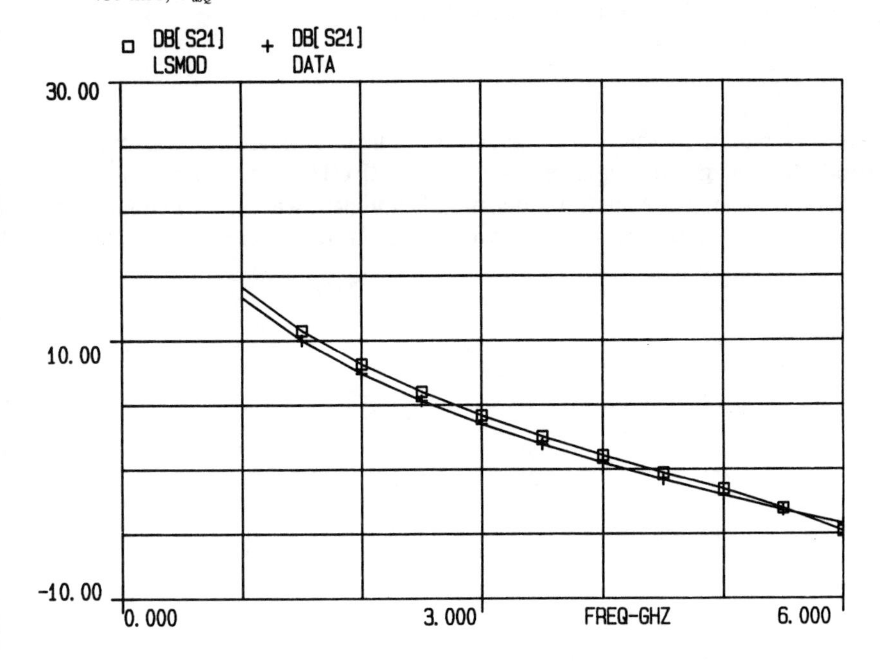


**Figure 3.21** Curves of $S_{21}$ in decibels for data and the large-signal model from 1 to 6 GHz where $I_{dsQ}$ = 480 mA; $V_{dsQ}$ = 10V.

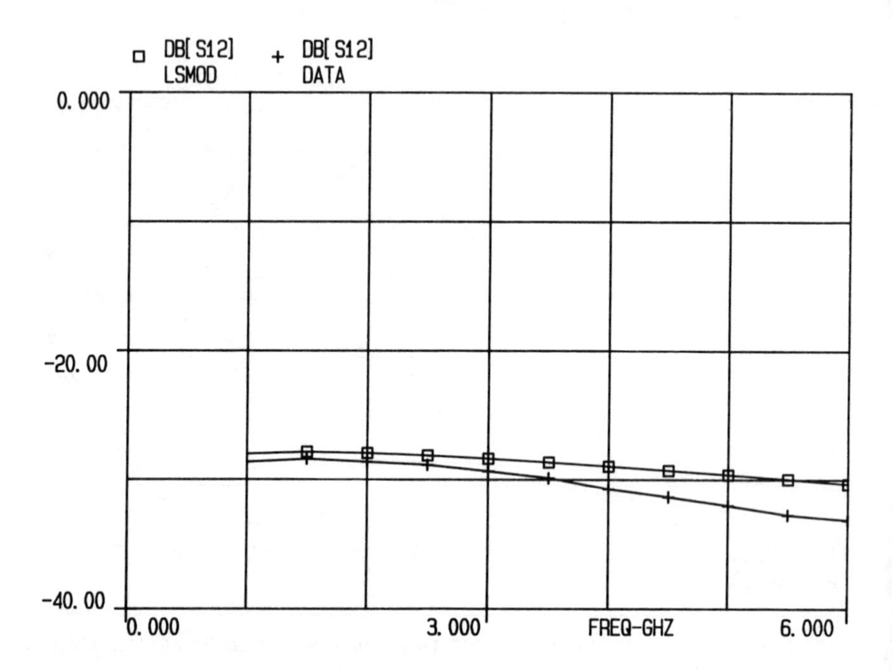


**Figure 3.22** Curves of $S_{12}$ in decibels for data and the large-signal model from 1 to 6 GHz where $I_{ds}$ = 480 mA; $V_{dsQ}$ = 10V.

shorted by means of the 33-pF capacitors indicated. The drain dc bias feed is a shorted stub, a quarter-wavelength long at 4.0 GHz. This stub provides a near open circuit at the fundamental operating frequency, while producing a near short circuit at even harmonics for improving dc-to-RF conversion efficiency [18]. The

**Table 3.1**
Large-Signal Model Parameters for the FLK202XV

| *Symbol* | *Keyword* | *Value* | *Symbol* | *Keyword* | *Value* |
|---|---|---|---|---|---|
| $R_g$ | RG | 0.035Ω | $\beta$ | BETA | 0.282 |
| $R_s$ | RS | 0.10Ω | $b$ | THETA | 0.536 |
| $R_d$ | RD | 0.19Ω | $\alpha$ | ALPHA | 1.3 |
| $R_i$ | RIN | 0.035Ω | $\lambda$ | LAMBDA | 0.0 |
| $R_{ds}$ | RC | 48Ω | $V_{TO}$ | VTO | −2.72V |
| | CRF | 2.2 $\mu F$ | | | |
| $C_{gs}$ | CGSO | 8.0 pF | | | |
| $C_{gd}$ | CGDO | 1.4 pF | | | |
| $C_{ds}$ | CDS | 1.3 pF | | | |

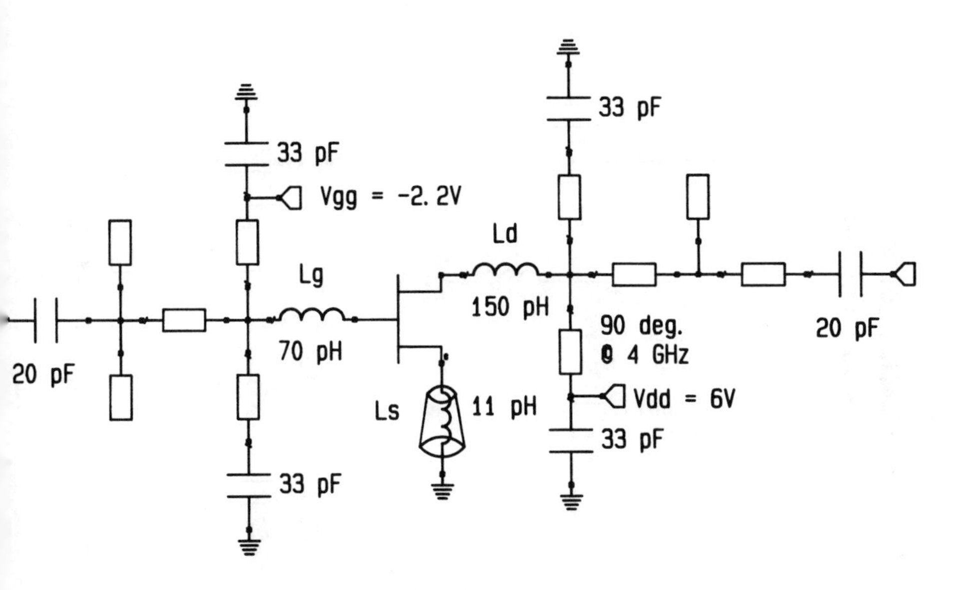


gure 3.23 The schematic for the 3.7- to 4.2-GHz, 1-W, Class AB amplifier.

her shorted stub at the drain is resonant with the output capacitance of the K202XV. According to Geller and Goettle, the remaining T-network was signed to transform the 50-$\Omega$ load to a 15.5-$\Omega$ load line for the transistor. At e input, a pair of shorted stubs at the gate resonates with the transistor input pacitance at 4.2 GHz. The remaining lines are designed to form a low-pass ipedance match to 50-$\Omega$ at 4.2 GHz. The dimensions for the microstrip lines on 0-$\mu$m-thick alumina for the input and output matching networks were extracted om the photograph of the circuit in Geller and Goettle's paper. The microstrip mensions were then transferred into the identical CAD microstrip layout shown Figure 3.24 for simulation.

The source-to-ground via inductance was fixed at 11 pH. This was the value tracted in the $S$ parameter model fitting. With the microstrip circuit design fixed, e only variables left for adjustment in the simulation were the bond wire inducnces at the gate and drain, $L_g$ and $L_d$, respectively. Geller and Goettle describe ing a Class AB quiescent bias of approximately 10% of $I_{dss}$. Referring to the $V$ curves of Figure 3.18, this corresponds to setting $I_{dsQ} = 80$ mA. Geller and oettle presented measurements of $S_{11}$ and $S_{21}$ of the amplifier from 2.5 to 5.0 Hz at this quiescent bias. Using $S$ parameters generated from the linearized largenal model at this quiescent bias, a linear simulation of $S_{11}$ and $S_{21}$ was compared th Geller and Goettle's measurements. Optimization was used to adjust only the known bond wire inductances, $L_g$ and $L_d$, for the best agreement. The plotted

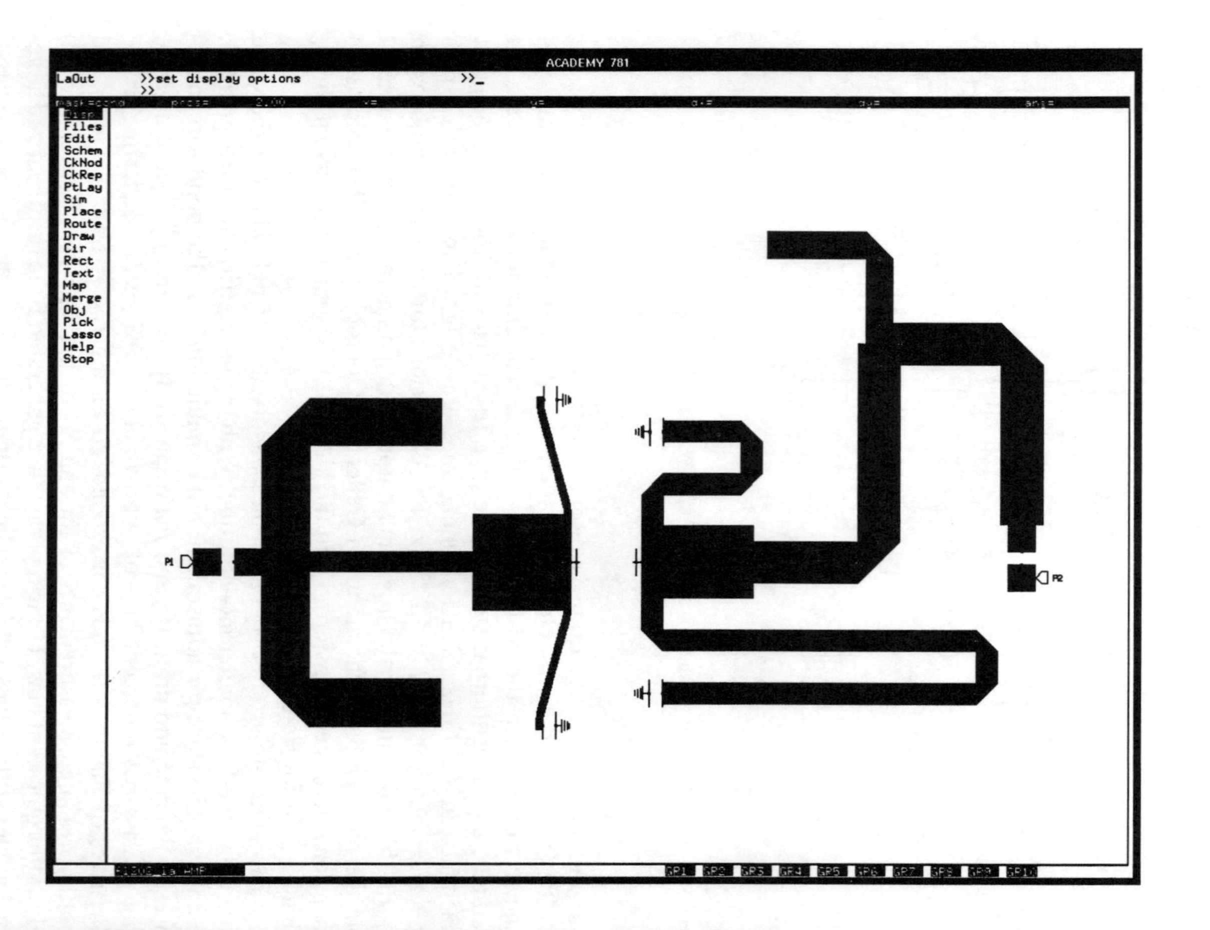


[illegible] 3.7 to 4.2 GHz 1 W Class AB amplifier

omparison of the simulated and measured results is shown in Figure 3.25. The eatest deviation between measured and modeled $S_{11}$ is <1.5 dB, and for $S_{22}$ is 1 dB.

A comment should be made regarding the simulations that follow; Geller and oettle did not specify the dc drain supply voltage for operation of the amplifier. itially, a supply voltage of 10V was assumed in the simulations since the manfacturer's data [17] illustrated transistor application at this operating voltage. owever, it was experimentally found during the simulations that by decreasing e supply voltage to 6V the simulations of output power and power-added efficiency agreed more closely with the measurement data.[1]

Comparisons of simulated and actual output power and power-added efficiency versus input power at 4.1 GHz are shown in Figures 3.26 and 3.27, respecvely. In Figure 3.26 the agreement for output power is excellent—within 1 dB

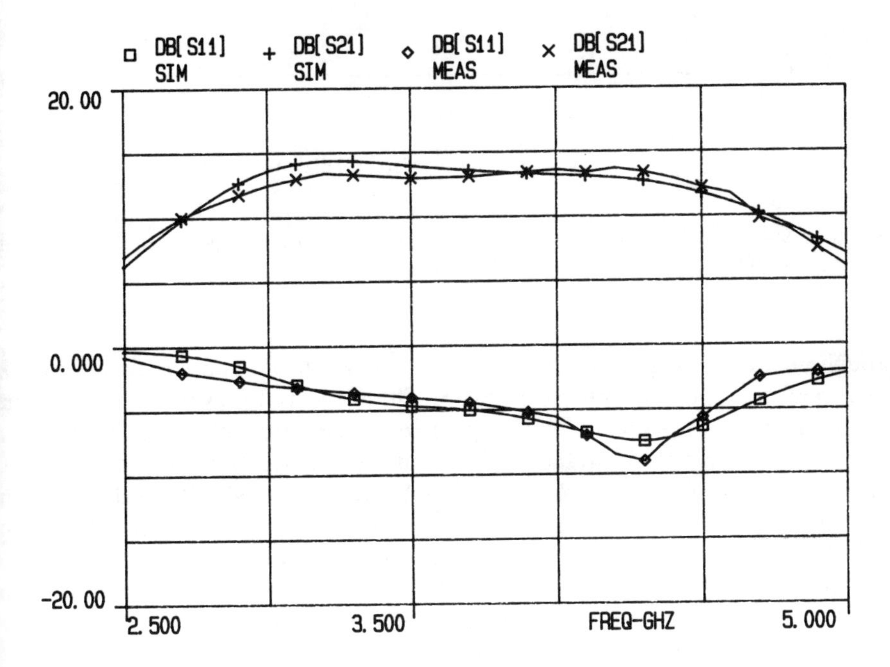


gure 3.25 Measured and simulated curves of $S_{11}$ and $S_{21}$ for the 3.7- to 4.2-GHz, 1-W, Class AB amplifier.

[1] checking with Geller during the simulations of power-added efficiency as a function of supply voltage, eller confirmed that his measurements had been made using a 6-V supply voltage. All large-signal nulations presented here have been made using this drain supply voltage.

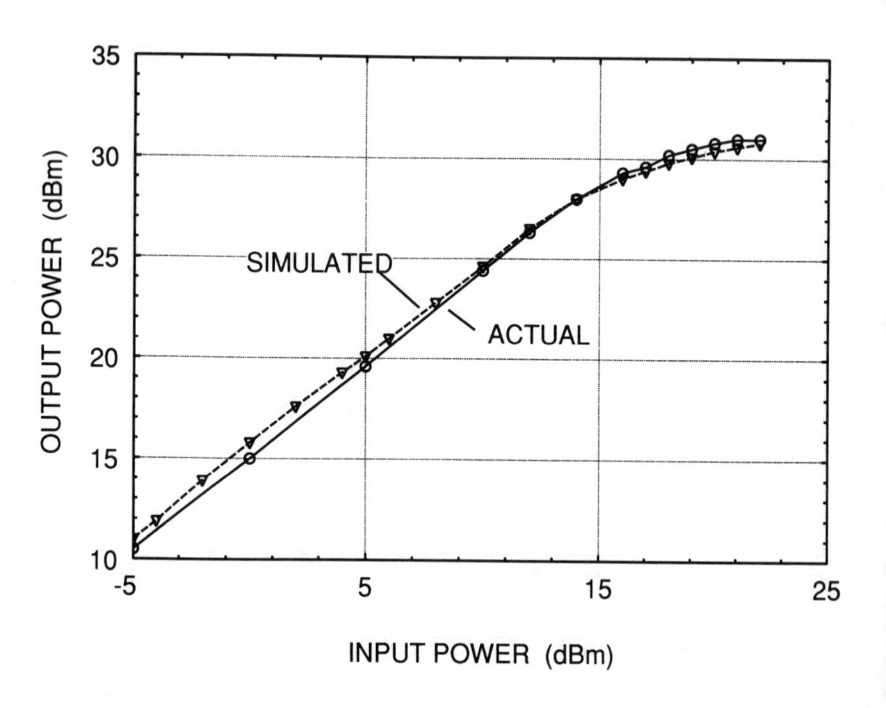


**Figure 3.26** Measured and simulated curves of output power versus input power at 4.1 GHz for t amplifier.

throughout the input power range from −5 to 22 dBm. The simulated and th actual power output saturate at 31 dBm. The measured power-added efficienc peaked at 65%, while the simulation peaked at approximately 61%. These simu lations were made with $R_{ds}$ in the model set at 1 k$\Omega$ rather than the extracted valu of $R_{ds}$ = 48$\Omega$. As mentioned in Section 3.2.1, the popular SPICE model of Figu 3.2 does not validly represent the effect of the actual $R_{ds}$ on the output power an dc-to-RF conversion efficiency for Class B operation with $R_{ds}$ modeled as a pu resistance. The simulated curves in Figure 3.28 compare the output power an efficiency for $R_{ds}$ values of 1 k$\Omega$ (upper curves) and for $R_{ds}$ of 48$\Omega$ (lower curves The differences in power-added efficiency are especially large.

The curves in Figure 3.29 show simulated output power at the 4.1-GHz fu damental and at the second and third harmonic frequencies versus input powe The second harmonic is suppressed below the third harmonic throughout becau of the short-circuit termination of the even harmonics.

The multiple-amplitude waveform sets presented in Figures 3.30 through 3. correspond to stepped input powers of 1, 13, and 19 dBm at 4.1 GHz. The voltag at the 50-$\Omega$ load is shown in Figure 3.30. The low-pass output matching netwo

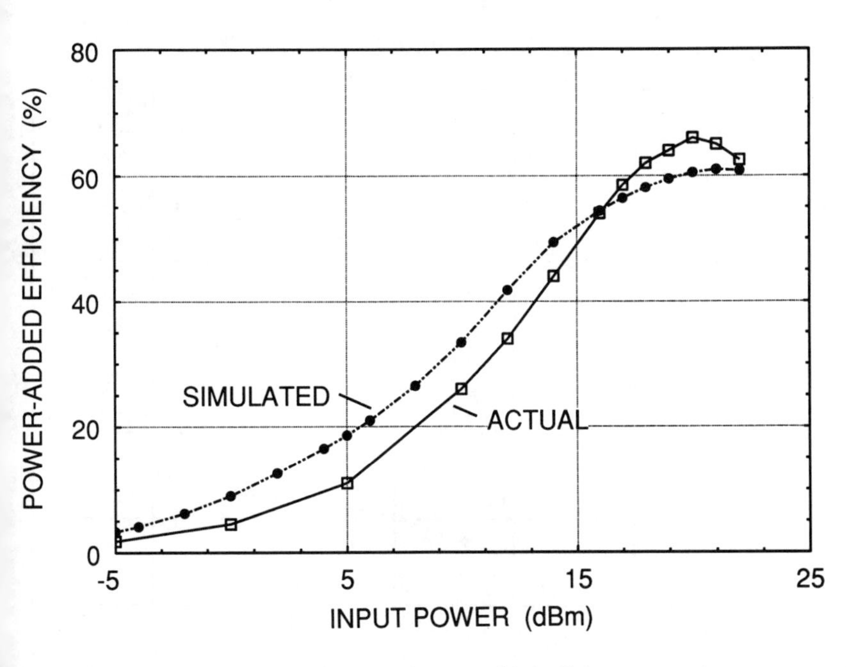


**Figure 3.27** Measured and simulated curves of power-added efficiency versus input power at 4.1 GHz for the amplifier.

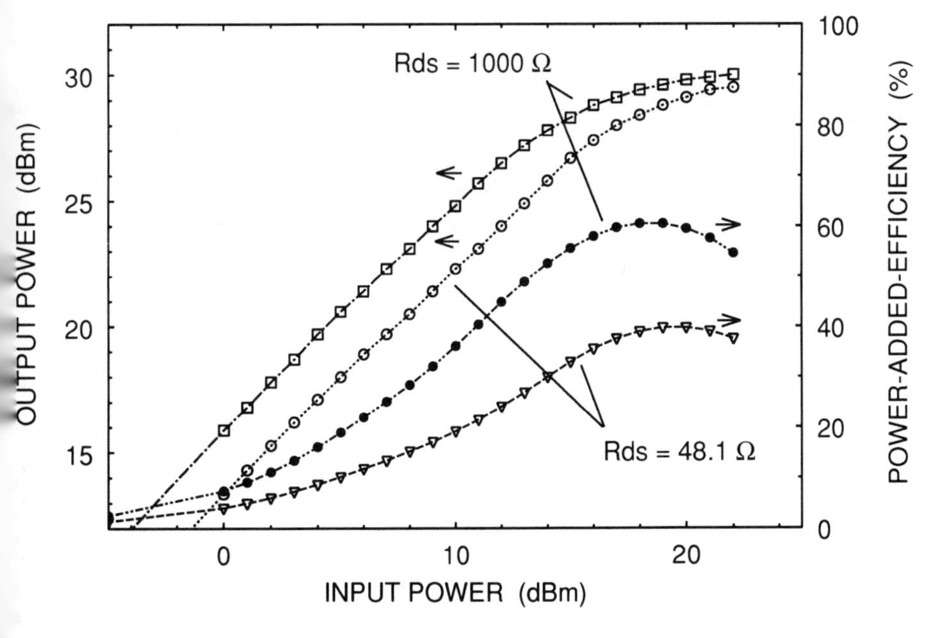


**Figure 3.28** Simulated output power and power-added efficiency versus input power at 4.1 GHz for $R_{ds} = 1000\Omega$ and $R_{ds} = 48\Omega$.

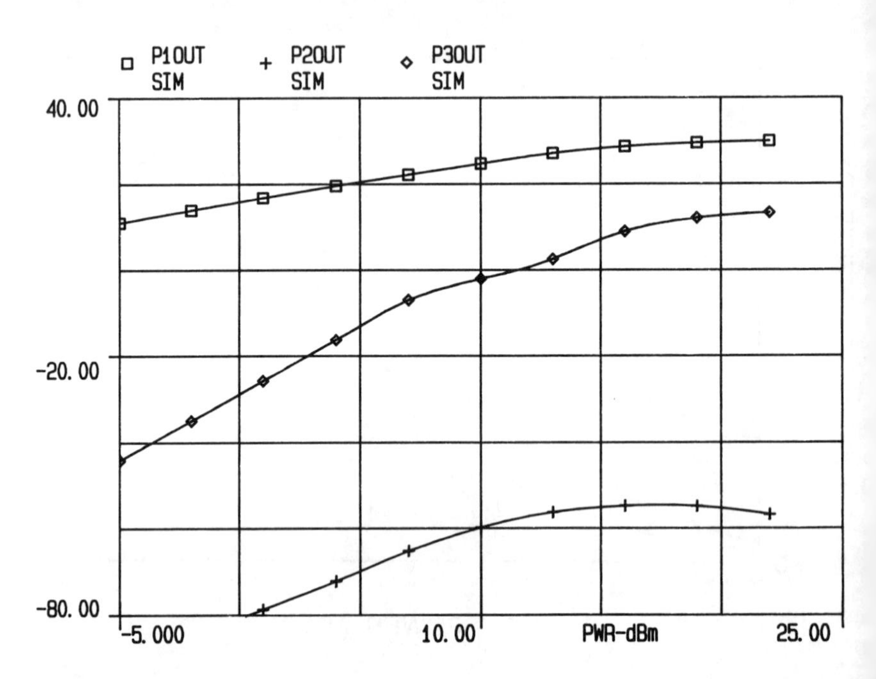


**Figure 3.29** Simulated output power versus input power of the 4.1-GHz fundamental, and second a third harmonics.

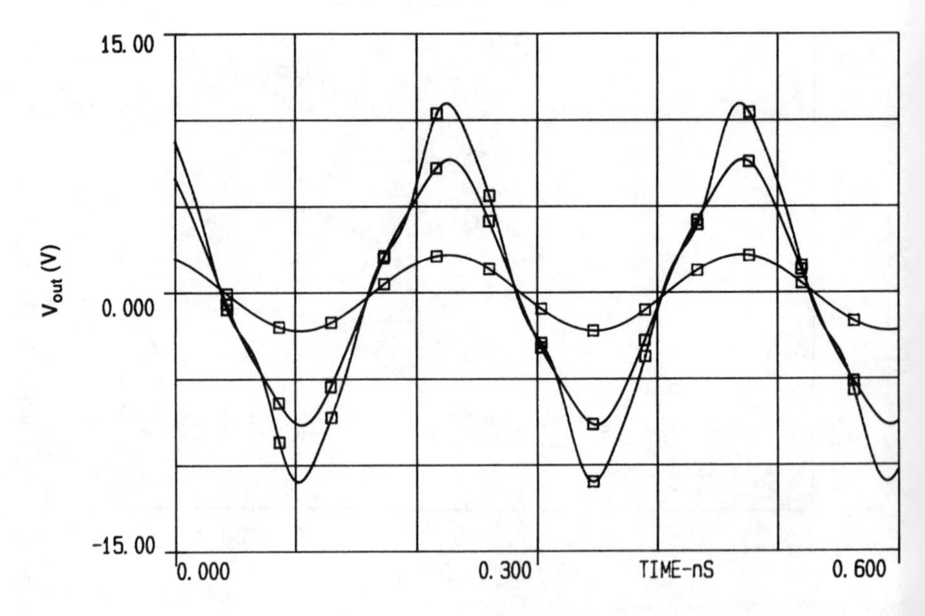


**Figure 3.30** Simulated voltage across the 50-$\Omega$ load termination for input power levels of 1, 13, and dBm at 4.1 GHz.

as modified this voltage from that of $v_{ds}$ at the drain terminal of the MESFET. Figure 3.31 shows $v_{ds}$, which is seen to be sinusoidal at the 1-dBm drive power and becoming trapezoidal (almost a square wave) at 19 and 22 dBm. As predicted by Snider [18], this is the waveshape that should occur in output power saturation when the even harmonics are short-circuit terminated. The $i_{ds}$ waveforms of Figure 3.32 show unclipped, Class A behavior at 1-dBm drive. As the drive increases, however, the amplifier operation progresses into the classic, Class B, half-sine waveshape for $i_{ds}$.

The input voltage to the MESFET, $v_{gs}$, is illustrated in Figure 3.33. As is typical for Class B MESFET operation, the minimum instantaneous gate-source voltage of $-6.5$V is more than twice the pinch-off voltage of $-2.7$V. Since this negative peak of $v_{gs}$ occurs in synchronism with the positive peak of $v_{ds}$, the voltage between the drain and gate can reach surprisingly large peak values. In this circuit the simulations indicate $v_{dg}$ peaking at 18V in Figure 3.34.

The information in the $v_{ds}$ and $i_{ds}$ waveforms can be used to describe the dynamic load line trajectory on the static *I–V* characteristics as shown in Figure 3.35 for input drives of 1, 7, 13, and 22 dBm. Rather than following a purely resistive ideal load line, there is a reactive component that causes the loops at lower drive powers. For clarification of saturated load line behavior, Figure 3.36 indicates the output *i–v* contour for the 22-dBm input drive alone. The self-biasing

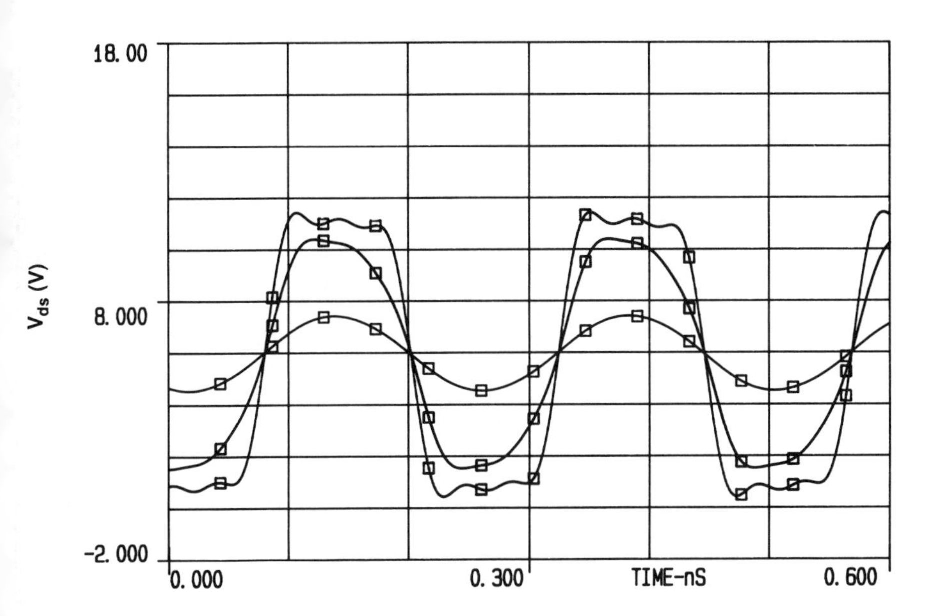


**Figure 3.31** Simulated drain-source voltage for input power levels of 1, 13, and 19 dBm at 4.1 GHz.

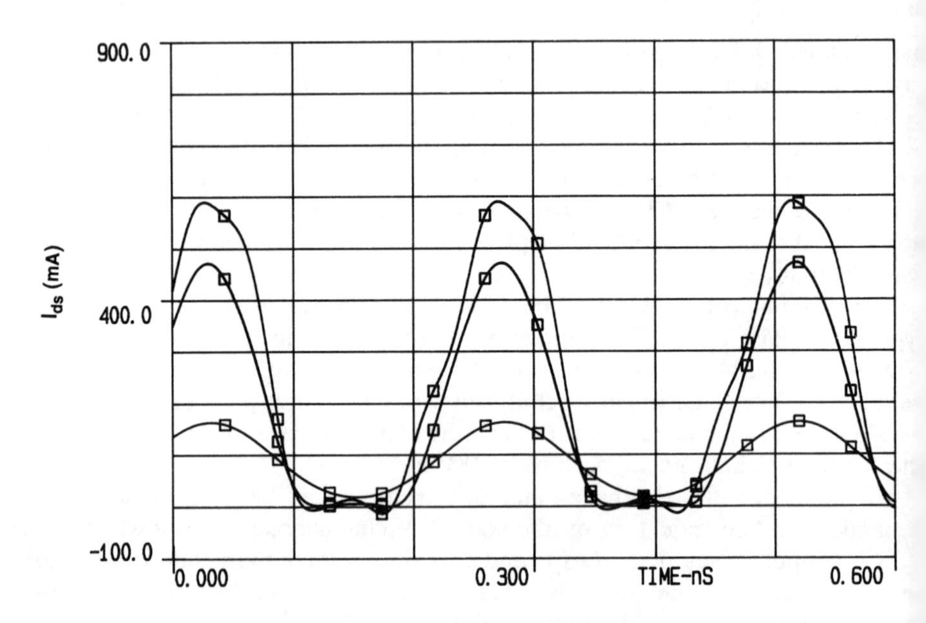


**Figure 3.32** Simulated drain-current waveform for input power levels of 1, 13, and 19 dBm at 4.1 GHz

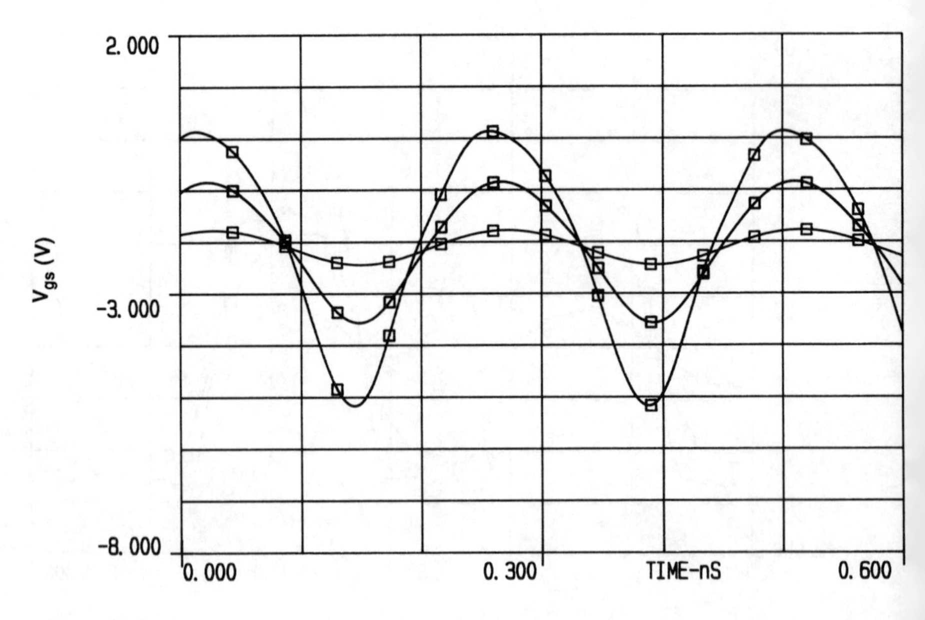


**Figure 3.33** Simulated gate-source voltage for input power levels of 1, 13, and 19 dBm at 4.1 GHz

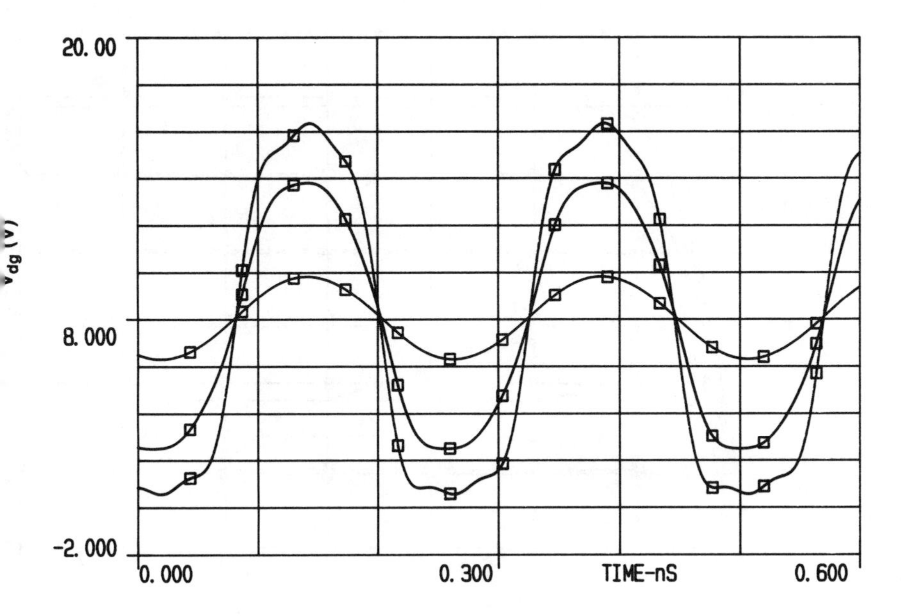


ure 3.34 Simulated drain-gate voltage for input power levels of 1, 13, and 19 dBm at 4.1 GHz.

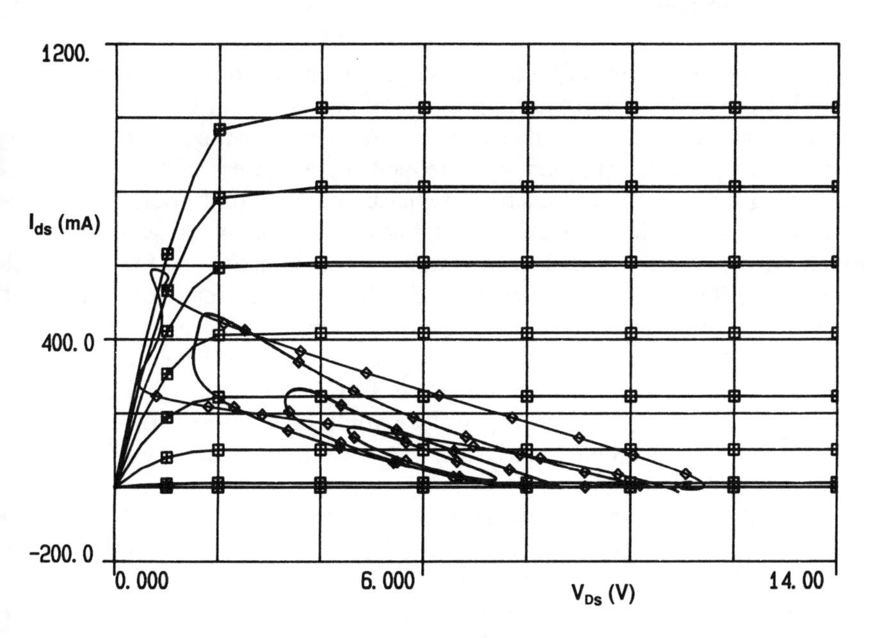


ure 3.35 Simulated dynamic load-line trajectory at 4.1 GHz for input power levels of 1, 7, 13, and 22 dBm.

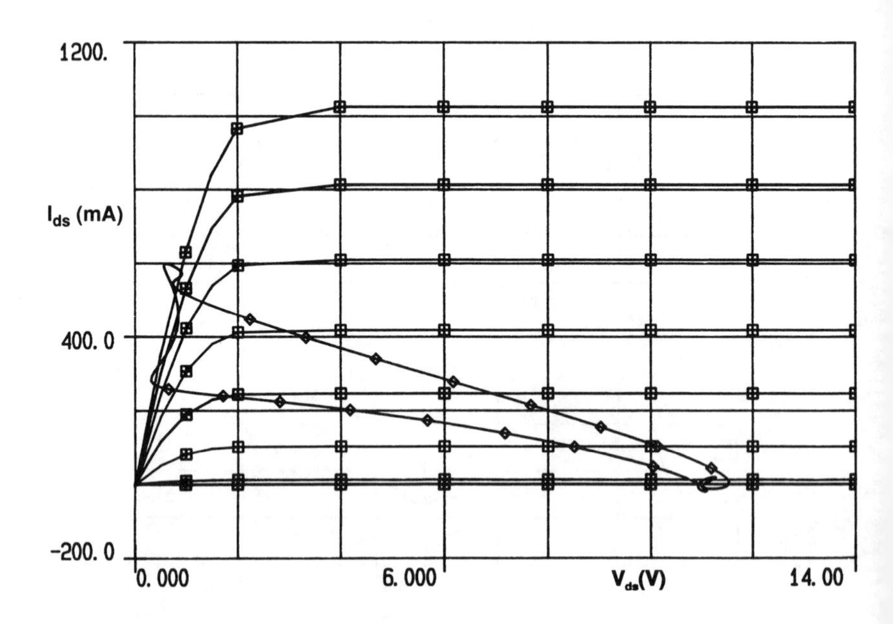


**Figure 3.36** Simulated dynamic load line trajectory at saturation at 4.1 GHz for an input power le of 22 dBm.

in Class B operation can be seen to have raised the average dc drain current $V_{ds}$ = 6V from $I_{dsQ}$ = 80 mA to approximately $I_{ds}$ = 230 mA.

Geller and Goettle provided some measurements of output power and pow added efficiency versus frequency at stepped input powers of 5, 10, 14, and dBm. In Figure 3.37 the output power measurements (solid lines) are compar with output power simulations (dotted lines). Agreement is well within 1 throughout the frequency and power ranges. The comparisons for power-add efficiency are illustrated in Figure 3.38. At the saturated input drive of 19 d the actual power-added efficiency is higher than simulated at most frequencies differing by 5% or less. At the lower power, the simulated efficiency is consisten higher. This is largely a result of setting $R_{ds}$ to 1 k$\Omega$ for better modeling of efficie in saturated operation, whereas $R_{ds}$ = 48$\Omega$ would be a better choice at the lov drive powers. This example clearly illustrates a major deficiency of most of models in popular usage.

Overall, it is apparent from the preceding example that simulation can r licate with reasonably good accuracy many important aspects of electrical perf mance such as power gain, output power, and conversion efficiency. In additi simulation can illuminate large-signal amplifier operational characteristics that c not be explored by any other means at microwave frequencies at this time.

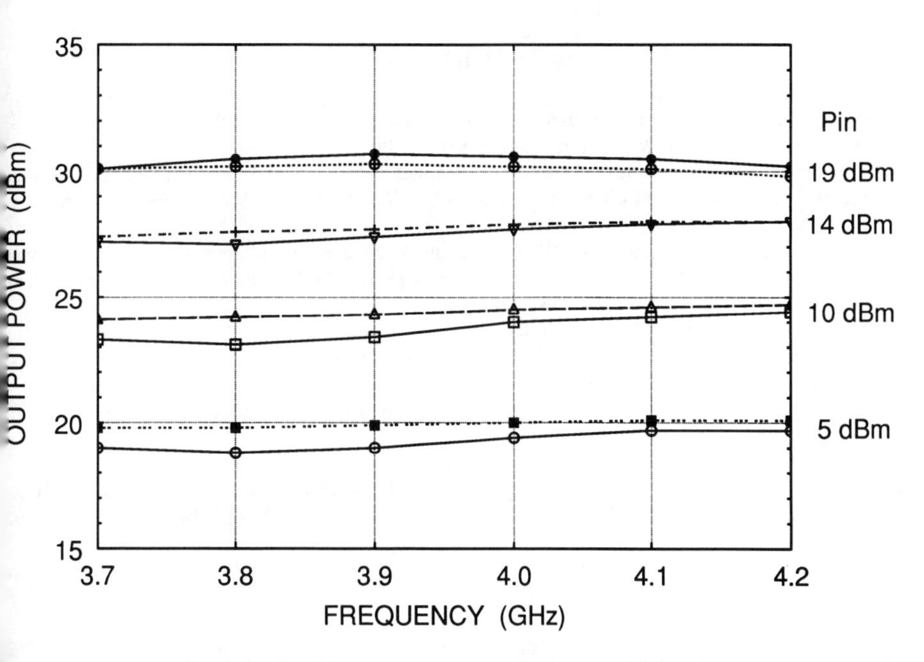


gure 3.37 Measured and simulated output power versus frequency with input power as a parameter.

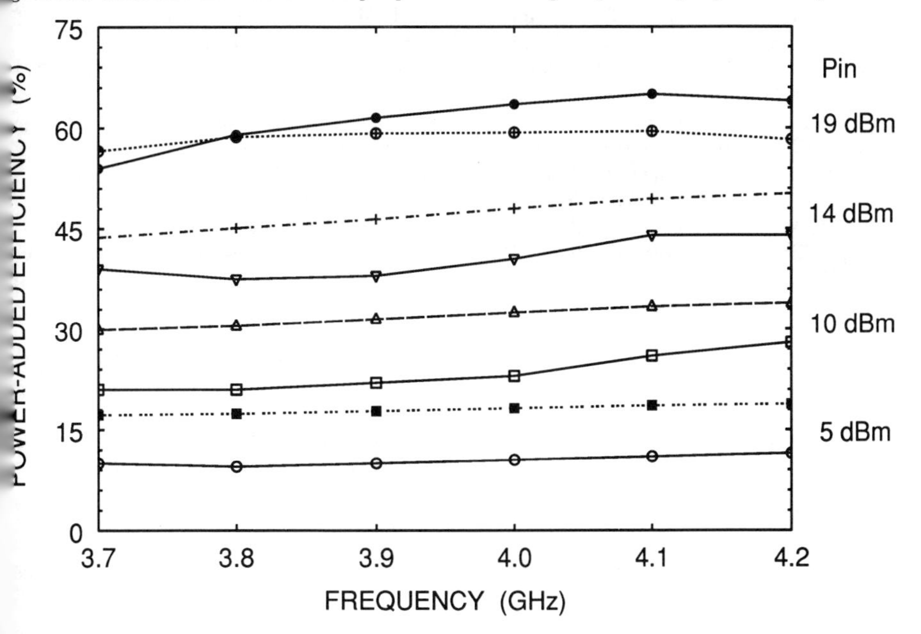


gure 3.38 Measured and simulated power-added efficiency versus frequency with input power as a parameter.

## REFERENCES


1. Holmes, C., O. Pitzalis, and Y. Yuan, "Harmonic Balance Software Simulates Nonlinear Circuits, *Microwaves & RF*, Vol. 26, October 1987, pp. 141–149.
2. Cammacho-Penalosa, C., and C. S. Aitchison, "Modeling Frequency Dependence of Output Impe ance of a Microwave MESFET at Low Frequencies," *Electron. Lett.*, Vol. 21, June 1985, pp. 528 529.
3. Platzker, A., A. Palevsky, S. Nash, W. Struble, and Y. Tajima, "Characterization of GaAs Device by a Versatile Pulsed I-V Measurement System," *IEEE MTT-S Int. Microwave Symp. Diges* Dallas, TX, May 8–10, 1990, pp. 1137–1140.
4. Struble, W., S.L.G. Chu, M. J. Schindler, Y. Tajima, and J. Huang, "Modeling Intermodulatio Distortion in GaAs MESFETs Using Pulsed I-V Characteristics," *1991 GaAs IC Symp. Digest*, p 179–182.
5. Sango, M., O. Pitzalis, L. Lerner, C. McGuire, P. Wang, and W. Childs, "A GaAs MESFE Large-Signal Circuit Model for Nonlinear Analysis," *IEEE MTT-S Int. Microwave Symp. Diges* New York, NY, May 25–27, 1988, pp. 1053–1056.
6. Statz, H., P. Newman, I. W. Smith, R. A. Pucel, and H. A. Haus, "GaAs FET Device and Circu Simulation in SPICE," *IEEE Trans. on Electron Devices*, Vol. ED-34, February 1987, pp. 160 169.
7. Scheinberg, N., R. Bayruns, P. Wallace, and R. Goyal, "An Accurate MESFET Model for Line and Microwave Circuit Design," *IEEE J. Solid-State Circuits*, Vol. SC-24, April 1989, pp. 532 539.
8. Maas, S. A., "A GaAs MESFET Balanced Mixer with Very Low Intermodulation," *IEEE MT S Int. Microwave Symp. Digest*, Las Vegas, NV, June 9–11, 1987, pp. 895–898.
9. Curtice, W. R., "A MESFET Model for Use in the Design of GaAs Integrated Circuits," *IEE Trans. on Microwave Theory and Techniques*, Vol. MTT-28, May 1980, pp. 448–456.
10. Shichman, H., and D. A. Hodges, "Modeling and Simulation of Insulated-Gate Field Effect Tra sistor Switching Circuits," *IEEE J. Solid-State Circuits*, Vol. SC-3, September 1968, pp. 285–28
11. Taki, T., "Approximation of Junction Field-Effect Transistor Characteristics by a Hyperbolic Fun tion," *IEEE. J. Solid-State Circuits*, Vol. SC-13, October 1978, pp. 724–726.
12. Kacprzak, T., and A. Materka, "Compact dc Model of GaAs FET's Large-Signal Computer Ca culation," *IEEE. J. Solid-State Circuits*, Vol. SC-18, April 1983, pp. 211–213.
13. Curtice, W. R., and M. Ettenberg, "A Nonlinear GaAs FET Model for Use in the Design Output Circuits for Power Amplifiers," *IEEE. Trans. on Microwave Theory and Techniques*, Vo MTT-33, December 1985, pp. 1383–1394.
14. Rehn, B., and R. Mitterer, "Ein MOS-Model mit Pseudophysikalischen Parametern," *NTG Fac berichte*, Band 51, March 1975, pp. 162–167.
15. McCamant, A. J., G. D. McCormack, and D. H. Smith, "An Improved GaAs MESFET Mod for SPICE," *IEEE Trans. on Microwave Theory and Techniques*, Vol. MTT-38, June 1990, p 822–824.
16. Geller, B. D., and P. E. Goettle, "Quasi-Monolithic 4-GHz Power Amplifiers with 65-Perce Power-Added-Efficiency," *IEEE MTT-S Int. Microwave Symp. Digest*, New York, NY, May 2 27, 1988, pp. 835–838.
17. Fujitsu Microwave Semiconductors 1989/90 Data Book, FLK202XV Data, pp. 48–49.
18. Snider, D. M., "A Theoretical Analysis and Experimental Confirmation of the Optimally Load and Overdriven RF Power Amplifier," *IEEE Trans. on Electron Devices*, Vol. ED-14, Decembe 1967, pp. 851–857.

# Chapter 4
# High-Power GaAs FET Amplifier Design

**R. B. Culbertson and R. E. Lehmann**

*Texas Instruments*

## .1 INTRODUCTION

n this chapter we present an elementary, but practical, introduction to power mplifier design. The essential ingredients are threefold, namely, FET character- zation for power and efficiency, small-signal modeling, and a linear load line pproximation technique. We assume the reader is familiar with small-signal *S* arameter amplifier design techniques, which are widely discussed in microwave extbooks. Bias conditions, output circuit loss, load impedance, and gain are major esign considerations to achieve maximum amplifier performance. Although the pproach applies to narrowband and broadband amplifiers alike, the focus here is n radar applications with bandwidths of the order of 30%. The chapter concludes ith a dual-gate FET 3-W amplifier example.

## .2 BUDGETING TRANSMITTING CHAIN RF PERFORMANCE

efore delving into amplifier design details it is useful to consider a module-level ransmitter block diagram performance budget such as that shown in Figure 4.1. n this example a multistage amplifier is driving a balanced power amplifier. An itial allocation of FET gate periphery is usually made prior to detailed amplifier esign so that an overall transmitting chain can be optimized. For example, we ay size the last two stages for operation at the maximum efficiency point, which pically occurs 2 dB into gain compression. We may want to saturate the driver mplifier third stage to provide some tolerance to variations in the input signal ower level, while the driver amplifier second stage might be sized for best effi- iency, and the first stage operated in the small-signal mode for high gain. Drain

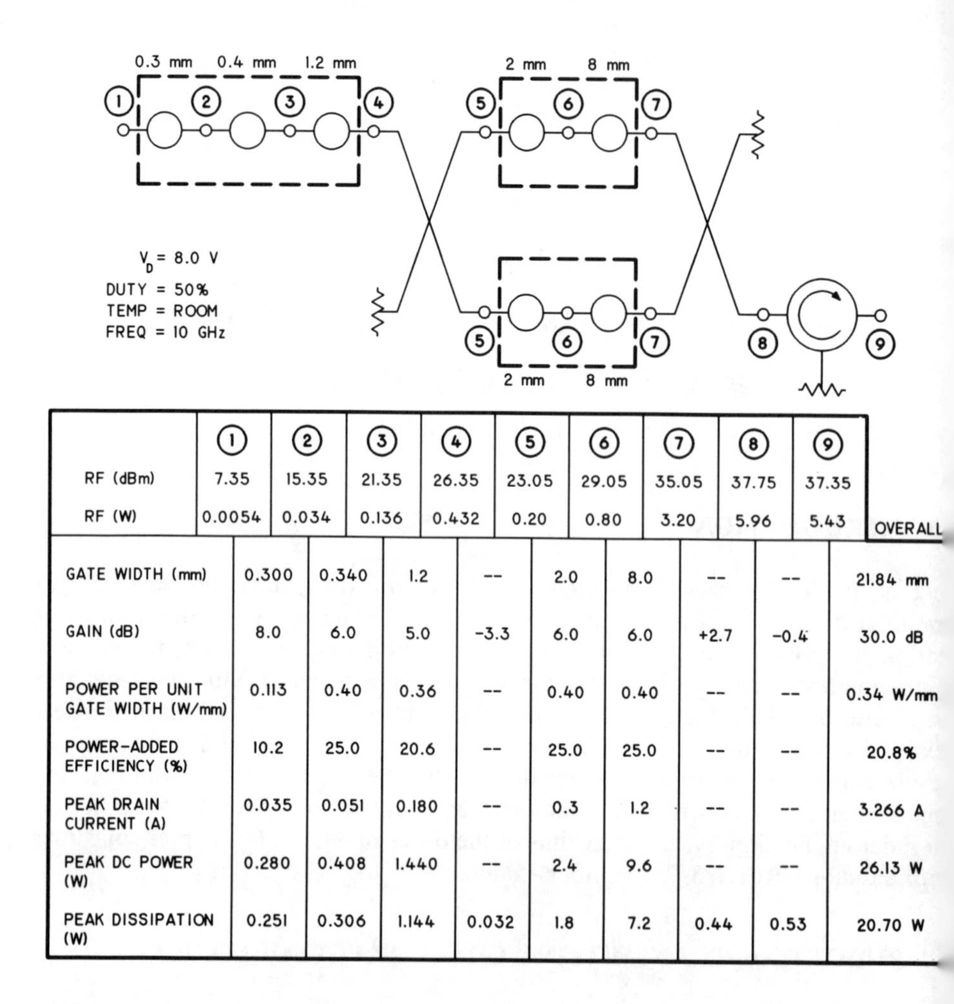


| | 1 | 2 | 3 | 4 | 5 | 6 | 7 | 8 | 9 |
|---|---|---|---|---|---|---|---|---|---|
| RF (dBm) | 7.35 | 15.35 | 21.35 | 26.35 | 23.05 | 29.05 | 35.05 | 37.75 | 37.35 |
| RF (W) | 0.0054 | 0.034 | 0.136 | 0.432 | 0.20 | 0.80 | 3.20 | 5.96 | 5.43 |

| | | | | | | | | | OVERALL |
|---|---|---|---|---|---|---|---|---|---|
| GATE WIDTH (mm) | 0.300 | 0.340 | 1.2 | -- | 2.0 | 8.0 | -- | -- | 21.84 mm |
| GAIN (dB) | 8.0 | 6.0 | 5.0 | -3.3 | 6.0 | 6.0 | +2.7 | -0.4 | 30.0 dB |
| POWER PER UNIT GATE WIDTH (W/mm) | 0.113 | 0.40 | 0.36 | -- | 0.40 | 0.40 | -- | -- | 0.34 W/mm |
| POWER-ADDED EFFICIENCY (%) | 10.2 | 25.0 | 20.6 | -- | 25.0 | 25.0 | -- | -- | 20.8% |
| PEAK DRAIN CURRENT (A) | 0.035 | 0.051 | 0.180 | -- | 0.3 | 1.2 | -- | -- | 3.266 A |
| PEAK DC POWER (W) | 0.280 | 0.408 | 1.440 | -- | 2.4 | 9.6 | -- | -- | 26.13 W |
| PEAK DISSIPATION (W) | 0.251 | 0.306 | 1.144 | 0.032 | 1.8 | 7.2 | 0.44 | 0.53 | 20.70 W |

**Figure 4.1** Example of a transmitting chain block diagram and performance budget.

RF impedance-matching networks can be designated as either power matching f the compressed stages or small-signal matching for the small-signal stages. Arm with a knowledge of the small-signal and large-signal performance capabilities a nominal FET at the highest intended frequency of operation, the required ga periphery and associated performance can be estimated stage by stage, starti from the output. After the amplifiers are built and measured, the performan budget is a useful tool to understand performance shortfalls, revise the chain, a modify the initial performance assumptions.

Power-added efficiency ($\eta$) is given by

$$\eta = \frac{P_{\text{out}} - P_{\text{in}}}{V_{ds} I_d} 100\% \tag{4.1}$$

here $P_{\text{out}}$ = RF output power, $P_{\text{in}}$ = RF input power, $V_{ds}$ = drain voltage, and = drain current. Any dc or RF power entering the stage that is not transmitted RF output power will be dissipated as heat. The dissipation ($P_{\text{diss}}$) is thus given by

$$P_{\text{diss}} = V_{ds} I_d + P_{\text{in}} - P_{\text{out}} \tag{4.2}$$

he performance budget table could be expanded to separate the FETs, impedance atching networks, and the interconnection losses.

A few other efficiency-related expressions are useful. Since the RF gain $G$, by definition, given by $G = P_{\text{out}}/P_{\text{in}}$ and the applied dc power is $P_{\text{dc}} = V_{ds} I_d$, e can rearrange (4.1) to yield

$$\eta = \frac{P_{\text{out}}}{P_{\text{dc}}}\left(1 - \frac{1}{G}\right) 100\% \tag{4.3}$$

here the ratio $P_{\text{out}}/P_{\text{dc}}$ is called the *drain efficiency*. If the output circuit loss $G_{\text{out}}$ included, then the efficiency becomes

$$\eta = \frac{P_{\text{out}}}{P_{\text{dc}}}\left(G_{\text{out}} - \frac{1}{G}\right) 100\% \tag{4.4}$$

dicating that output loss directly affects the efficiency. Similarly, if the input rcuit loss is included, then the efficiency becomes

$$\eta = \frac{P_{\text{out}}}{P_{\text{dc}}}\left(G_{\text{out}} - \frac{1}{G_{\text{in}} G}\right) 100\% \tag{4.5}$$

dicating that the input loss is buffered by the amplifier gain. Finally, two cascaded nplifiers have an efficiency given by

$$\eta = \frac{\eta_1 \eta_2 \left[1 - \dfrac{1}{G_1 G_2}\right]}{\eta_1 \left[1 - \dfrac{1}{G_2}\right] + \eta_2 \left[\dfrac{1}{G_2} - \dfrac{1}{G_1 G_2}\right]} \tag{4.6}$$

## 4.3 PERFORMANCE CHARACTERIZATION AND MODELING

### 4.3.1 Pulsed RF Testing

Power characterization of FETs for bias, output power, efficiency, and gain at given frequency provides the basis for transmitting chain performance budgetin and precedes detailed amplifier design. Figure 4.2 is a block diagram for a wave guide phase bridge that, despite its tedious point-by-point manual measuremen is very useful for pulsed characterization of FETs and amplifiers. The measurement are of a substitution type. A power meter is used to establish a reference level a the test ports, and subsequent power levels are determined by adjusting the pre cision waveguide attenuators and recording the attenuator settings. Relative inser tion phase measurements are similarly made by subtracting subsequent precisio phase shifter settings required to null the output of the phase detector. Since th

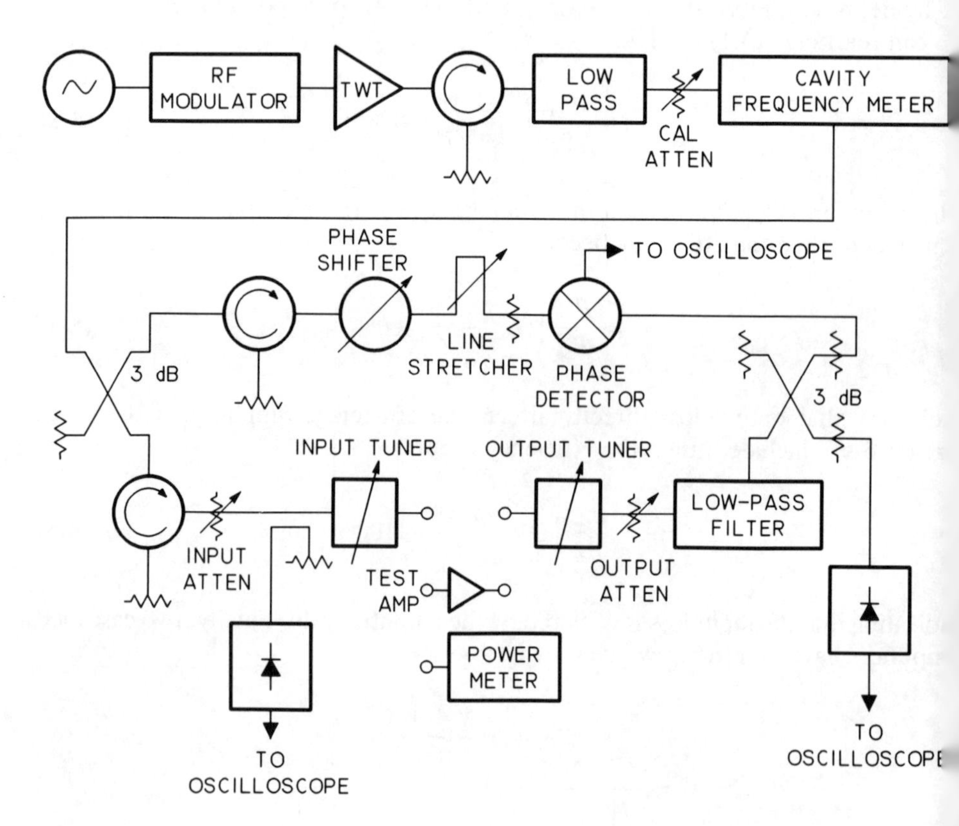


**Figure 4.2** Waveguide phase bridge for pulsed amplitude and phase characterization.

test set uses detectors and oscilloscopes, the amplitude and phase fidelity of pulsed amplifiers can be directly observed. An ac-coupled current probe senses the pulsed drain current.

Gain compression curves under a variety of conditions comprise the bulk of power characterization. Figure 4.3 shows the typical effect of small-signal versus large-signal output tuning for a 1200-$\mu$m-gatewidth FET. When the output is tuned for a small-signal conjugate match, then the device has 1 dB more small-signal gain but 2 dB less saturated output power than when tuned for maximum saturated output power. If the output tuner is adjusted at every drive level for maximum power, then we find the transition between large-signal and small-signal optimum tuning occurs over a narrow interval near the 20-dBm input level. Because the transition is relatively small, output matching networks are designed for either a large-signal match or a small-signal match. A compromise of both is generally not desired. Input tuning does not change significantly for either output tuning condition.

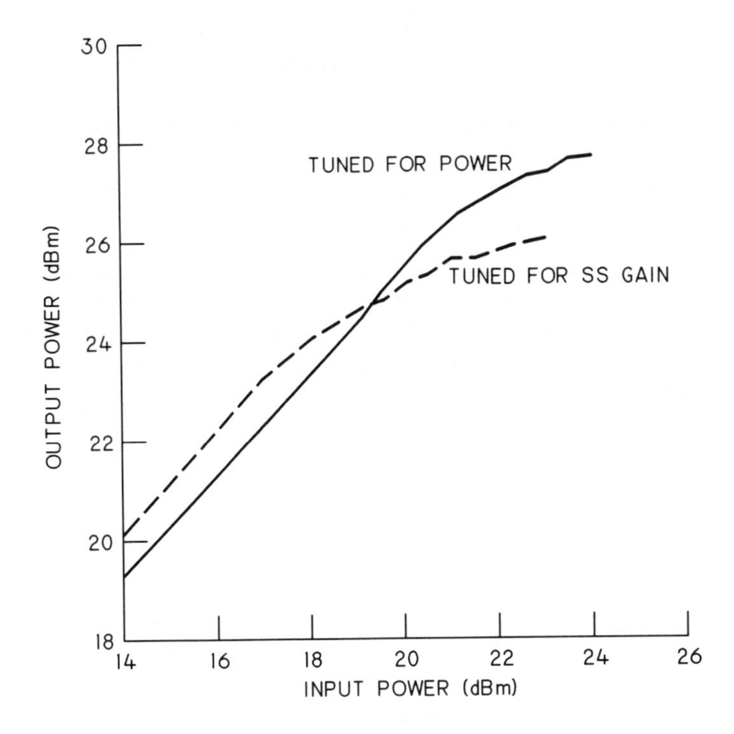


**Figure 4.3** Gain compression curves for a 1200-$\mu$m-gatewidth FET at 10 GHz for large-signal and small-signal tuning.

The following sequence is a useful guideline for lowering the risk of amplifier failure during initial biasing. If damage does occur, it is often minimized sufficiently to perform useful failure analysis.

1. Set the gate voltage power supply to 0V and connect to the device under test. Increase the RF input level and observe the rectification current (ensuring gate bias and RF input are indeed connected).
2. Increase the gate voltage to an operational level and check that no gates are dc shorted to ground. (Limit the supply current to roughly 10 mA/mm of device gatewidth.)
3. Using a pulsewidth and duty factor of the order of 2 $\mu$s and 1%, respectively increase the drain voltage to about 1V and vary the gate voltage to verify gate control of the drain current and that the device can be pinched off.
4. Use the input tuner to minimize the reflected RF power.
5. Raise the short-pulse, low-duty-cycle drain voltage to the operational level while observing pulsed current and detected RF output (ensuring no low frequency oscillations or missing RF output connections).
6. Use the output tuner to minimize output mismatches.
7. Ensure the drain voltage pulse is reasonably square without ringing or transients (often caused by excessive inductance between the drain modulator and the RF amplifier).
8. Increase the duty cycle to 50%. A drop in RF output power, typically of the order of 0.5 dB, will be observed, which indicates that no major thermal conduction problems are present due to poor die attachment or an air gap between the carrier plate and the test fixture.

In addition, follow basic static precautions by using antistatic work surfaces, grounded equipment, antistatic storage enclosures, and antistatic wrist straps.

Finally, a curve tracer measurement of gate-drain breakdown just prior to RF testing is recommended. This simple test is an effective screen to separate mysterious failures during RF turn-on. Some device failures are due to oscillation or transients while other failures are due to devices that have been weakened during fabrication or that have encountered static damage.

### 4.3.2 Bias Points and Class of Operation

The terminology for the class of operation has been carried over from vacuum tube and bipolar transistor amplifier design into the GaAs FET literature. The operating class (Class A, B, AB, C) generally denotes the quiescent bias point and resulting conduction angle irrespective of whether the amplifier is tuned to present a short circuit, open circuit, resistance, or a combination of these to the device at th harmonic frequencies. A variety of switched-mode operations (Classes D, E, F, G, H, S and more!) also exist. An excellent review of these conditions can be found in [1]. Many of the modes of operation characteristic of lower frequency designs are not presently obtainable in the upper microwave bands due to the relatively low gain at the fundamental and harmonic frequencies. For this reason most high-efficiency designs are Class AB, and so it is more meaningful to denote the bias point as a percentage of $I_{dss}$, usually between 10% and 40% (with no RF applied).

The drain voltage is usually selected to achieve the best efficiency or a compromise between efficiency and output power. Gain compression curves are shown in Figure 4.4 for a 4-mm-gatewidth single-stage MMIC that has been tuned for optimum performance at three different drain voltages. In this example, the highest efficiency is obtained with 5V and the highest power at 9V. Note that the power-added efficiency "rolls over," indicating that the input power level needs to be maintained within a 1- to 2-dB window, which is centered roughly at the 2-dB gain compression point.

Similarly, Figure 4.5 shows the effect of varying the gate voltage on the power output and efficiency after the output has been retuned for the best efficiency at each setting. In this example the best efficiency is obtained by biasing at 33% of $I_{dss}$. If the device is biased at 50% of $I_{dss}$, then a familiar Class A characteristic is observed. For Class A, the drain current does not change with RF input drive level and, as a result, the dissipation is greater for small-signal input levels than for large-signal inputs.

For a fixed drain voltage, the optimum gate voltage and output tuning is often easily found by observing the pulsed detected output power and drain current on the same oscilloscope. The voltage and tuning are adjusted to maximize the power and the difference between the power and current waveforms, i.e., achieving the most power for the least current.

### 4.3.3 Small-Signal Modeling

Small-signal $S$ parameters are measured over a broad bandwidth and the values in a lumped-element model are then adjusted to produce an equivalent circuit. Traditionally, CAD circuit optimizers are used for the parameter fitting, but due to the large number of variables, it is desirable to hold constant as many parameters as possible. Drain resistance $R_d$, gate resistance $R_g$, and source resistance $R_s$, for example, can be determined from curve tracer measurements. Transconductance $g_m$ and drain-source resistance $R_{ds}$ can be established from measuring $S$ parameters at 100 MHz where the device equivalent network can be simplified. Then $R_{ds}$ and $g_m$ are given by the following equations:

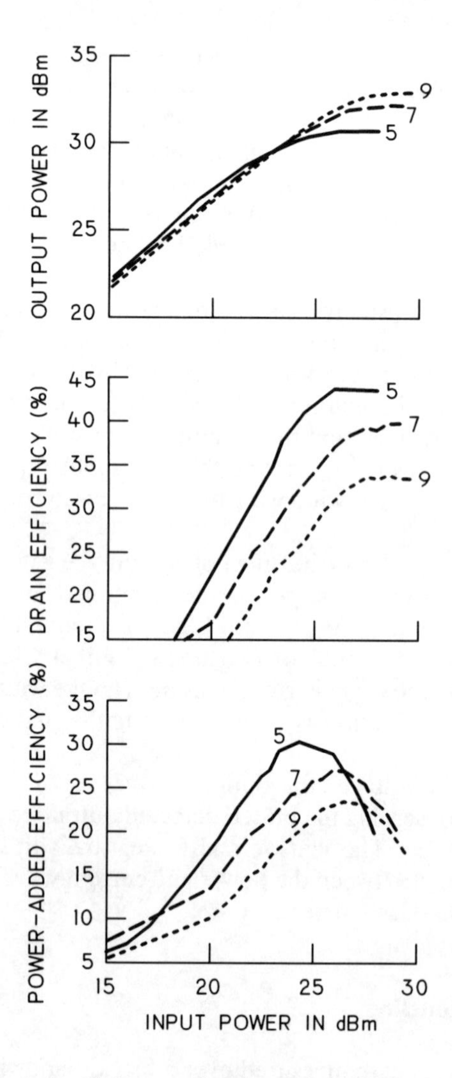


**Figure 4.4** Output power, drain efficiency, and power-added efficiency for a 4.0-mm amplifier stag at 10 GHz and drain voltages of 5, 7, and 9V.

$$R_{ds} = \frac{C - ABR_s}{1 + BR_s} \tag{4.7}$$

$$g_m = \frac{B(R_{ds} + A)}{R_{ds}} \tag{4.8}$$

where

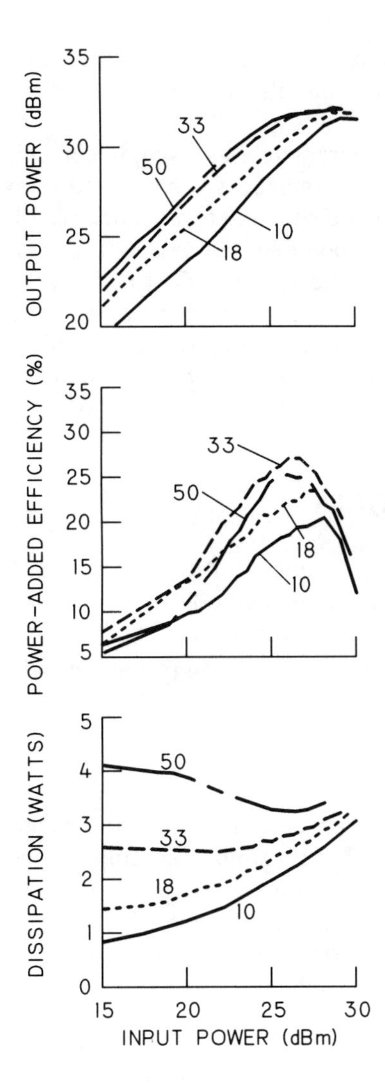


**Figure 4.5** Output power, power-added efficiency, and dissipation for a 4.0-mm amplifier stage at 10 GHz and a drain voltage of 7V with quiescent currents of 10, 18, 33, and 50% $I_{dss}$.

$$A = R_s + R_d + Z_0$$

$$B = \frac{|S_{21}|}{2Z_0 - R_s|S_{21}|}$$

$$C = \frac{Z_0(1 + |S_{22}|)}{1 - |S_{22}|} - R_s - R_d$$

and $Z_0$ is the measurement system impedance (usually $50\Omega$). Recent progress in parameter extraction has all but eliminated the need for a brute force optimization approach [2].

Finally, since the $S$ parameters are weakly dependent on drain voltage, but strongly dependent on gate voltage, it is natural to ask what gate voltage should be used for the small-signal characterization. For example, if power and efficiency tests indicate that the bias point should be set at 25% of $I_{dss}$ with no RF present but the operating current increases to 50% of $I_{dss}$ with RF applied, then should the small-signal measurements correspond to 25 or 50% of $I_{dss}$? Because $S_{11}$ changes a rather small amount with increasing drive, we believe that the gate voltage corresponding to the condition with no RF applied is more correct for the $S$ parameter characterization.

## 4.4 DESIGN TECHNIQUES

Power amplifier design techniques can be divided into three categories, namely simple load line approximation methods, designs based on measured load-pull characteristics, and nonlinear circuit simulators. The design technique emphasized here is the load line approach or the Cripps method [3].

### 4.4.1 Load-Pull

Design by load-pull is the most empirical of the three approaches. In its basic form an output tuner is adjusted to maximize the output power, the tuner is removed and then its input impedance is measured on a network analyzer. The tuner impedance and associated output power for enough additional tuner settings are then measured to plot constant power contours on a Smith chart. The test fixture electrical length and bond wire inductance must be deembedded before the data are plotted.

The resulting contours form the region of input impedance required for the impedance matching network. With small-signal power levels, the contours correspond to constant gain circles generated from measured $S$ parameters. As the input level is increased to saturate the FET, the contours become closer together and less circular. Without automation the method quickly becomes tedious as test conditions such as input signal level, bias settings, and frequency are changed. With automation a large quantity of data is recorded and reduced into a manageable subset useful for amplifier design. Automation can include a stepping motor-driven mechanical slide-screw tuner for which the input impedances for all positions have been stored. An alternative approach is to replace the passive tuner with an acti

r electronic equivalent [4]. A signal that is derived from the same source as the
put signal is injected into the output arm. An attenuator and phase shifter are
sed to vary the magnitude and angle of the "reflected" wave, which is analogous
varying the depth and position of a slide-screw tuner. Again, the adjustments
an be automated with switched attenuators and stepping motor-driven phase shif-
rs.

The disadvantages of load-pull design can include entering regions of insta-
ility, tedium if manual, significant investment if automated, uncontrolled imped-
nces at harmonic frequencies, FET periphery limits imposed by the maximum
assive tuner VSWR, and, in the active tuner case, component power limitations.
n the other hand, when dealing with a new, and as yet unmodeled, device or
vestigating performance sensitivity to impedance mismatches, carefully taken
ad-pull measurements can provide the most convincing evidence for amplifier
npedance-related behavior.

## .4.2 Nonlinear CAD

icrowave nonlinear circuit simulators can provide insight into circuit behavior
hich cannot be predicted by overly simplified quasilinear approximations or which
nnot be directly measured. The steady-state solutions of harmonic balance and
e older time-domain transient response simulators have advantages and disad-
ntages that make both techniques useful for circuit analysis. As analysis tools
ey can provide proof-of-concept, predict performance limitations, and explain
herwise anomalous behavior.

## 4.3 Modified Cripps Method

he utility of the loadline approach when used with a small-signal model in opti-
ization software makes a simple, but surprisingly successful, design tool. Because
is method incorporates the primary limitations to output power, namely, limi-
tions on RF voltage and current swing, the projected output power is often very
curate.

Recall that the average power delivered from a sinusoidal current generator
ith $i(t) = i_{peak} \cos \omega t$ to a network having an input impedance $Z = |Z| \angle \theta$ can
found by integrating the instantaneous power over a period. The voltage across
e load, $v(t)$, is given by

$$v(t) = i_{peak} |Z| \cos(\omega t + \theta) \tag{4.9}$$

and hence the instantaneous power dissipated in the load, $p(t)$, is given by

$$p(t) = v(t)i(t) = i_{\text{peak}}^2 |Z| \cos(\omega t + \theta) \cos \omega t \qquad (4.10$$

After using two common trigonometric identities, (4.10) can be rewritten as

$$p(t) = i_{\text{peak}}^2 |Z| \left( \frac{1}{2} \cos \theta + \frac{1}{2} \cos \theta \cos 2\omega t - \frac{1}{2} \sin \theta \sin 2\omega t \right) \qquad (4.11$$

which, when integrated over a period, gives

$$P_{\text{RF}} = \frac{1}{2} i_{\text{peak}}^2 |Z| \cos \theta = \frac{1}{2} i_{\text{peak}} v_{\text{peak}} \cos \theta \qquad (4.1$$

Now suppose that $v_{\text{peak}}$ and $i_{\text{peak}}$ are limited to $v_{\text{peak,max}}$ and $i_{\text{peak,max}}$, respectively [Fi
4.6(b)]. Then the optimum load impedance is a pure resistance of

$$R_{\text{opt}} = \frac{v_{\text{peak,max}}}{i_{\text{peak,max}}} \qquad (4.1$$

For this condition $\theta = 0$ and the maximum RF output power is

$$P_{\text{RF,max}} = \frac{1}{2} v_{\text{peak,max}} i_{\text{peak,max}} = \frac{1}{2} i_{\text{peak,max}}^2 R_{\text{opt}} \qquad (4.1$$

We are interested in the maximum power delivered to the load (without clippin
if $Z$, which may be complex, is not equal to $R_{\text{opt}}$. Assume the output current reach
$i_{\text{peak,max}}$. Then the power delivered to the load is

$$P_{\text{RF}} = \frac{1}{2} i_{\text{peak,max}}^2 |Z| \cos \theta = \frac{1}{2} i_{\text{peak,max}}^2 R_L \qquad (4.1$$

where $R_L$ is the real part of the load impedance. Dividing both sides by (4.1
yields

$$\frac{P_{\text{RF}}}{P_{\text{RF,max}}} = \frac{R_L}{R_{\text{opt}}} \qquad (4.1$$

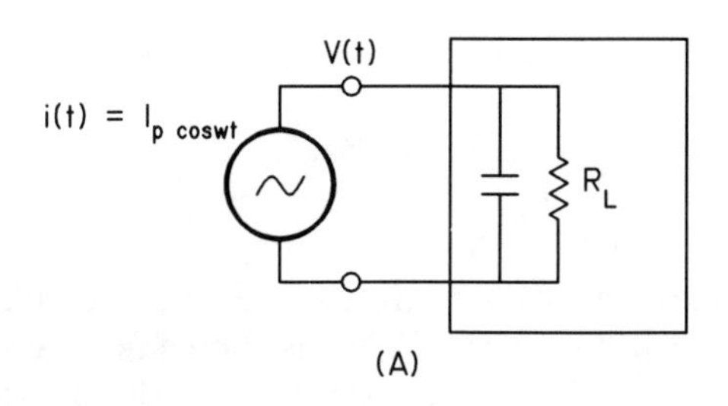


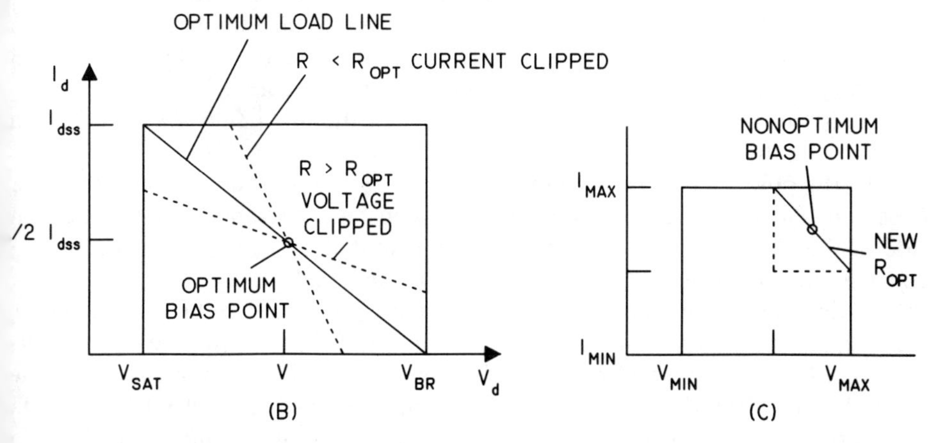


igure 4.6 Output load line effects: (a) current generator with a complex load, (b) load lines with an optimum bias point, and (c) load line with a nonoptimum bias point.

ut, $v_{\text{peak}} = i_{\text{peak,max}}|Z| = v_{\text{peak,max}}|Z|/R_{\text{opt}} < v_{\text{peak,max}}$. So

$$|Z| < R_{\text{opt}} \tag{4.17}$$

the condition for which (4.16) holds. If, on the other hand, the output voltage aches $v_{\text{peak,max}}$, then the power delivered to the load is given by

$$P_{\text{RF}} = \frac{1}{2}\frac{v^2_{\text{peak,max}}}{|Z|\cos\theta} = \frac{1}{2}\frac{v^2_{\text{peak,max}}}{R_L} \tag{4.18}$$

gain dividing (4.18) by (4.14) yields

$$\frac{P_{\text{RF}}}{P_{\text{RF,max}}} = \frac{R_{\text{opt}}}{R_L} \tag{4.19}$$

But, $i_{\text{peak}} = v_{\text{peak,max}}/|Z| = i_{\text{peak,max}} R_{\text{opt}}/|Z| < i_{\text{peak,max}}$. So

$$|Z| > R_{\text{opt}} \tag{4.20}$$

is the condition for which (4.19) holds.

When the load is no longer a pure resistance, then the load line is no longe
a straight line; instead, it becomes an ellipse as shown in Figure 4.7. (Remembe
that the load line is simply a plot of the instantaneous drain-source voltage versu
drain current.) The major axis of the ellipse lies along the load line for maximu
RF output power. Conditions (4.17) and (4.20) ensure that the ellipse is bounde
by $0 \le I_{ds} \le I_F$, $V_K \le V_{ds} \le V_{dgB} - |V_P|$ as shown in Figure 4.7.

If a Smith chart is normalized to $R_{\text{opt}}$, then the line $|Z| = R_{\text{opt}}$ is a vertica
line dividing the chart, as shown in Figure 4.8. On the left half of the chart, $|Z| <$
$R_{\text{opt}}$, so the power to the load is limited by $i_{\text{peak,max}}$, and contours of constant powe
(less than $P_{\text{RF,max}}$) correspond to contours of constant resistance, $R_L$. On the righ
side of the chart, $|Z| > R_{\text{opt}}$, so the power is limited by $v_{\text{peak,max}}$, and contours o
constant power correspond to contours of constant conductance, $1/R_L$. Notice tha
the power contours in Figure 4.8 are more tightly grouped than the correspondin
small-signal mismatch contours. A reflection coefficient magnitude of 0.33, fo
example, has a small-signal mismatch loss of only 0.5 dB, but for a load line th
power loss is 1 dB if the reflection coefficient angle is ±90 deg and 3 dB if it shoul
fall on the resistance axis! Hence, it is very important for the impedance locus o
output matching networks to fit within a noncircular power contour instead o

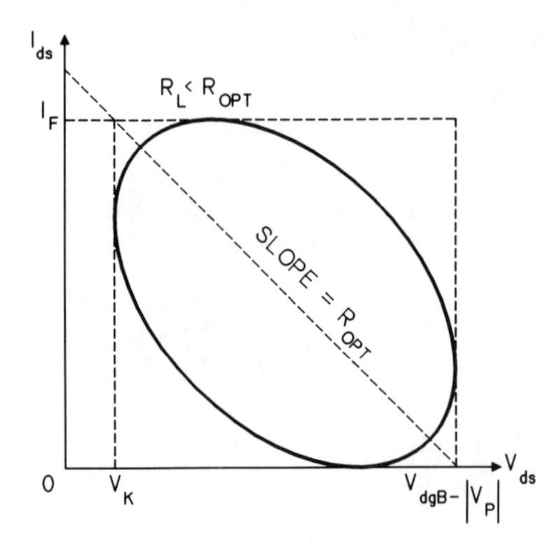


**Figure 4.7** Elliptical load line resulting from a complex load impedance.

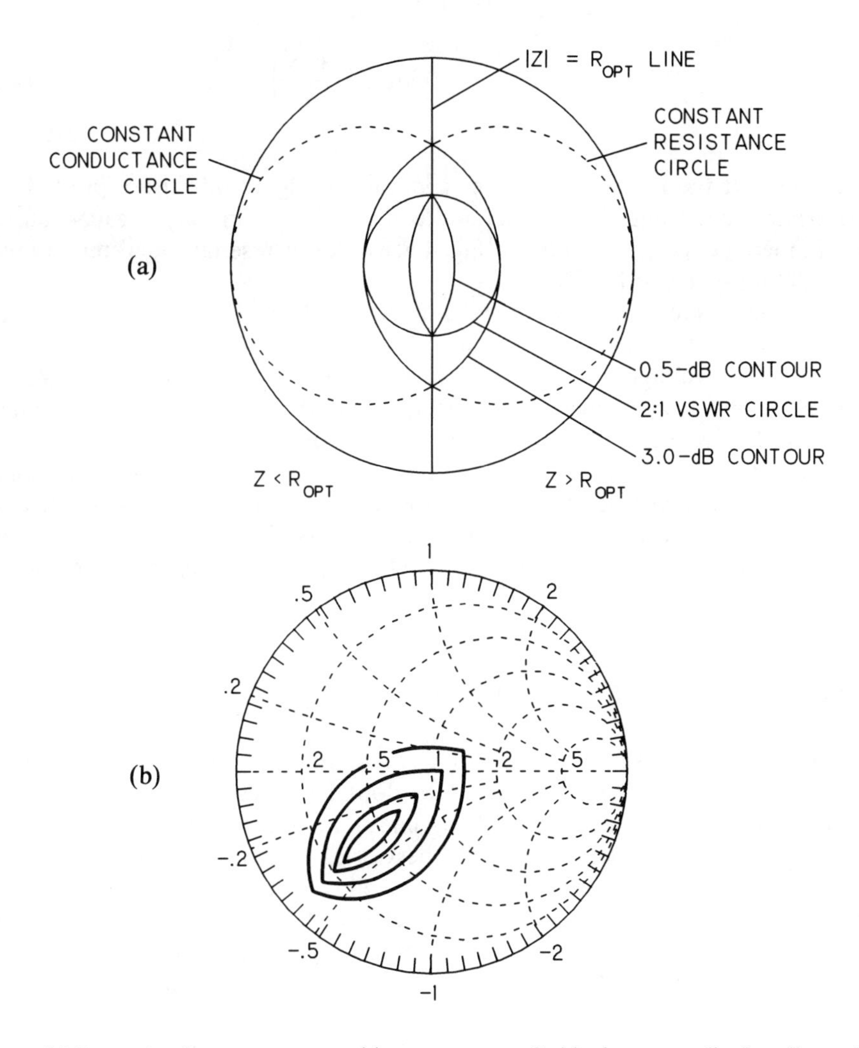


**igure 4.8** Large-signal power contours: (a) contours on a Smith chart normalized to $R_{opt}$ and (b) contours on a 50-$\Omega$ Smith chart for $R_{opt} = 35\Omega$ in parallel with 0.35 pF.

erely within a circular VSWR contour. The maximum and minimum load line ismatch loss ($L_{clip}$) can be expressed in terms of VSWR, $S$, as

$$\max L_{clip} = 10 \log_{10} S \tag{4.21}$$

nd

$$\min L_{clip} = 10 \log_{10}\left(\frac{1 + S^2}{2S}\right) \qquad (4.22$$

respectively. If the normalized power contours of Figure 4.8(a) are plotted on
50-$\Omega$ Smith chart using $R_{opt} = 35\Omega$ and combined with an output capacitance c
0.35 pF, then the resulting contours [Fig. 4.8(b)] closely resemble load-pull contour
for a 1200-$\mu$m-gatewidth FET.

The drain-source bias voltage $V_{ds}$ is chosen from FET characterization o
other requirements and $v_{peak,max}$ is then considered to be $V_{ds} - V_K$, where $V_K$ is th
saturation knee voltage in the FET drain *I-V* curves (typically 1 to 2V). We the
want breakdown to be above $2(V_{ds} - V_K)$, so that the load voltage swings equall
above and below the bias point. In the absence of a load-pull measurement fo
$R_{opt}$, $i_{peak}$ can be assumed to be $I_{dss}/2$. This bounded linear approach also applie
to a nonoptimum bias [Fig. 4.6(c)] by considering the maximum linear swing abo
the new bias point, or new voltage or current limits. The revised limits give ne
values for $R_{opt}$ and $P_{RF}$, which can be used in the above equations for power clippin
loss.

## 4.5 SCALING

Scaling concepts are used throughout power amplifier development. Increasing th
gate periphery of a particular stage can be done by (1) connecting several prove
cells in parallel (lowest risk), (2) adding additional fingers to a given cell, or (
extending the length of gate fingers (highest risk). If additional fingers are adde
to a cell, the RF performance can be predicted by analyzing the gain of lumpe
element models for each finger interconnected with appropriate lengths of tran
mission line as shown in Figure 4.9. As more fingers are added, resistive loss
and phase imbalance will decrease the gain with respect to an entirely lumpe
element model.

The trends encountered when extending gate finger length can be predict
using distributed models for the gate finger as shown in Figure 4.10. Coupled-mo
theory [5, 6] is used to analyze the wave propagation effects along the gate strip
Again, resistance loss and phasing effects are apparent. The typical series resistan
for a 6000-Å-thick, 0.5-$\mu$m gate stripe is 300 $\Omega$/mm. Figure 4.11 shows the predict
maximum gain for gate metal resistances of 0, 100, and 400 $\Omega$/mm versus frequen
and finger length.

Finally, source grounding inductance and thermal effects must be estimat
when altering the FET geometry. Ideally, one would choose from a family
standard cell devices, but custom MMIC designs invariably dictate minor alteratio
in the FET layout for sensible compatability with the matching networks and

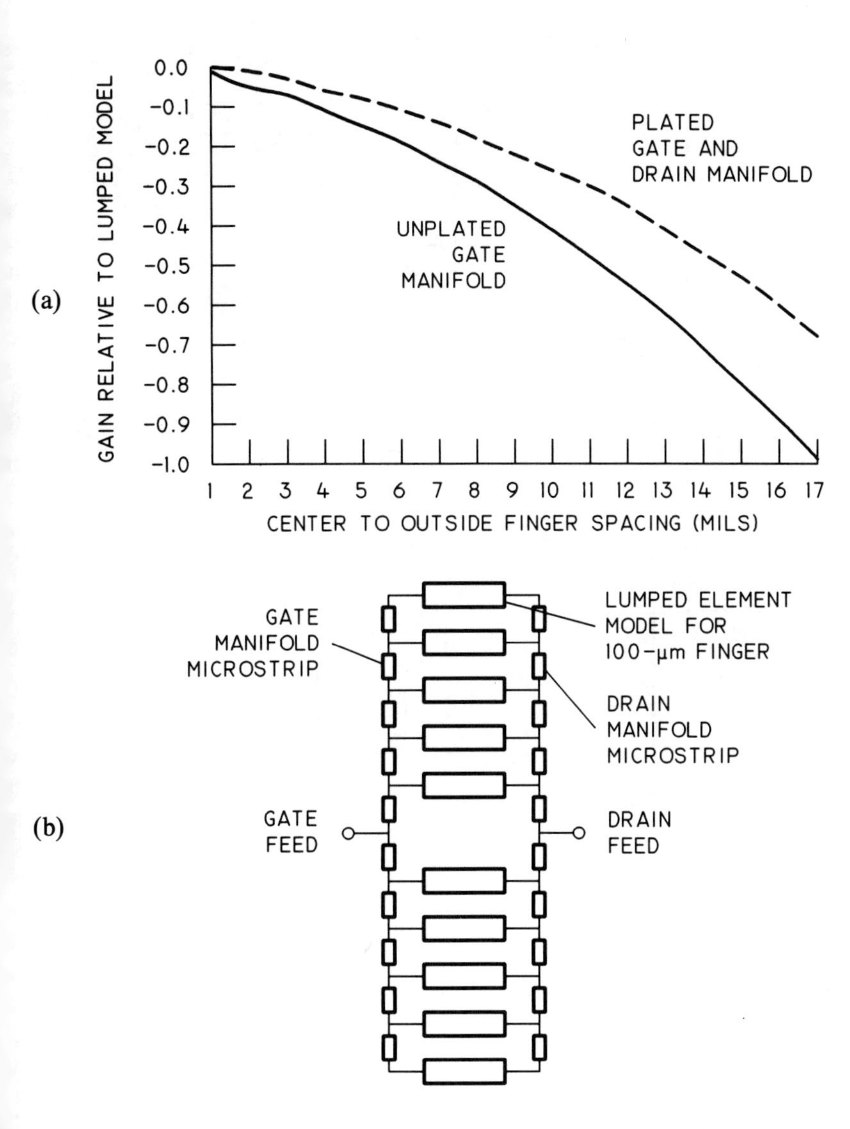


**Figure 4.9** Modelling the effect of adding additional FET fingers in parallel: (a) calculated gain degradation at 10 GHz and (b) model schematic diagram.

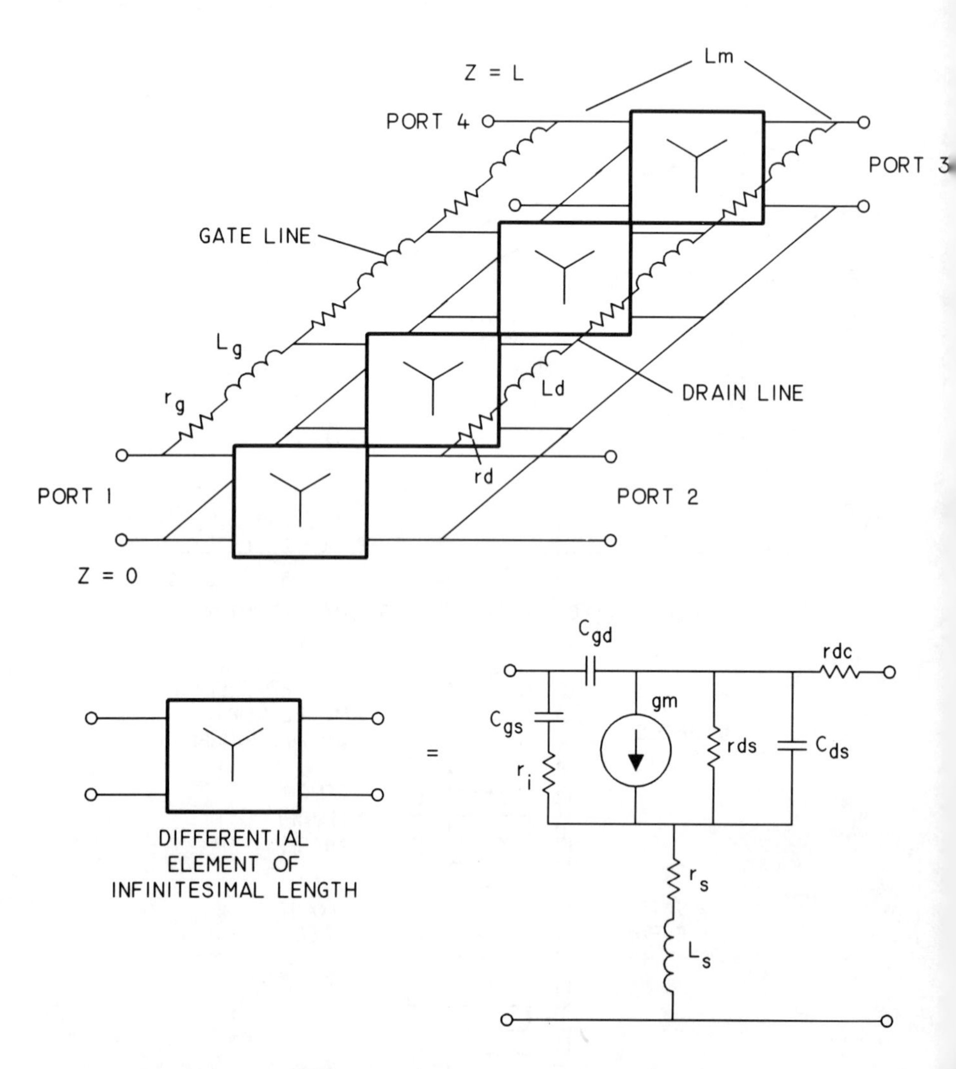


**Figure 4.10** Distributed model of a single FET finger using coupled-mode analysis. (After LaRue et al. [6]).

achieve the total required gate periphery. For example, it may be preferable to use eight slightly modified cells instead of nine standard cells.

Low-loss power combining of cells depends on good amplitude and phase balance. Phase is particularly important. The loss in combining two signals with amplitude and phase imbalance is given by

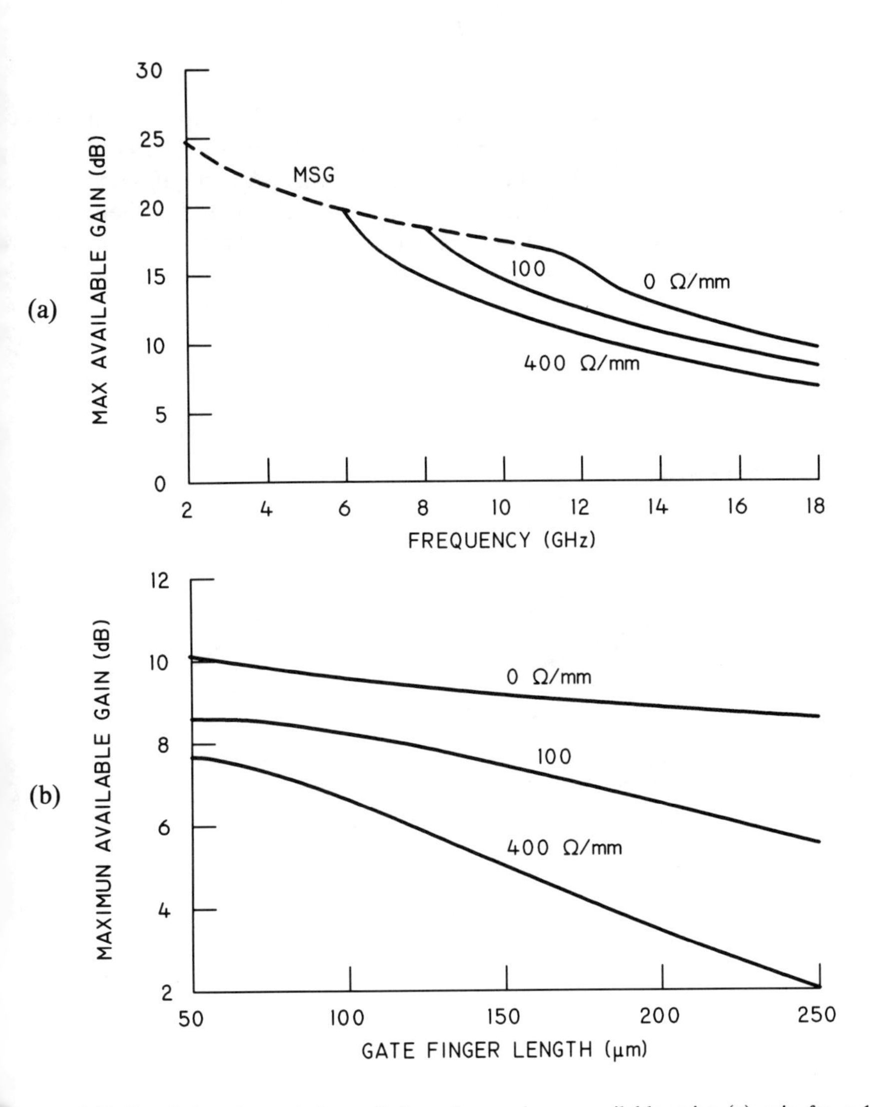


**Figure 4.11** Coupled-mode analysis predictions for maximum available gain: (a) gain for a 100-$\mu$m finger with series gate resistance of 0, 100, and 400 $\Omega$/mm and (b) gain at 18 GHz versus finger length and gate resistance.

$$IL_{ex} = 10 \log_{10}\left(\frac{1}{2} + \frac{\Delta^{1/2}}{1 + \Delta} \cos \theta\right) \quad (4.23$$

where $\theta$ is the phase imbalance in degrees and $10 \log_{10} \Delta$ is the amplitude imbalanc in decibels. Using this expression, combining loss contours are plotted in Figur 4.12, which emphasize the importance of phase balance over amplitude balance These computations assume that the signal sources are isolated. When reactiv combining is used, phase and amplitude imbalance causes additional losses due t the unavoidable mistuning of one source or the other.

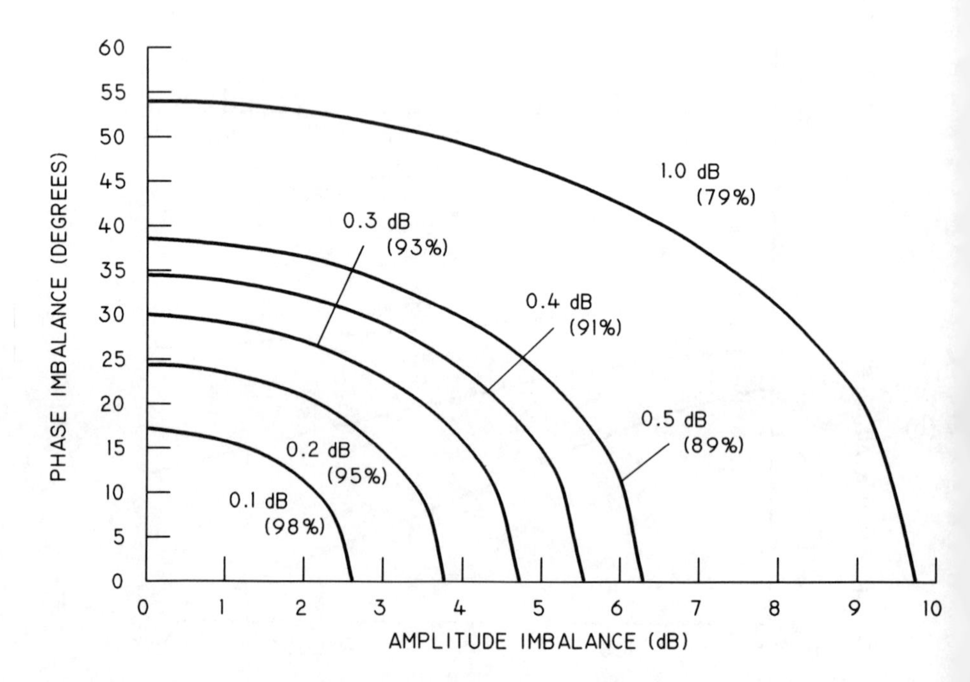


**Figure 4.12** Combiner loss contours for phase and amplitude imbalance.

## 4.6 MATCHING NETWORK DESIGN

Matching network design is perhaps the major task in any amplifier design. Unfo tunately, it is beyond the scope of this chapter, so we refer the reader to literatu n the subject [7–13] and present a few points particularly relevant to power mplifier design.

Impedance matching is often divided into parasitic absorption and resistance vel transformation. The parasitic absorption part of the problem is measured in , the resistance-to-reactance ratio at band center. Figure 4.13 shows a simplified iagram of the impedance-matching problem for a two-stage amplifier. The gate as a higher $Q$, representing a more difficult matching problem, than the drain. ote that the $Q$, however, does not scale with gatewidth. Hence, from a parasitic bsorption point of view, large periphery FETs pose no more difficulty than smaller ETs. The resistance transformation, however, is increasingly difficult with larger ETs. The impedance levels for the example shown in the figure require trans- rmation ratios of 14:1 for the output network, 25:1 for the interstage, and 57:1 r the input network. These ratios are much less than the 143:1 transformation at would be required if the second-stage input were matched directly to $50\Omega$. npedance transformation is usually obtained with either a frequency-independent orton transformer [8, 9] if bandpass networks are used, or with a pseudo low- ass transforming network [12, 13]. The results of gain bandwidth calculations for w-pass transformers of various complexities and transformation ratios are shown Figure 4.14. Very high transformation ratios using low-order networks are pos- ble for narrow bandwidths, while for decade bandwidths even a transformation 2:1 can be impractical.

## 6.1 Output and Interstage Network Load Line Analysis

he FET load impedance can be calculated from the predicted $S$ parameters for e output matching network, including a capacitor representing the FET output pacitance. The value of the FET parasitic capacitance can come from load-pull easurements, or it can be represented simply as the shunt capacitance of the FET tput when the input is conjugately matched (slightly larger than $C_{ds}$). The output ower and clipping loss can be calculated using the equations of Section 4.4. The ssipation of the output network is given by

$$\text{dissipation} = \frac{|S_{21}|^2}{1 - |S_{11}|^2} \tag{4.24}$$

he final output power is determined by the load line maximum power minus the tput network dissipation. Using this simple approach during the design empha- zes the importance of a close match to the optimum power impedance and favors riations in load reactance over variations in resistance.

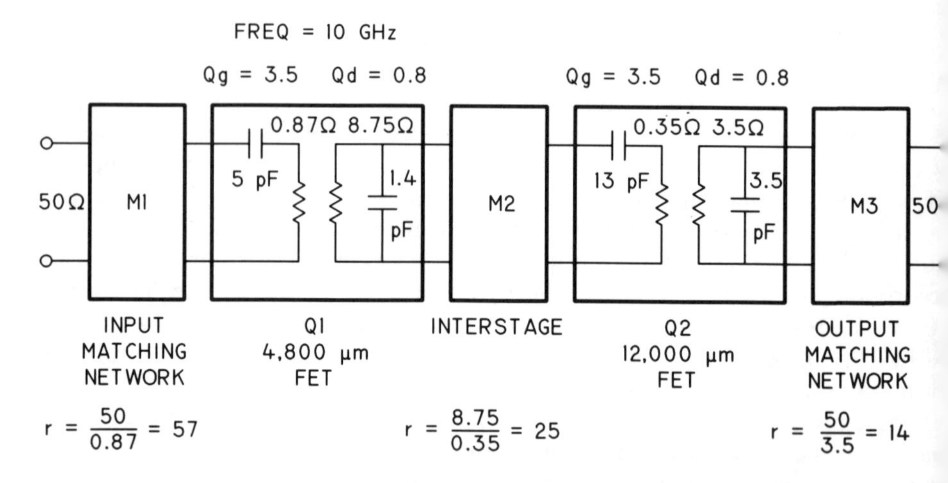


**Figure 4.13** Impedance-matching considerations for a sample two-stage power amplifier.

Equally important is the load line evaluation of the interstage. The comput *S* parameters for the cascade of the next-to-last FET output capacitance, intersta network, output FET, and output matching network are used to compute the lo impedance to the driving FET. The load line clipping loss or saturated power this FET is obtained. Equation (4.24) in this case represents the dissipation of t interstage and output networks and the small-signal gain of the output FET. T gain, added to the saturated output power of the driving FET, must equal or exce the previously predicted saturated output power of the output FET; otherwise, t driving FET will not be able to drive the output FET into saturation.

Some design margin should be applied at this point. For example, a 20 lower value for $g_m$ should be used in the FET model to ensure the small-sig projection still exceeds the saturated output power.

### 4.6.2 Harmonic Termination Effects

The preceding design method emphasized a linear, or bounded linear, approa that neglected harmonics. Even if the amplifier is biased class A, however, i expected that the FET will be driven into compression and hence the wavefor will be clipped symmetrically with some harmonics generated. If biased tow class B, then the current waveform is, again, intentionally clipped. The result power will be degraded if the harmonics are improperly terminated. Therefore common practice is to strive for the output capacitance and matching network present a short-circuit impedance at the harmonic frequencies. Choosing a lc

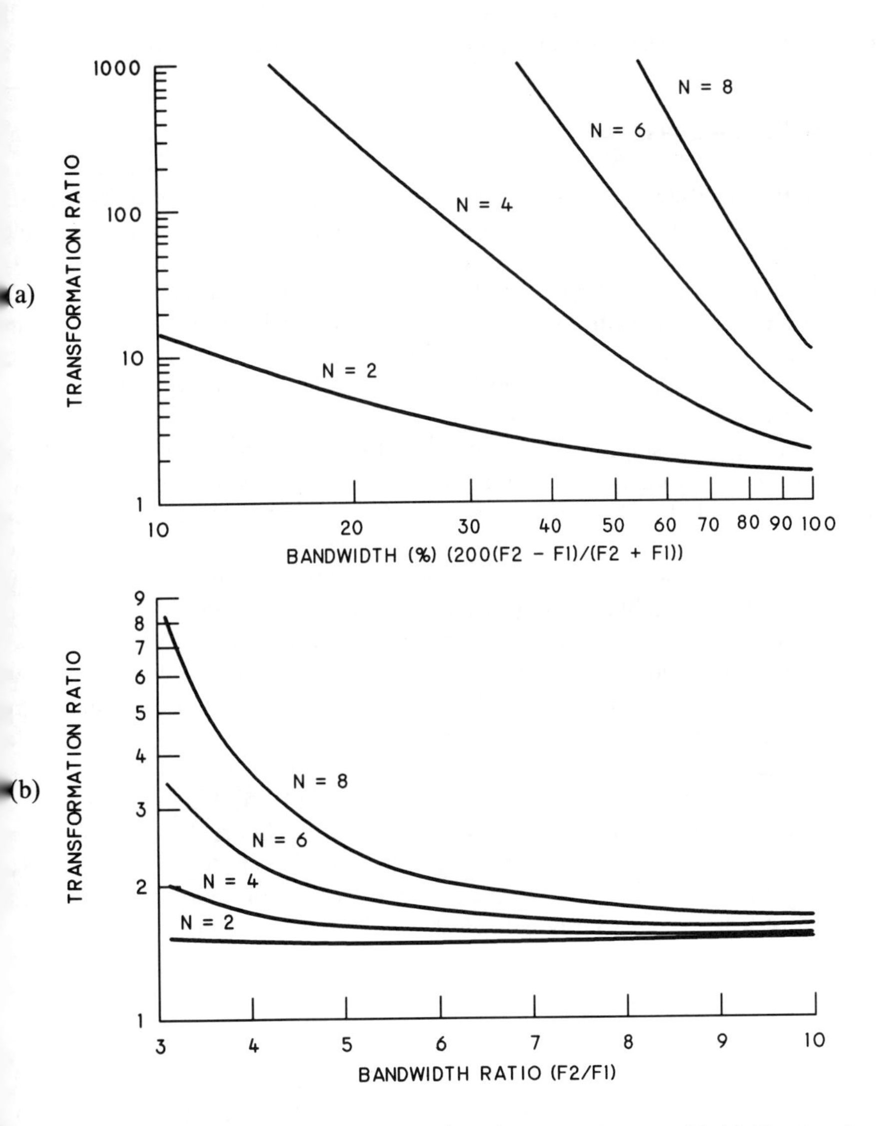


**Figure 4.14** Gain bandwidth results for low-pass impedance transformers with 15-dB return loss: (a) narrowband results and (b) broadband results.

pass topology for the matching network tends to promote this. If a bandpas network is used, then additional network elements may be added to provide th short circuit at the expense of additional dissipation and circuit complexity.

### 4.6.3 Stability Considerations

Leaving stability circle analysis to others, we will merely mention a few consid erations that are sometimes overlooked. First, it is insufficient to analyze a mu tistage amplifier as a whole. Each stage must be individually examined for stability Second, it is important to include the bias networks and the loops formed by bu bars joining the bias of several stages together. Figure 4.15 suggests a partitionin for the stability analysis of a two-stage amplifier, which does not rigorously chec the effects of both loops simultaneously, but does allow conventional two-port parameter analysis.

The FET can be stabilized by controlling the resistance on the gate or drai but to maintain high efficiency, any resistance in the drain is undesirable. Ga resistance is preferable if the value is low enough so that reasonable gate curre will not result in a significant change in bias voltage. High values of resistanc sometimes used for stage-to-stage bias network isolation in small-signal amplifie are therefore unacceptable for high-efficiency power amplifiers. Low-frequen instability due to bias loops can be addressed by lowering the low-frequency loo gain using a small value interstage blocking capacitor. Because device $g_m$ can va 20% due to fabrication tolerances, and because it varies inversely with temperatur the stability analysis should be performed with a high value of $g_m$ adjusted fo process tolerance and the lowest temperature of operation.

A differential mode of oscillation can also occur [14, 15] that can be alleviate with isolation resistors or *mode strapping*. In hybrid amplifiers this requires shor length bond wires connecting the gate or drain bond pads together. In MMI designs, large periphery stages composed of many cells can be joined by gate ar drain manifolds to create a large gatewidth contiguous structure.

## 4.7 THERMAL CONSIDERATIONS

FET saturated output power and small-signal gain decrease with temperature t between 0.01 and 0.015 dB/°C. A transmitting chain that can maintain the outp stage in compression over temperature will exhibit only the power degradation a single stage as the temperature is varied. However, the small-signal gain w show the cumulative change of every stage. A useful estimate for the temperatu dependence of $g_m$ is

$$g_m(T) = g_m(300)\sqrt{300/T} \quad (4.2$$

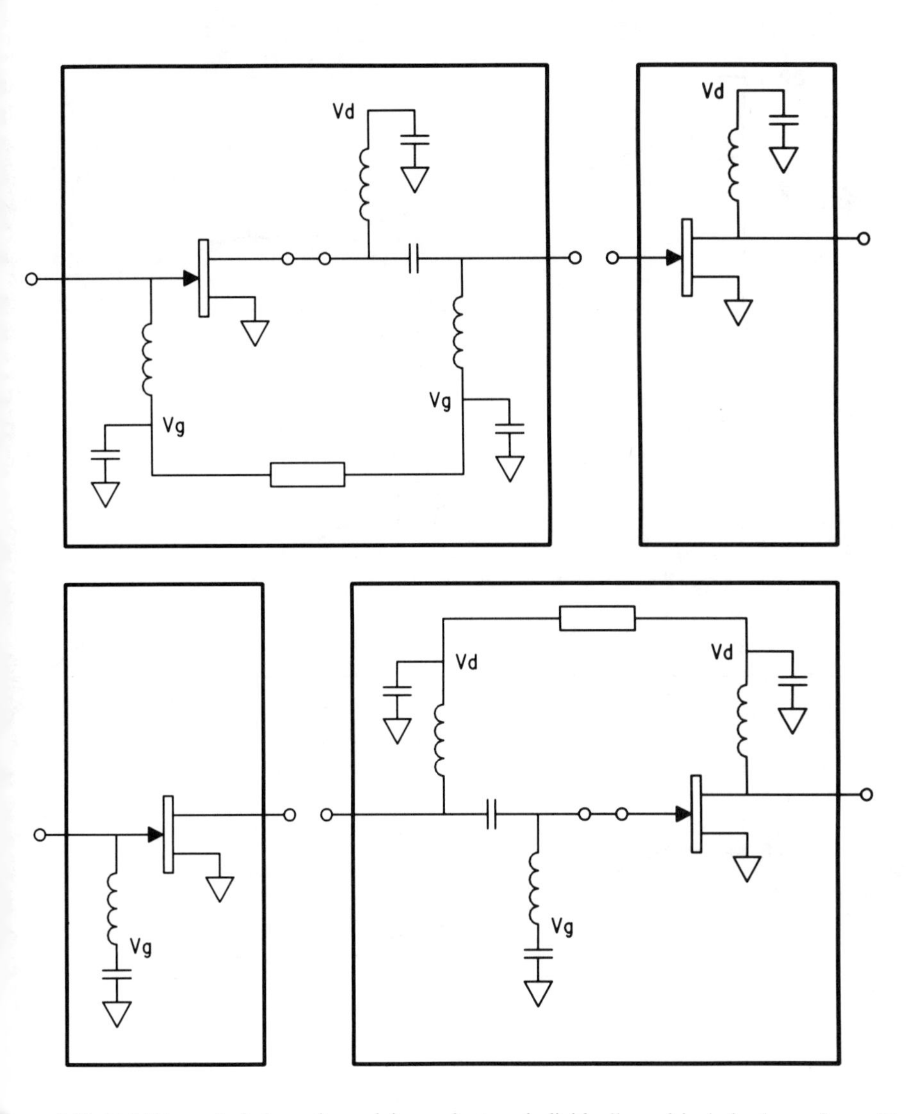


**gure 4.15** Stability analysis is performed for each stage individually and includes loops formed by bias interconnections.

here $T$ is the temperature in Kelvin. This corresponds to a 0.015 dB change in e gain per degree Celsius.

Figure 4.16 shows the power and efficiency of a 4-mm total gatewidth single-age amplifier operated with a 5-$\mu$s pulse as a function of duty cycle. Typically, e output power drops by as much as 1 dB between 1% and 100% duty factor.

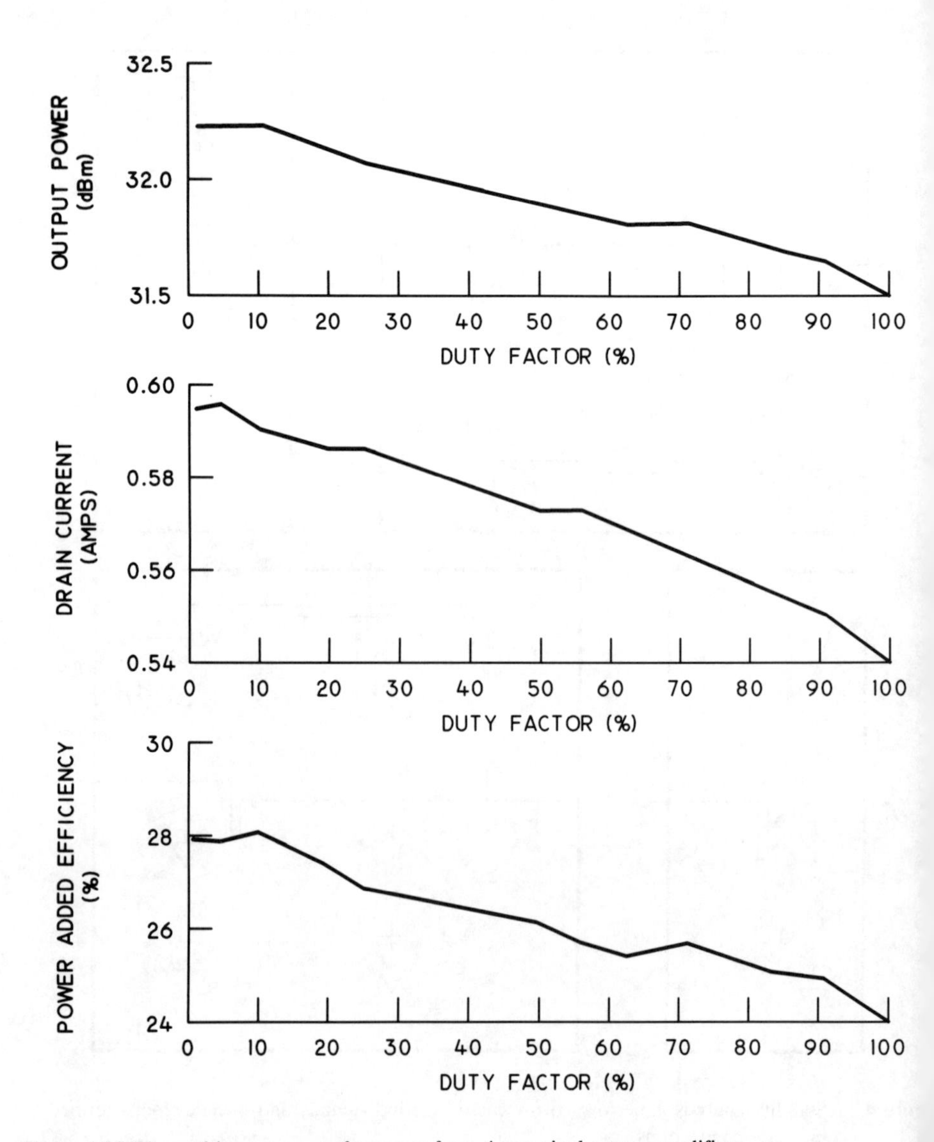


**Figure 4.16** Thermal impact on performance for a 4-mm single-stage amplifier.

Figure 4.17(a) shows the measured output power as a function of pulse length low duty cycle for the same device, while Figure 4.17(b) shows the correspondi predicted increase in channel temperature (see Sec. 5.4). Typically, the droop power output is as large as 1 dB.

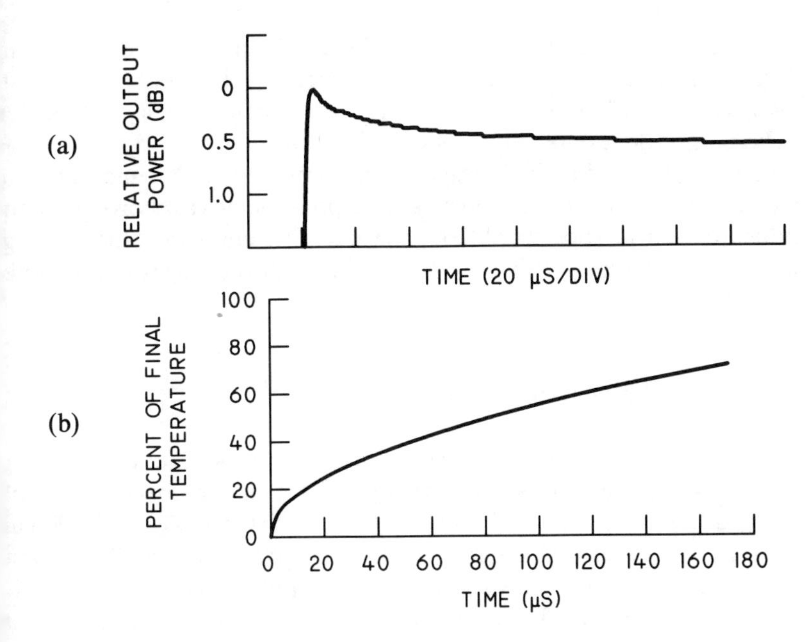


'igure 4.17 Thermal effect of a step function dissipation: (a) measured output power droop for a 4-mm single-stage amplifier and (b) predicted increase in temperature as a percentage of final steady-state value.

## .8 GATE CURRENT AND INSERTION PHASE

Gate current of the order of 1 to 2 mA/mm of gate periphery is normally encoun-ered when stages are operated at the maximum efficiency and power condition orresponding to 2 to 3 dB into gain compression. When the drain voltage is chosen ɔ maximize power, the resulting RF voltage and current swings can extend from reakdown at one end of the load line to rectification at the other. The two currents ıay cancel, or the variations in the load line from one frequency to another may ary in such a way as to exhibit rectification current at one frequency and breakdown urrent at another. Even in designs that have current flowing only in one direction, ɔt-to-lot FET processing variations in $I_{dss}$ or breakdown voltage can change the ate current behavior of a given design.

During FET and amplifier characterization it is quite possible to bias or tune ɔr good performance, but with excessive gate current. Excessive gate current ıcreases the requirements on bias voltage regulation networks and could influence ET reliability. Metal migration is a function of materials and current density, but ɔ estimate a limit, consider a 0.1-mm gate finger of gold with a 0.5-$\mu$m gate length and a metal thickness of 0.5 $\mu$m. A current density limit of $0.7 \times 10^6$ A/cm translates into a maximum of 1.75 mA for the finger or 17.5 mA/mm of gat periphery. A gate current limit of 2 mA/mm seems adequate for good RF perfor mance and should not be a reliability problem.

The insertion phase of a single-stage power amplifier at X-band generall varies by 5 to 10° between the linear and saturated power states as shown in Figur 4.18. In multistage transmitting amplifiers, the overall chain can exhibit a larg phase variation when the input drive level is varied if all the stages are operate in compression.

## 4.9 DUAL-GATE FET POWER AMPLIFIER

This section discusses various aspects of the design of a two-stage MMIC powe amplifier [16], shown in Figure 4.19, using dual-gate FETs and represents an appl cation of the load line design method presented in this chapter. The variable gai dual-gate FET is one of several approaches used in providing amplitude tape across the aperture of an active phased array. In this application it is desirable fc the amplifier to exhibit a minimal change in insertion phase when varying the gai

The design approach uses the bias-dependent small-signal model presente in [17]. Small-valued capacitors are used to terminate the RF port of the secon gate. The matching networks are designed using the small-signal model associate with the maximum gain bias values and the load line approach outlined in Sectio 4.6. Finally, using the bias-dependent model, the small-signal insertion phase vers gain control can be predicted.

The lumped-element FET model consists of two symmetrical FETs connecte as shown in Figure 4.20. The drain bias is fixed and broadband $S$ parameters a measured for eight values of the second gate bias. Eight corresponding sets model parameter values are mathematically fitted to the data. Finally, a composi model is obtained by fitting polynomials to the bias-dependent model parameter The major second gate voltage-dependent parameters are the drain resistance, $g$ and feedback capacitance of the first FET as shown in Figure 4.21.

The RF configuration of the dual-gate FET is equivalent to a cascode stag i.e., a common-source FET followed by a common-gate FET. The ideal RF impe ance terminating the second gate for minimal insertion phase variation is a sho circuit. Rather than using a large bypass capacitor, which resonates with the v and connecting line length to produce an inductive termination, capacitors (0.3 p mm of gate periphery) were used that resonate above the operational band. Figu 4.22 illustrates two cells having a gate periphery of 2 mm, which was used in t first stage of the amplifier. Four such cells are placed in parallel to form the 8-m output stage.

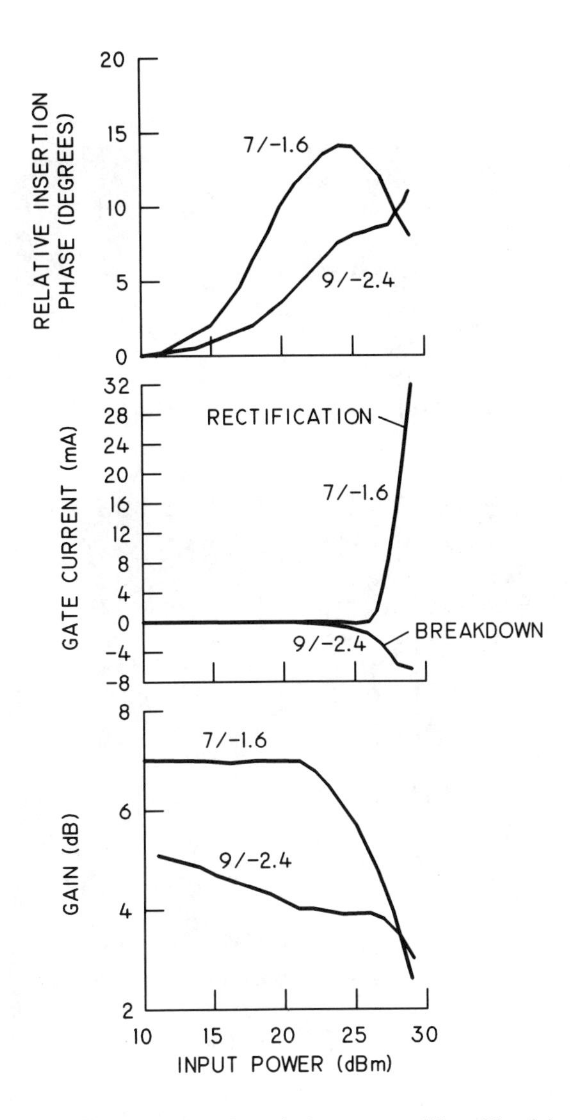


**Figure 4.18** Gate current and insertion phase for a single-stage amplifier with original bias $V_{gs} = -1.6$V, $V_{ds} = 7$V and when biased at $V_{gs} = -2.4$V, $V_{ds} = 9$V to induce breakdown current.

The output matching network is a low-pass transformer to match the 6.8-$\Omega$ load line resistance to 50$\Omega$. A high-pass $\pi$-section network is inserted in the middle of the network to provide an injection point for the drain bias, as shown in Figure 4.23. The final output network (including the FET capacitance) results in a 35-

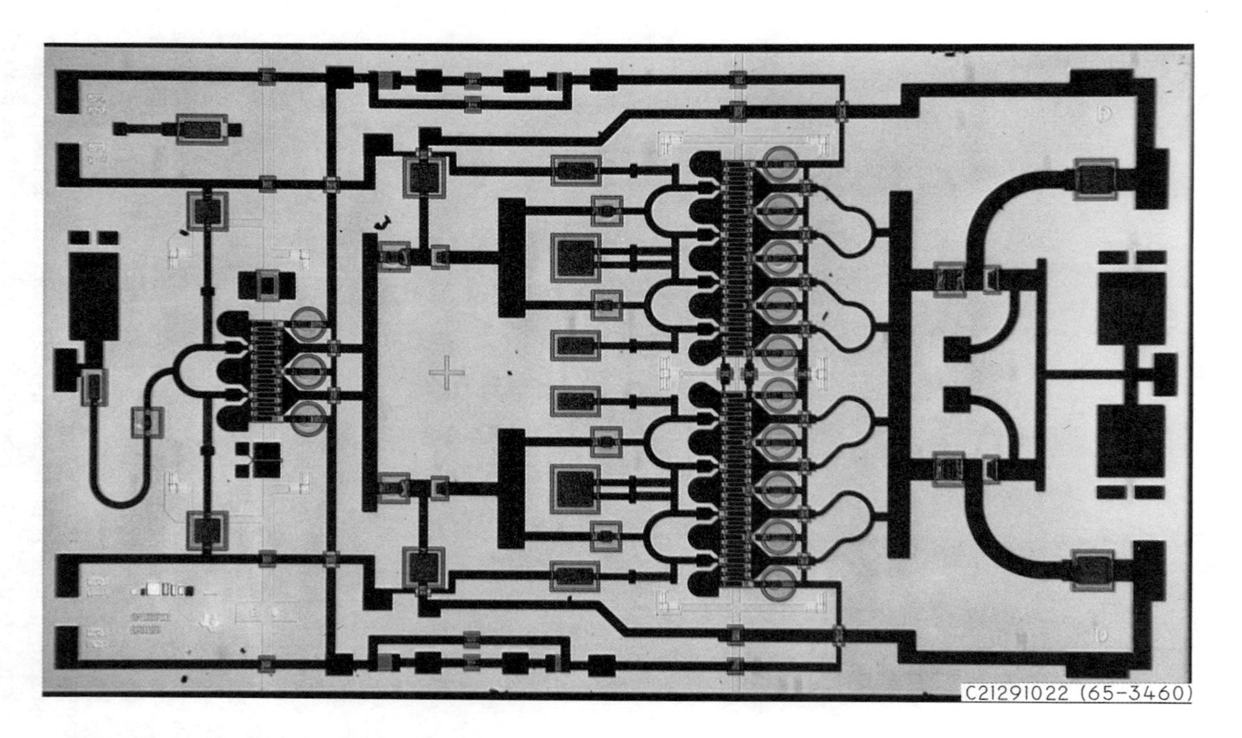


**Figure 4.19** A 3-W two-stage dual-gate amplifier (6.6 by 3.8 by 0.1 mm).

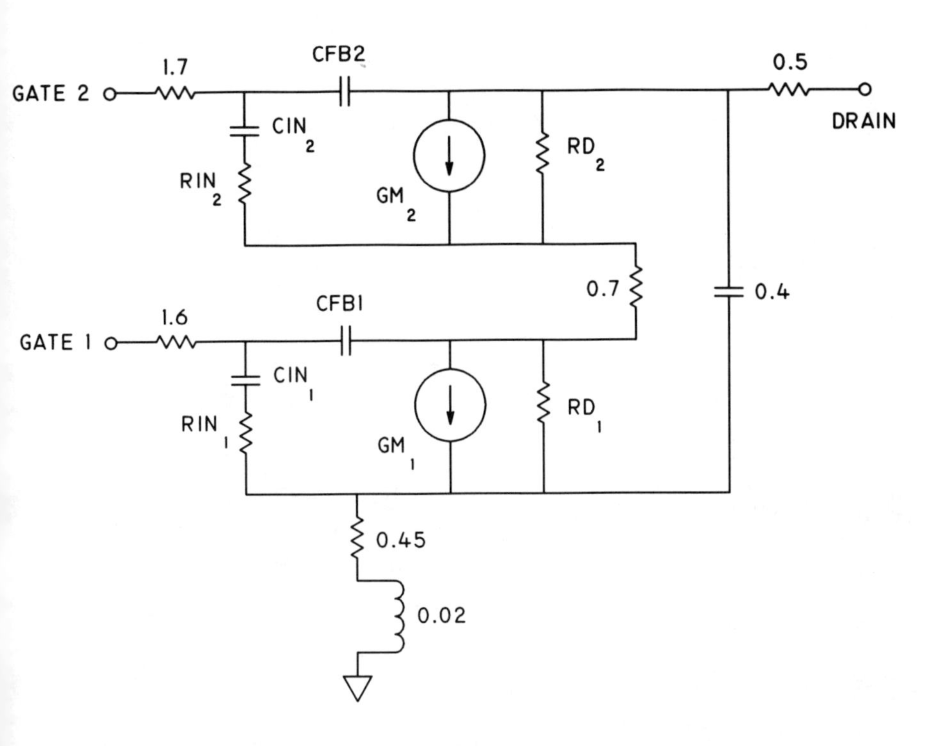


gure **4.20** Lumped-element model for a 2-mm dual-gate FET showing constant and second gate bias-dependent elements.

Bm maximum load line power curve, shown in Figure 4.24, using the equations f Section 4.4.3.

The interstage network design begins with a lossy subnetwork (Fig. 4.25). As own in Figure 4.26, the lossy network flattens the gain slope of the FET, improves ability, lowers the effective $Q$ of the FET input, and provides a convenient point r applying gate bias. Load line analysis using the interstage input impedance ($\Gamma_3$ Fig. 4.25) predicts a 29-dBm maximum first-stage output power (Fig. 4.24). hen this power is combined with the small-signal gain of the second stage, e projected power is greater than the maximum output power. This indicates at the first stage will be able to compress the second stage everywhere across e band.

The amplifier typically produces 3W with 23% power-added efficiency (Fig. 27). Figure 4.28 shows the change in insertion phase at eight frequencies while rying the gain control over a 17-dB range. The phase remains constant within 6°.

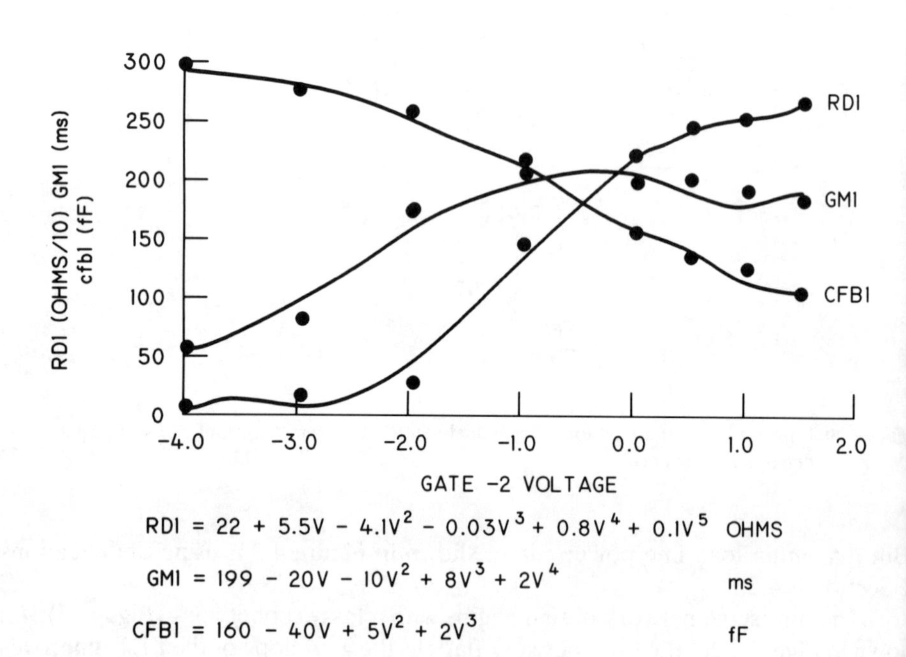


$$\mathrm{RDI} = 22 + 5.5V - 4.1V^2 - 0.03V^3 + 0.8V^4 + 0.1V^5 \quad \mathrm{OHMS}$$

$$\mathrm{GMI} = 199 - 20V - 10V^2 + 8V^3 + 2V^4 \quad \mathrm{ms}$$

$$\mathrm{CFBI} = 160 - 40V + 5V^2 + 2V^3 \quad \mathrm{fF}$$

**Figure 4.21** Polynomials fitting the three major second gate bias-dependent elements.

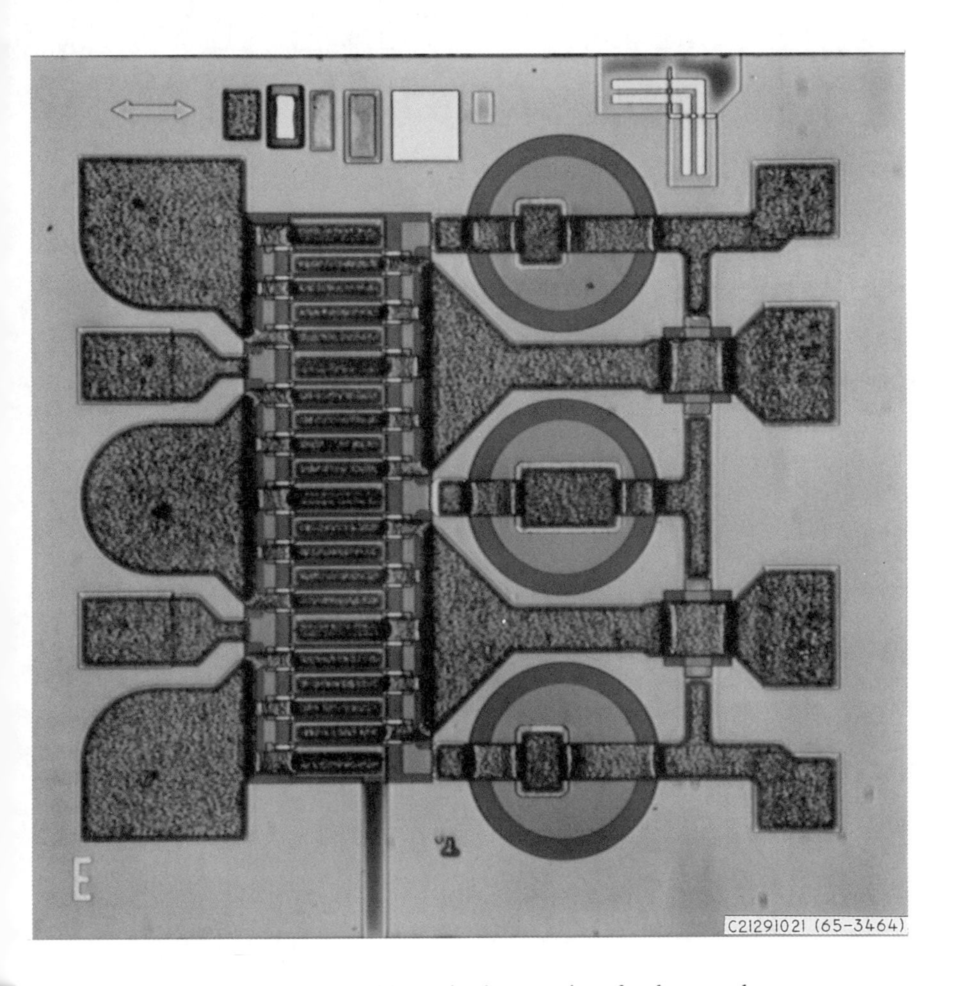


**Figure 4.22** A 2000-μm dual-gate FET with termination capacitors for the second gate.

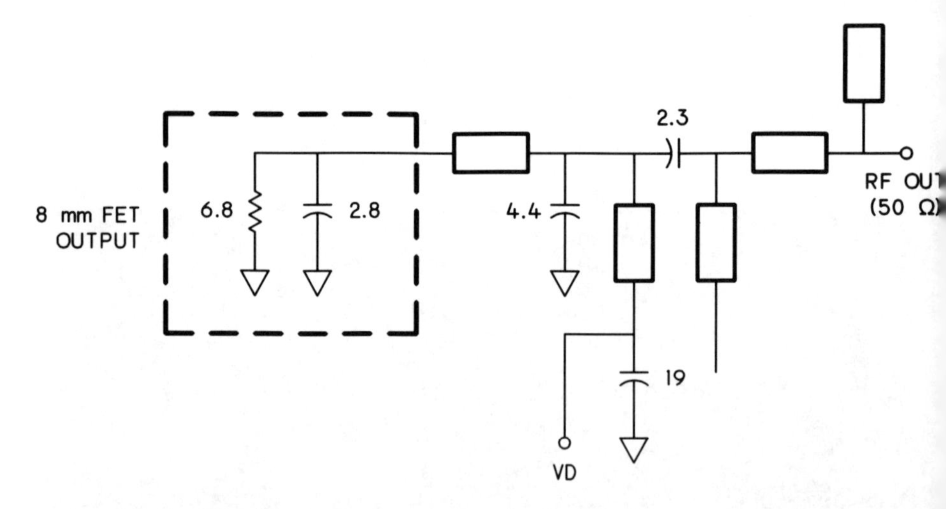


**Figure 4.23** Output matching network.

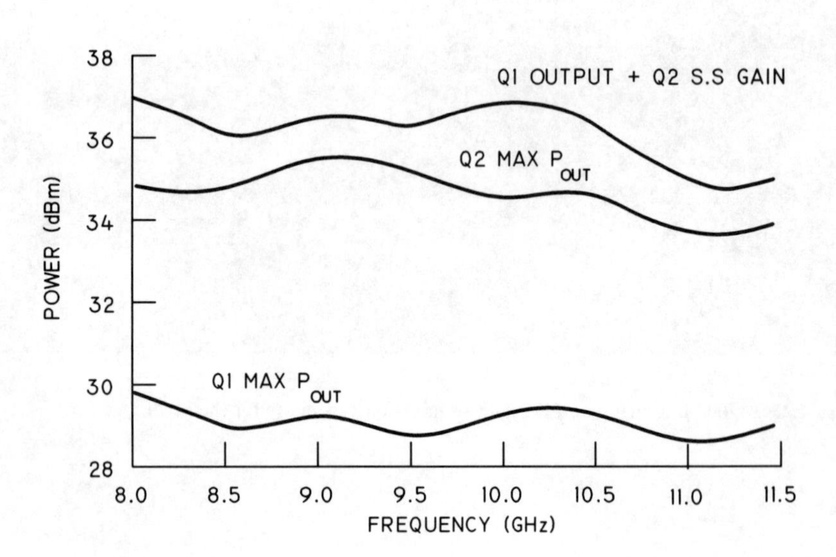


**Figure 4.24** Predicted large-signal response.

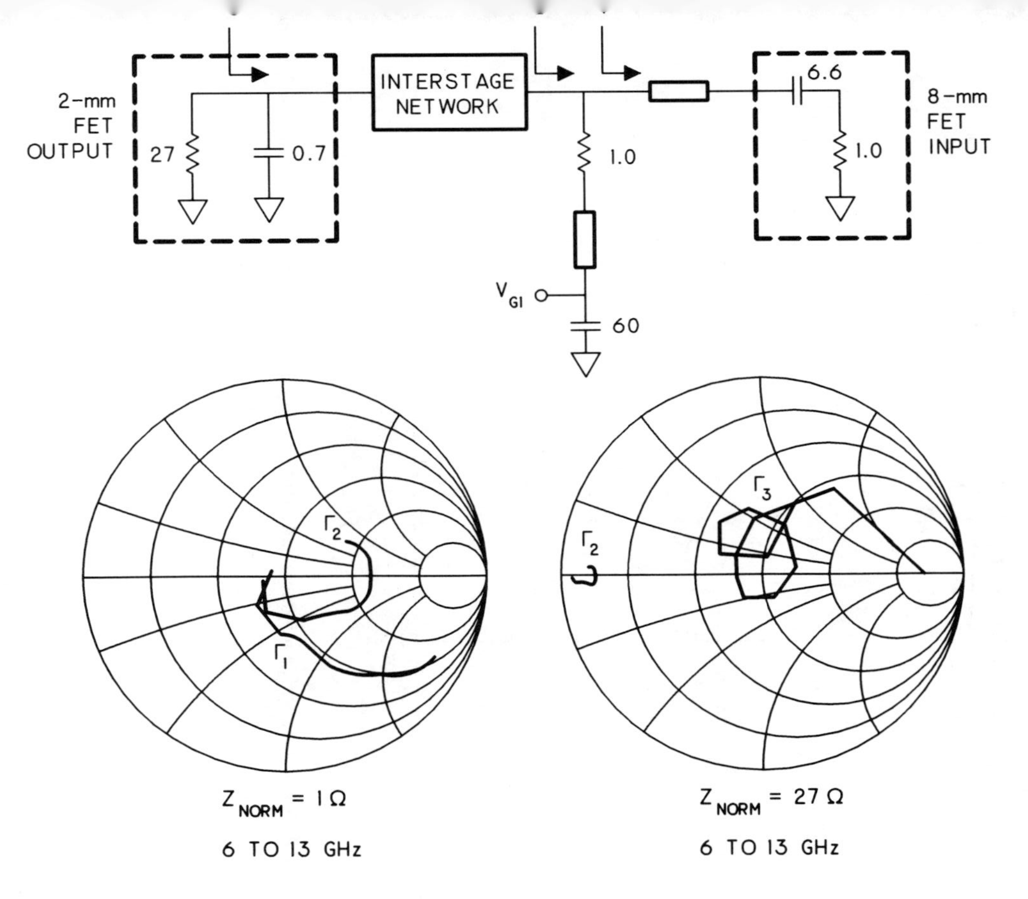


**Figure 4.25** Lossy matching network for interstage impedance matching.

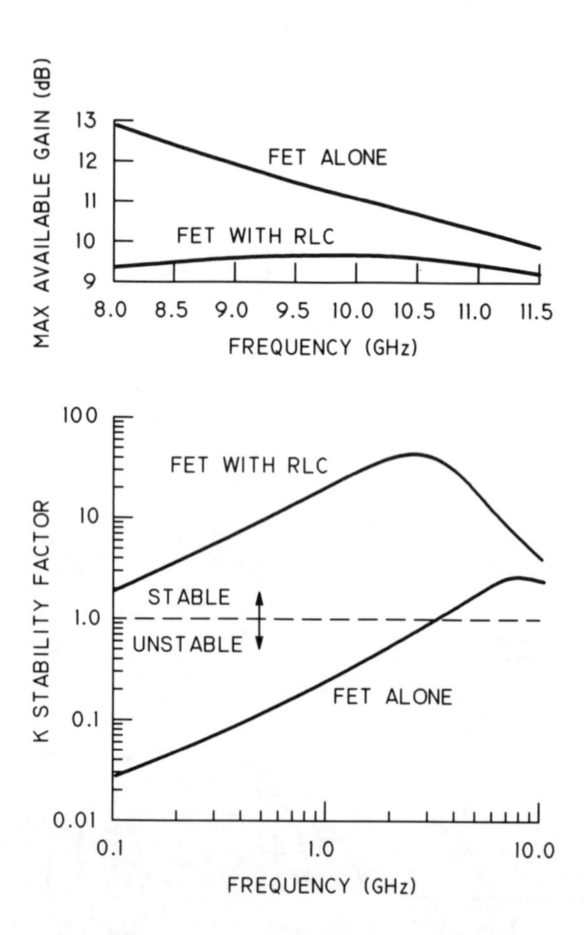


**Figure 4.26** Second-stage gain flatness and stability before and after including lossy matching eleme

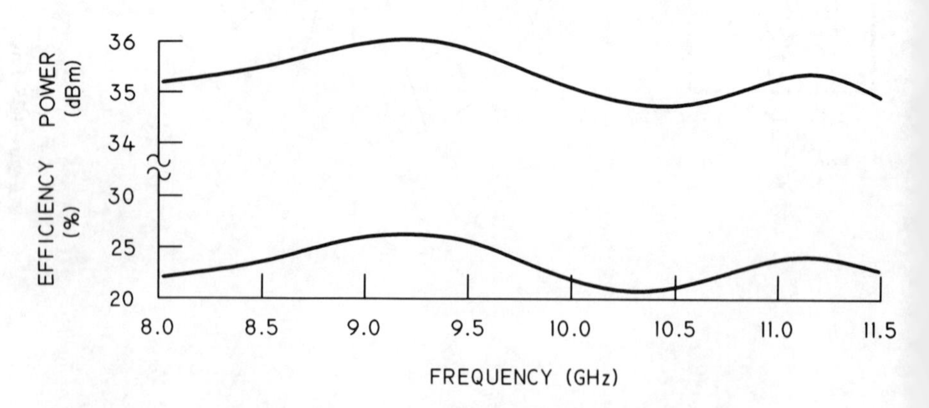


**Figure 4.27** Measured output power and efficiency of two-stage dual-gate FET MMIC.

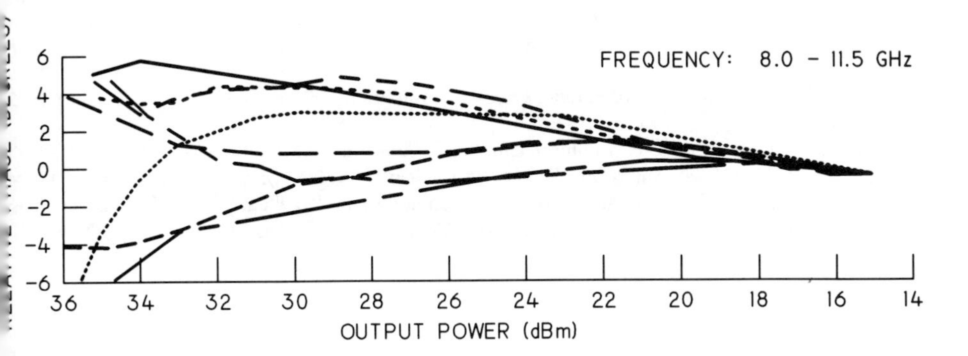


**Figure 4.28** Measured large-signal insertion phase variation with second gate gain control.

## REFERENCES

1. Krauss, H. L., C. W. Bostian, and F. H. Raab, *Solid State Radio Engineering*, New York: John Wiley & Sons, 1980, Chaps. 12–14.
2. Dambrine, G., A. Cappy, F. Heliodore, and E. Playez, "A New Method for Determining the FET Small-Signal Equivalent Circuit," *IEEE Trans. on Microwave Theory and Techniques*, Vol. MTT-36, July 1988, pp. 1151–1159.
3. Cripps, S. C., "A Theory for the Prediction of GaAs FET Load-Pull Power Contours," *IEEE MTT-S Int. Microwave Symp. Digest*, Boston, MA, May 31–June 3, 1983, pp. 221–223.
4. Takayama, Y., "A New Load-Pull Characterization Method for Microwave Power Transistors," *IEEE MTT-S Int. Microwave Symp. Digest*, Cherry Hill, NJ, June 14–16, 1976, pp. 218–220.
5. Kretschmer, K. H., P. Grambow, and T. Sigulla, "Coupled-Mode Analysis of Travelling-Wave MESFETs," *Int. J. Electronics*, Vol. 58, No. 4, April 1985, pp. 639–648.
6. LaRue, R., C. Yuen, and G. Zdasiuk, "Distributed GaAs FET Circuit Model for Broadband and Millimeter Wave Applications," *IEEE MTT-S Int. Microwave Symp. Digest*, San Francisco, CA, May 30–June 1, 1984, pp. 164–166.
7. Abrie, P. L. D., *The Design of Impedance-matching Networks for Radio Frequency and Microwave Amplifiers*, Norwood, MA: Artech House, 1985.
8. Cuthbert, T. R., Jr., *Circuit Design Using Personal Computers*, New York: John Wiley & Sons, 1983.
9. Levy, R., "Explicit Formulas for Chebyshev Impedance-matching Networks," *Proc. IEE*, Vol. 111, No. 6, June 1964, pp. 1099–1106.
10. Ku, W. H., and W. C. Peterson, "Optimum Gain-bandwidth Limitations for Transistor Amplifiers as Reactively Constrained Active Two-port Networks," *IEEE Trans. on Circuits and Systems*, Vol. CAS-22, June 1975, pp. 523–533.
11. Apel, T. R., "Bandpass Matching Networks Can Be Simplified By Maximizing Available Transformation," *Microwave System News*, December 1983, pp. 105–117.
12. Matthaei, G. L., "Tables of Chebyshev Impedance Transforming Networks of Low-Pass Filter Form," *Proc. IRE*, August 1964, pp. 939–943.
13. Matthaei, G. L., "Short Step Chebyschev Impedance Transformers," *IEEE Trans. on Microwave Theory and Techniques*, Vol. MTT-14, August 1966, pp. 372–383.

14. Kassakian, J. G., and D. Lau, "An Analysis and Experimental Verification of Parasitic Oscillation in Paralleled Power MOSFETs," *IEEE Trans. on Electron Devices*, Vol. ED-31, July 1984, pp 959–963.
15. Nelson, J. E., "Stabilization of Parallel Transistor Amplifiers," *Proc. IEEE*, Vol. 60, January 1972 p. 140.
16. Culbertson, R. B., and D. C. Zimmermann, "A 3-Watt X-Band Monolithic Variable Gain Ampli fier," *IEEE MTT-S Int. Microwave Symp. Digest*, New York, NY, May 25–27, 1988, pp. 171–174
17. Drury, D. M., D. C. Zimmermann, and D. E. Zimmerman, "A Dual-gate FET Constant Phase Variable Power Amplifier," *IEEE MTT-S Int. Microwave Symp. Digest*, St. Louis, MO, June 4-6, 1985, pp. 219–222.

# Chapter 5
# Thermal Effects and Reliability

**H. F. Cooke**
*Consultant*

## 5.1 INTRODUCTION

Thermal design and reliability are closely related since the ultimate failure modes in power FETs are all determined by the operating temperature of the channel, which also affects the dc and RF performance of the FET. This chapter will begin by describing the fundamentals of heat flow and will then use these to estimate the channel temperature of some real devices. Examples of the effect of geometry on thermal resistance will be given. The later section on reliability will start with a review of the mathematics necessary to make reliability calculations. Next, reliability tests will be described including a completely automated test setup.

GaAs power FETs are most often used fairly close to their useful frequency limit. This is because the upper limit is constantly being pushed to fill the need for efficient, higher frequency devices for commercial and military applications. As a result, device geometry has become extremely small and the internal spacings shorter in order to maintain RF gain at the higher frequencies, but the overall efficiency has tended to remain in the 20% to 50% range. State-of-the-art devices are often reported with even lower numbers. With efficiencies in this range, the disposal of heat becomes a serious problem, since too high a channel temperature will degrade performance and shorten the life of a semiconductor device. Thus several of the chip mounting techniques that are common for low-power devices are usually unacceptable for high-power amplifiers, such as die mounting on alumina substrates or to Kovar carriers and die attachment using conductive epoxy.

Heat leaves the die by convection, radiation, or conduction. The first two are usually insignificant and thus a good thermal design must consider how the heat is to be conducted away from the chip or MMIC. The next section explain the fundamentals of heat flow by conduction.

## 5.2 THERMAL FUNDAMENTALS

Figure 5.1(a) shows a block of semiconductor mounted on a large surface calle an *infinite heat sink*. Assume that power is flowing uniformly into the upper surfac and that the upper and lower surfaces are isothermal. As indicated in Figure 5.1(a the heat flow in this case is always parallel to the edges of the chip and this termed *columnar heat flow*. The difference in temperature between the top an bottom surfaces, $\Delta T$, will be proportional to the power $P$ and die thickness $h$ an inversely proportional to the area $A$, i.e.,

$$\Delta T = \frac{Ph}{\kappa A} \quad (5.1)$$

where the constant of proportionality $\kappa$ is termed the *thermal conductivity*. We ca see from (5.1) that $\kappa$ has the units of W $cm^{-1}$ $°C^{-1}$ if power is in watts, thickne in centimeters, and area in square centimeters. The thermal conductivity of mate rials used in semiconductor manufacture varies a great deal, from 6.31 W cm $°C^{-1}$ for the type 2A diamond, to 0.0088 W $cm^{-1}$ $°C^{-1}$ for $SiO_2$. Table 5.1 lists th thermal conductivity of a range of relevant materials. Note that the thermal co ductivity of GaAs, like other semiconductors, is not a constant, but decreases a temperature increases and is about 0.45 W $cm^{-1}$ $°C^{-1}$ at 300K.

The thermal resistance $R_\theta$ of the structure shown in Figure 5.1(a) is define by

$$R_\theta = \frac{\Delta T}{P} = \frac{h}{\kappa A} \quad (5.2)$$

Thermal resistance is the thermal analogue of electrical resistance.

Figure 5.1(b) illustrates another type of heat flow, known as *spreading*. Th occurs when the heated area is much smaller than the die itself. In the extrem case of a semi-infinite piece of semiconductor with a constant heat flux applie over a circle of diameter $d$, the thermal resistance becomes

$$R_\theta = \frac{1}{\pi \kappa d} \quad (5.3)$$

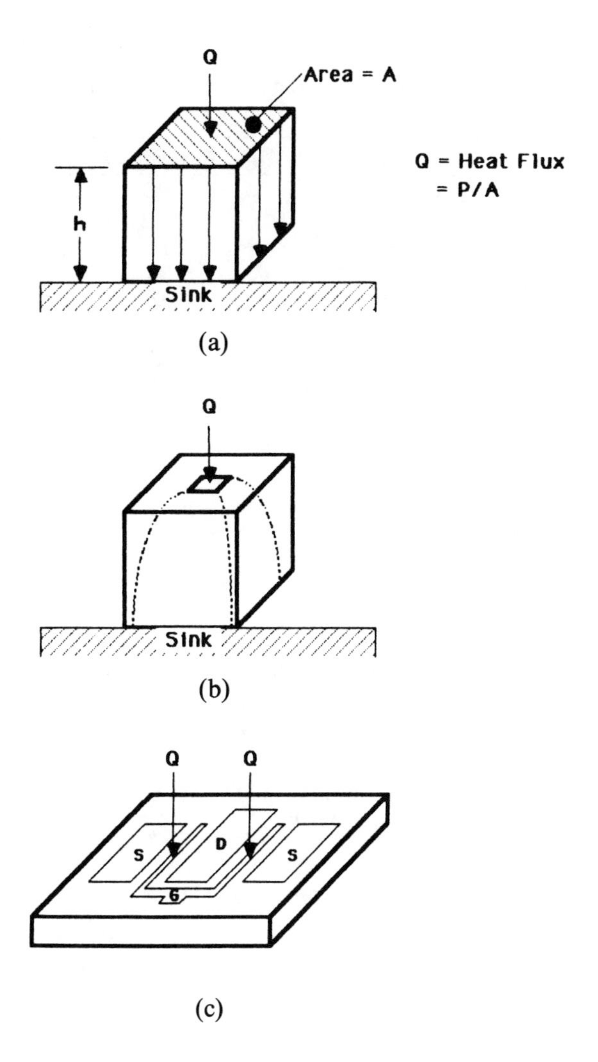


igure 5.1 Three examples of heat flow: (a) columnar, (b) spreading, and (c) combined (FET).

## .3 THERMAL CALCULATIONS FOR PRACTICAL FETs

leither pure columnar nor pure spreading are found in practical semiconductor evices such as the FET shown in Figure 5.1(c). The heat flow is a combination f both and the calculation of thermal resistance is relatively complex. The three asic methods of determining the temperature profile and thermal resistance are nalytical solution of the three-dimensional partial differential equation describing

**Table 5.1**
Thermal Conductivity of Some Materials Used in Semiconductors, Packages and Mounts in W $cm^{-1}$ $°C^{-1}$

| *Material* | *Conductivity* |
|---|---|
| Alloy 42 | 0.122 |
| Alumina, $Al_2O_3$ | 0.292 |
| Aluminium | 2.17 |
| Beryllium oxide | 2.34 |
| Copper | 3.97 |
| Diamond, type 2A | 6.31 |
| Elkonite or Thermkon | 1.69 |
| Epoxy, silver-filled | 0.0585 |
| Epoxy, gold-filled | 0.0064 |
| Glass, $SiO_2$ | 0.0088 |
| Gold | 2.93 |
| Kovar | 0.155 |
| Sapphire | 0.25 |
| Silver | 4.18 |
| Steel | 0.585 |

heat flow, numerical solution of the partial differential equation, and approxima analytical methods. These three techniques are now considered in turn.

The partial differential equation governing heat flow is [1]

$$\frac{\partial^2 T}{\partial x^2} + \frac{\partial^2 T}{\partial y^2} + \frac{\partial^2 T}{\partial z^2} = \frac{1}{\alpha}\frac{\partial T}{\partial t} \quad (5.$$

where $\alpha$ is the thermal diffusivity. For GaAs, $\alpha = 0.24$ $cm^2/s$ at room temperatur This equation has to be solved subject to the appropriate initial and bounda conditions. However, under steady-state CW conditions (5.4) reduces to the sta dard Laplace equation:

$$\frac{\partial^2 T}{\partial x^2} + \frac{\partial^2 T}{\partial y^2} + \frac{\partial^2 T}{\partial z^2} = 0 \quad (5.$$

which can be solved by the method of separation of variables with the appropria boundary conditions. Linsted and Surty [2] have used this method to obtain analytical solution for the structure shown in Figure 5.1(b). This solution has be programmed for use in a commercially available computer program [3] that ru on a PC. Culbertson and Lehmann[1] have extended Linsted and Surty's analysis

[1]This material was written by R. B. Culbertson and R. E. Lehmann.

clude multiple heat generation regions, which is much more representative of a pical interdigitated power FET. The temperature rise $\Delta T$ at any point on the rface for this case is given by

$$
\begin{aligned}
\frac{\Delta T}{Q} = {} & h \sum_{i=1}^{N} \frac{T_i L_i}{DW} \\
& + \sum_{m=1}^{\infty} \left( \sum_{i=1}^{N} \frac{4 T_i F}{m \pi D} \cos \frac{m \pi P_i}{W} \sin \frac{m \pi L_i}{2W} \right) \cos \frac{x m \pi}{W} \\
& + \sum_{n=1}^{\infty} \left( \sum_{i=1}^{N} \frac{4 L_i F}{n \pi W} \cos \frac{n \pi R_i}{D} \sin \frac{n \pi T_i}{2D} \right) \cos \frac{y n \pi}{D} \\
& + \sum_{m=1}^{\infty} \sum_{n=1}^{\infty} \left( \sum_{i=1}^{N} \frac{16F}{mn\pi^2} \cos \frac{m \pi P_i}{W} \sin \frac{m \pi L_i}{2W} \cos \frac{n \pi R_i}{D} \sin \frac{n \pi T_i}{2D} \right) \cos \frac{x m \pi}{W} \cos \frac{y n \pi}{D}
\end{aligned}
\tag{5.6}
$$

here

$$F = \frac{1}{G} \frac{1 - e^{-2Gh}}{1 + e^{-2Gh}}$$

$$G = \pi \left( \frac{m^2}{W^2} + \frac{n^2}{D^2} \right)^{1/2}$$

$N$ = number of gate fingers
$\kappa$ = thermal conductivity of GaAs
$Q$ = dissipated power density
$(W, D)$ = the $(x, y)$ dimensions of the chip
$(P_i, R_i)$ = the $(x, y)$ center location of the $i$th heat source
$(L_i, T_i)$ = the $(x, y)$ dimensions of the $i$th heat source
$h$ = thickness of the die.

major drawback to the series solution is the slow convergence for the geometries a typical FET. This slows the calculation and requires considerable memory to ore the Fourier coefficients when varying the $(x, y)$ location. Rather than using fixed upper limit for $m$ and $n$, which produces a rectangular array of coefficients, ch time $n$ is incremented a new upper limit for $m$ should be established by uncating the sum when terms contribute less than, perhaps, 0.0004°C to the mperature rise. The dimension associated with the channel length has a strong fluence on the rate of convergence. A convenient value for the width of the heat urce is the source-drain spacing. The variation in temperature and convergence summarized in Table 5.2 for a single 100-$\mu$m finger on a 100-$\mu$m-thick die.

The temperature rise given by (5.6) does not include the effect of thermal nductivity varying with temperature. However, this effect can be taken into

**Table 5.2**
Fourier Series Temperature Prediction

| *S-D Spacing* ($\mu$m) | $Q$ (W/mm$^2$) | $\Delta T$ (°C) | Number of Terms Required |
|---|---|---|---|
| 0.5 | 2000 | 57 | 183,000 |
| 1.0 | 1000 | 51 | 120,000 |
| 2.0 | 500 | 45 | 73,000 |
| 4.0 | 250 | 39 | 41,000 |
| 8.0 | 125 | 33 | 22,000 |

account in an exact analytical manner using Kirchhoff's transformation [4]. Th
temperature dependence of $\kappa$ can be approximated by

$$\kappa(T) = 0.45T_0/T \text{ W cm}^{-1} \text{ °C}^{-1} \qquad (5.7$$

where $T_0 = 300$K and $T$ is the temperature in Kelvin. Using (5.7) in conjunctio
with Kirchhoff's transformation results in the true FET surface temperature bein
given by

$$T_{\text{true}} = T_{\text{base}} \exp\left(\frac{\Delta T}{300}\right) \qquad (5.8$$

where $T_{\text{base}}$ is the heat sink temperature and $T_{\text{true}}$ and $T_{\text{base}}$ are both in Kelvin. A
an example, consider an FET with ten 100-$\mu$m fingers with a 25-$\mu$m gate-to-ga
spacing and a source-to-drain spacing of 4 $\mu$m centered on a 0.725- by 0.600- b
0.100-mm die with a power dissipation of 1W, i.e., $Q = 250$ W/mm$^2$. Then th
maximum temperature rise is $\Delta T = 67$°C, resulting in a true surface temperatu
of $T_{\text{true}} = 106$°C if the base is held at 30°C (i.e., the temperature dependence o
$\kappa$ adds an additional 9°C to the temperature rise in this case). If the gate-to-ga
spacing is increased to alternate between 29 and 35 $\mu$m, then the maximum surfa
temperature reduces to 90°C. This compares to 66°C for a single gate finger, whic
represents the lowest temperature that could be achieved. The temperature profi
for these two interdigitated structures as calculated by means of (5.6) is shown
Figure 5.2.

Equation (5.6) is an exact solution to the partial differential equation descri
ing heat flow, (5.5), under the assumption of a constant heat flux injected in
each source-drain region. An alternative method of calculating the thermal resi
tance or surface temperature is to make use of the analogy between electrostat
capacitance and thermal resistance [5]. Remember that electrostatic potential an

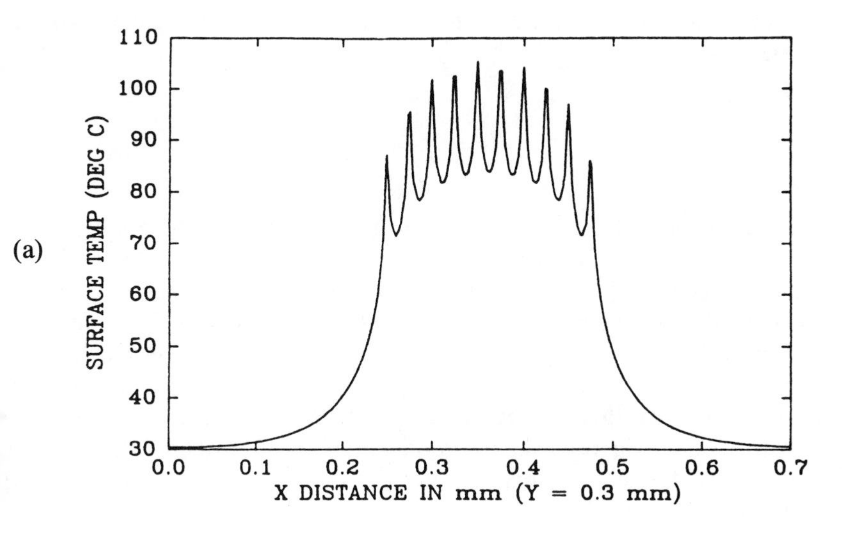


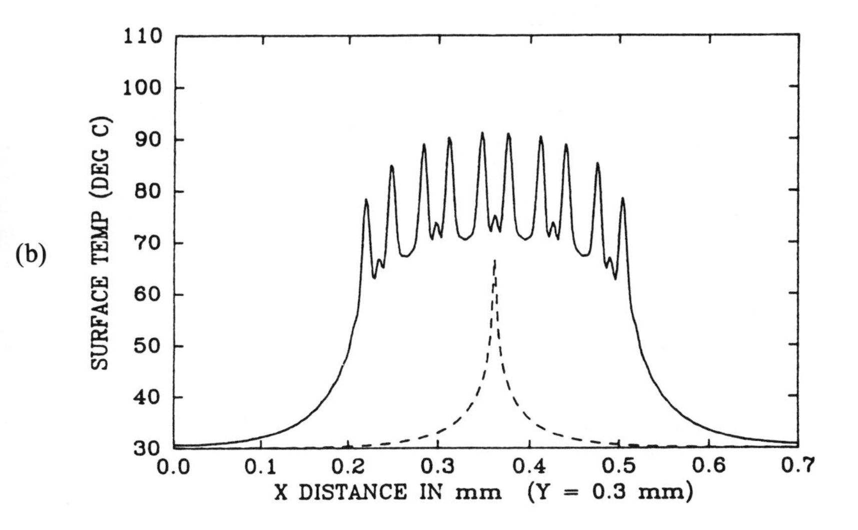


**Figure 5.2** Predicted thermal comparisons for two FET geometries: (a) Ten 100 micron fingers with 25 micron gate-to-gate spacing. (b) Spacing increased to 29/35 microns. The temperature for a single finger is shown for reference. (Courtesy of R. B. Culbertson and R. E. Lehmann).

temperature both satisfy Laplace's equation and that, therefore, the thermal resistance of a line heat source and the characteristic impedance of a microstrip line $Z_0$, are related by

$$R_\theta = \frac{\kappa}{\epsilon} Z_0 \tag{5.9}$$

The multiple-gate FET is the analog of the coupled transmission line shown in Figure 5.3. Since the equations for impedance and capacitance of the TEM line have already been solved [6], we can use them to solve for thermal resistance. In utilizing this analogy, the source of the heat must be defined first. The greater part of the heat in an FET is dissipated in the region between the gate and the drain where the field is at a maximum. Since the gate is very close to this region, it can be considered to be an isothermal source. While this assumption can be challenged it has to date resulted in remarkably good agreement between calculated and measured data [5]. The following expression gives the thermal resistance of an FET with any number of gates:

$$R_\theta = \frac{1}{w_g \kappa}\left[\frac{n}{2(n-1)\dfrac{\pi}{\ln M} - (n-2)\dfrac{\pi}{\ln P}}\right] \tag{5.10}$$

where

$$M = 2\,\frac{\sqrt{\dfrac{\cosh\left(\pi\dfrac{S+l_g}{4h}\right)}{\cosh\left(\pi\dfrac{S-l_g}{4h}\right)}+1}}{\sqrt{\dfrac{\cosh\left(\pi\dfrac{S+l_g}{4h}\right)}{\cosh\left(\pi\dfrac{S-l_g}{4h}\right)}-1}}, \qquad P = 2\sqrt{\frac{1+\operatorname{sech}\dfrac{\pi l_g}{4h}}{1-\operatorname{sech}\dfrac{\pi l_g}{4h}}}$$

The dimensions are defined in Figure 5.3. Note that $S$ is defined here as the distance between the centers of the gates. The essential difference between this method and that of Culbertson and Lehmann is that the latter solves Laplace's equation subje

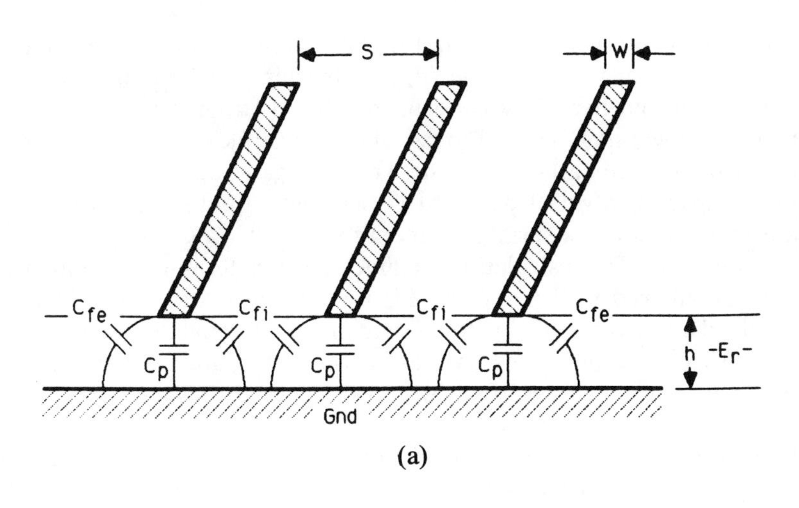


(a)

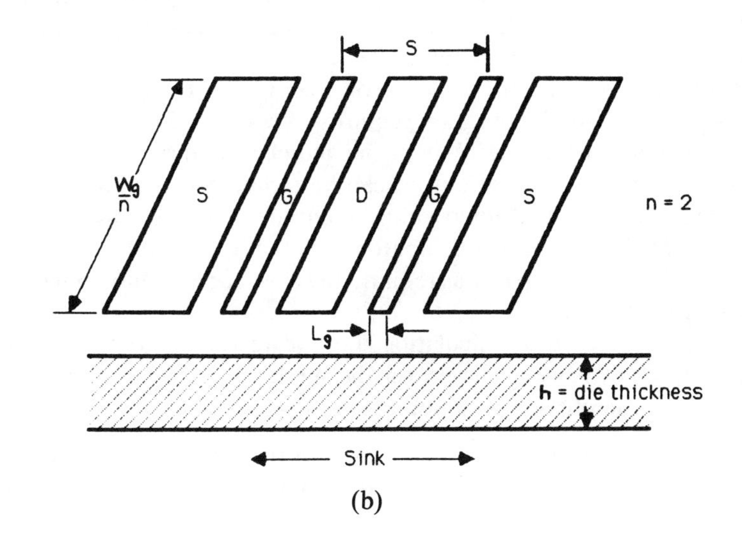


(b)

**gure 5.3** Microstrip—FET analogy: (a) microstrip and (b) FET.

the boundary condition of a constant heat flux input between each source-drain gion, while this method assumes the boundary condition is that all gates are at e same temperature. This permits a very simple expression for the thermal resis- nce or temperature rise. As can be seen from Figure 5.2, the assumption that all tes are at the same temperature is not in total agreement with the results of ulbertson and Lehmann's calculations. Note also that (5.10) implicitly assumes that either the die is much larger than the active area or that the die is sufficientl thin so that heat flow away from the edge of the chip parallel to the gate is alway by spreading, but no heat flow is assumed to occur from the ends of the gate Equation (5.10) will now be used to calculate the surface temperature rise of th power FET considered earlier, i.e., one having 10 gates each 100 $\mu$m long, 25-$\mu$ gate-to-gate spacing, a gate length of 0.5 $\mu$m, and a die thickness of 100 $\mu$m. I this case (5.10) predicts a surface temperature rise of 74°C compared with 67° predicted by (5.6). This calculation excludes the temperature dependence of which can be allowed for using (5.8). Figure 5.4 shows the effect of varying th gate length, die thickness, and gate spacing on the thermal resistance of an FE having twenty 60-$\mu$m-long gates. As can be seen, the gate spacing $S$ and the di thickness $h$ are the most useful tools to reduce thermal resistance if it is too hig

Certain simplifying assumptions had to be made concerning the bounda conditions in order to obtain the preceding analytical solutions of the heat flo equation. If greater accuracy is required for the FET thermal resistance, then w must resort to solving the heat flow equation numerically. Numerical analysis als permits much more complex structures to be analyzed. An approach utilizing finite element model has been successful in depicting temperature as a function position [7]. This method utilizes a personal computer to carry out the calculation but in its present form the program requires a very skilled operator. The data ent is started by imposing a "mesh" over the geometry and typically the running tir for this program is 4 hr. The output can be a multicolor temperature plot of th FET surface with constant temperature contours. When the average temperatu calculated by this method was compared with that computed from the capacitan analog model, the agreement was remarkably close for a 1200$\mu$m power FET havi twenty 60-$\mu$m-long gates.

The final method of calculating thermal resistance to be considered[2] is approximate analytical one based on the methods of Higashisaka [8] and Aor et al. [9]. Four simplifying assumptions are made:

1. The channel thickness is much less than the thickness of the GaAs so that is not necessary to consider the distribution of temperature over the dep or length of the channel.
2. Heat will flow outward at an angle of 45 deg below the channel surface initia [10].
3. The FET is mounted on an infinite heat sink such as a large copper base.
4. The thermal resistance of the solder between the GaAs and the heat sink ignored.

[2]This contribution on the approximate calculation of thermal resistance was written by Y. Aoki.

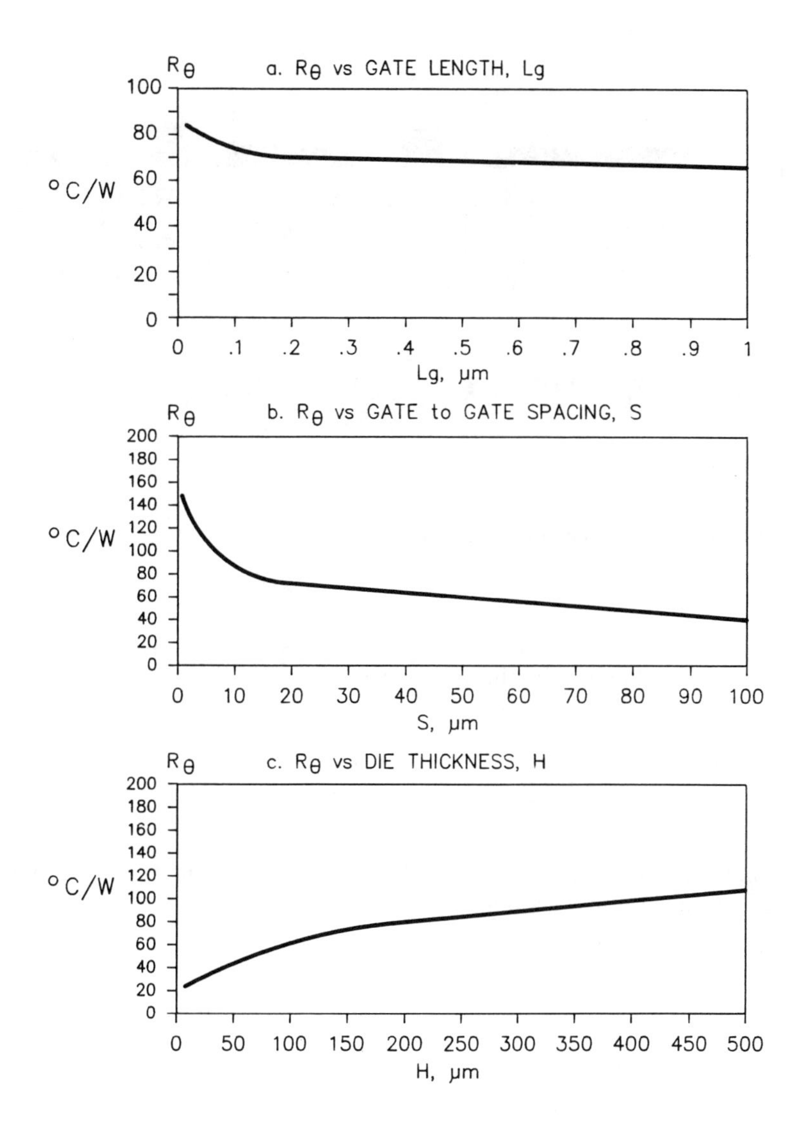


**ure 5.4** Effect of gate length, gate spacing, and die thickness on thermal resistance ($w_g$ = 1200 $\mu$m, $l_g$ = 0.3 $\mu$m, $S$ = 16 $\mu$m, $h$ = 100 $\mu$m, $N$ = 20, when not variables).

gure 5.5 shows a cross section of a typical interdigital FET where electrodes 1 d 2 represent the source and drain with respective widths $B_1$ and $B_2$, $P_1$ and $P_2$ intersection points in the heat flow path, and $R_1$ to $R_4$ are the thermal resistances the various layers. Let the $x$ axis be perpendicular to the channel surface with

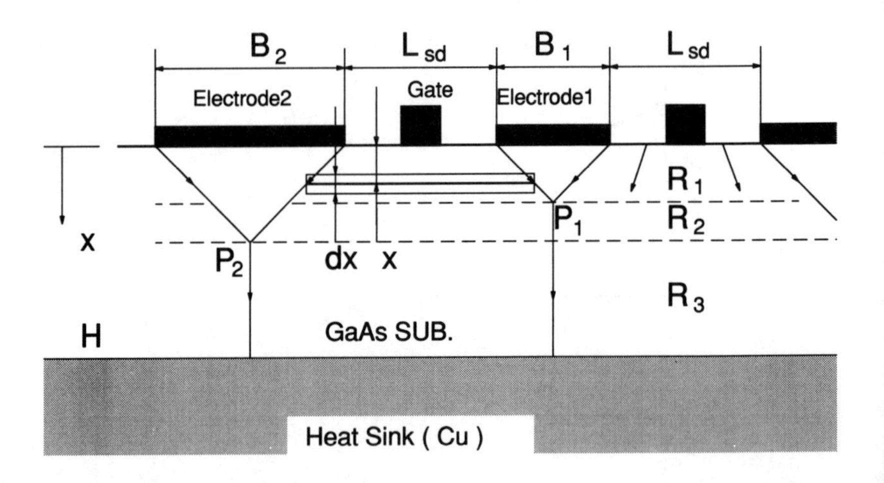


**Figure 5.5** Channel cross section of interdigital FET.

the origin at the surface. Consider a thin plate at point $x$ of thickness $\Delta x$; its are is $(l_{sd} + 2x)(w_g + 2x)$ and its thermal resistance is

$$\Delta R_1 = \frac{1}{\kappa} \frac{\Delta x}{(l_{sd} + 2x)(w_g + 2x)} \quad (5.1$$

The thermal resistance from the channel surface to the point $P_1$ corresponding $x = B_1/2$ is found by integrating and then dividing by the total number of gat $N$. Thus,

$$\begin{aligned} R_1 &= \frac{1}{N\kappa} \int_0^{B_1/2} \frac{dx}{(l_{sd} + 2x)(w_g + 2x)} \\ &= \frac{1}{N\kappa} \frac{1}{2(w_g - l_{sd})} \ln \frac{w_g(B_1 + l_{sd})}{l_{sd}(B_1 + w_g)} \end{aligned} \quad (5.1$$

Resistances $R_2$, $R_3$, and $R_4$ can be computed in a similar fashion:

$$R_2 = \frac{1}{N\kappa} \frac{1}{w_g - 2l_{sd} - B_1} \ln \frac{(w_g + B_1)\left(l_{sd} + \frac{B_1}{2} + \frac{B_2}{2}\right)}{(w_g + B_2)(l_{sd} + B_1)} \quad (5.1$$

$$R_3 = \frac{1}{2\kappa} \frac{1}{w_g - N\left(l_{sd} + \frac{B_1}{2} + \frac{B_2}{2}\right) + B_2} \times \ln \frac{\left[N\left(l_{sd} + \frac{B_1}{2} + \frac{B_2}{2}\right) - B_2 + 2h\right](w_g + B_2)}{\left[N\left(l_{sd} + \frac{B_1}{2} + \frac{B_2}{2}\right)\right](w_g + 2h)} \tag{5.14}$$

$$R_4 = \frac{1}{2\kappa_{Cu}} \frac{1}{N\left(l_{sd} + \frac{B_1}{2} + \frac{B_2}{2}\right) - B_2 - w_g} \ln \frac{N\left(l_{sd} + \frac{B_1}{2} + \frac{B_2}{2}\right) - B_2 + 2h}{w_g + 2h} \tag{5.15}$$

where $\kappa_{Cu}$ is the thermal conductivity of copper (assuming that the heat sink material is copper). Thus, $R_\theta$ is the sum of these four values, i.e.,

$$R_\theta = R_1 + R_2 + R_3 + R_4 \tag{5.16}$$

As an example, consider the FET analyzed previously: a device having 10 gates each 100 $\mu$m long with 25-$\mu$m gate-to-gate spacing and 4-$\mu$m source-to-drain spacing, then $R_\theta$ = 50.7 °C/W excluding the effect of the temperature dependence of thermal conductivity, which can be allowed for as previously, and assuming that $B_1 = B_2 = B$. This is 16 °C/W lower than the value predicted by the Culbertson and Lehmann method even though the boundary conditions are the same in both cases. This discrepancy is caused partly by the fact that the approximate technique calculates an average value for thermal resistance rather than the peak value given by the Culbertson and Lehmann method. Better agreement between the two methods of calculation may be possible by using a smaller value for $l_{sd}$ such as the gate length plus the gate-to-drain spacing rather than the source-to-drain spacing. Figure 5.6 shows the thermal resistance of power FETs calculated using this approximate technique assuming $B_1 = B_2 = B$. Also shown in Figure 5.6 are some experimental results, which, although slightly higher than the calculated values, are in reasonable agreement with the theory. Figure 5.7 shows measured and calculated values of thermal resistance as a function of the substrate thickness. As the substrate thickness is reduced, the predicted and measured values disagree to a greater extent, which is probably due to the effect of the solder layer between the GaAs substrate and the package, which has not been allowed for in these calculations.

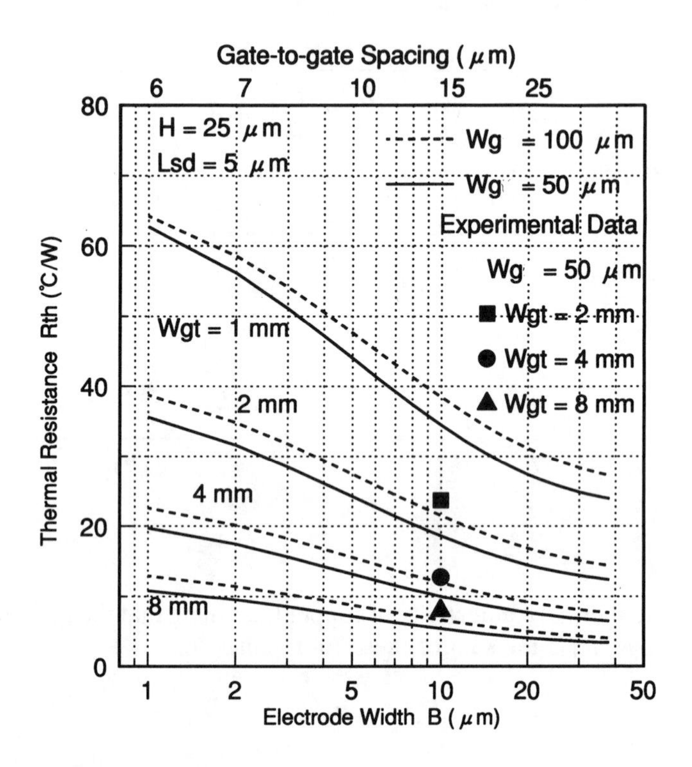


**Figure 5.6** Thermal resistance of interdigital FET as a function of gate-to-gate spacing where $W_{gt}$ is t total gate width. (Courtesy of Y. Aoki).

## 5.4 PULSED OPERATION

When an electronic device is thermally limited, that is, the dc input power cann be increased without decreasing the device life, pulsed operation can sometim be used to obtain a peak output power that is greater than the CW power outpu The vacuum tube is thermally limited by the plate dissipation, and the peak pow can be increased dramatically by raising the plate voltage and reducing the du cycle. For power FETs the gains from pulsed operation are modest at best a depend on the device design. If an FET is designed with a single gate or wi multiple gates so that the gate-to-gate spacing is relatively large (e.g., >20 μm it is likely that it can be operated at its maximum useful dc input. In this case t device is spread out so that the thermal resistance is relatively low and the pe pulse power will not be greater than the CW power. On the other hand, a ve compact geometry used to increase the RF gain may be thermally limited a

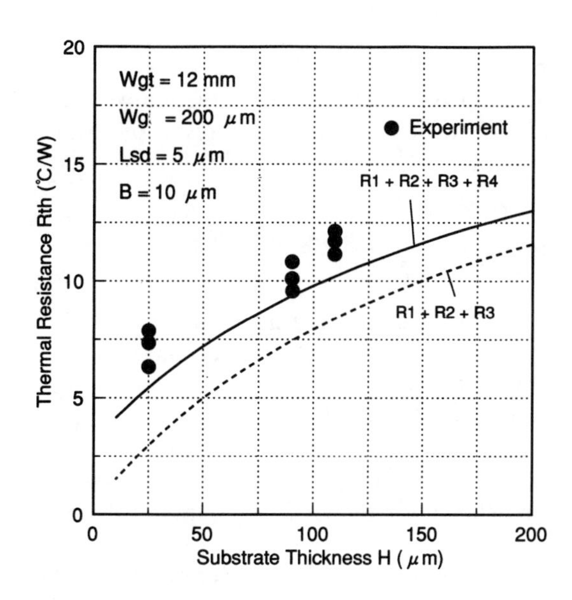


**gure 5.7** Experimental and calculated thermal resistance of an FET as a function of substrate thickness. (Courtesy of Y. Aoki).

lsed operation may be beneficial. To analyze this, it will be necessary to examine e thermal time constant.

*ermal Time Constant*

e surface temperature rise at any time after the dc power is switched on for the ucture shown in Figure 5.1(a) can be determined analytically by solving (5.4) bject to the appropriate initial and boundary conditions. The result is [11]

$$\Delta T = PR_\theta\left[1 - \frac{8}{\pi^2}\sum_{n=1,3,\ldots}^{\infty}\frac{\exp(-n^2 t/\tau)}{n^2}\right] \tag{5.17}$$

ere $R_\theta$ is the thermal resistance given by (5.2), $t$ is the time since the dc power s switched on, and $\tau$ is the thermal time constant given by

$$\tau = \left(\frac{2h}{\pi}\right)^2 \frac{1}{\alpha} \tag{5.18}$$

If $h = 100$ $\mu$m, then $\tau = 156$ $\mu$s for GaAs at room temperature since $\alpha = 0.$ cm²/s. Figure 5.8 is a graph of (5.17). As $t \rightarrow \infty$ the temperature rise given by (5.1 tends asymptotically to the steady-state value given by (5.1). For $t < \tau$ it can shown [12] that

$$\Delta T \approx PR_\theta \sqrt{t/2\tau} \qquad (5.1$$

and thus at $t = \tau$ the surface temperature will have reached about 70% of final steady-state value. The value of $\Delta T$ determined by (5.19) differs from t exact value given by (5.17) by less than 3% for $t < \tau$. While (5.19) was deriv for the structure shown in Figure 5.1(a), it can be shown [12] that this result a plies to the structure of Figure 5.1(b) and also to any power GaAs FET pr vided $t < 0.3\tau$. The thicker the die, the higher the steady-state surface temperatu rise, but the longer it takes for the surface temperature to reach its steady-sta value.

When an FET is operated under pulsed conditions, each pulse adds an inc ment of heat while it is on. As soon as that pulse is off, the heat diffuses into

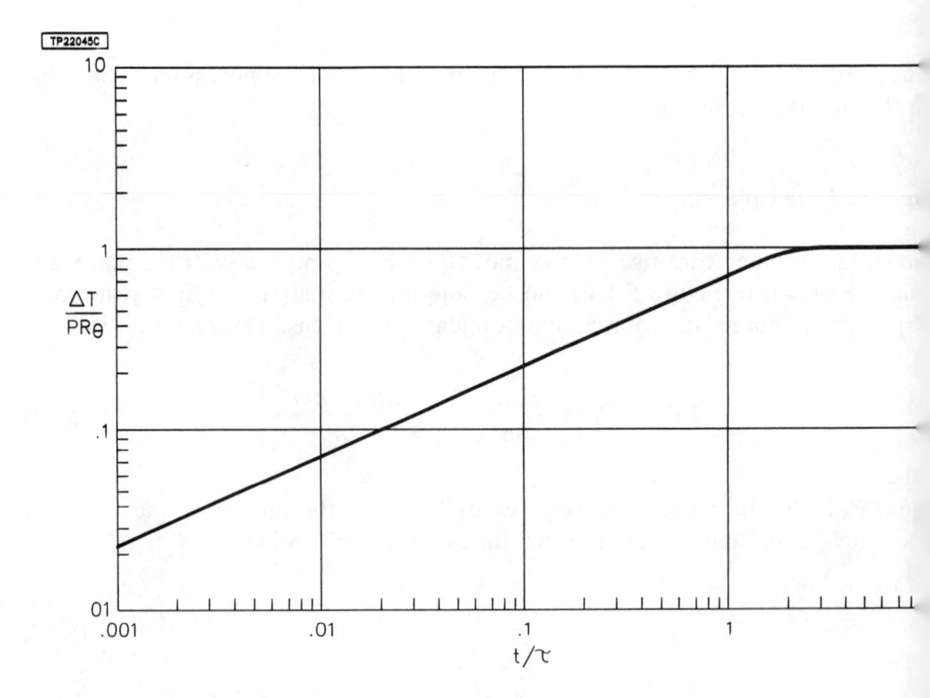


**Figure 5.8** Single-pulse transient temperature rise.

nk and the channel starts to cool down. The temperature at $t$ seconds after the art of the first pulse (at $t = t_0$) will be

$$\Delta T(t) = PR_\theta \sqrt{\frac{t - t_0}{2\tau}} \tag{5.20}$$

his holds until $t - t_0 = \tau$ or until the pulse goes off at $t = t_p$ for $t_p < \tau$. After e pulse goes off the die starts to cool according to

$$\Delta T(t) = T_{max}\left(1 - \sqrt{\frac{t - t_p}{2\tau}}\right) \tag{5.21}$$

here $T_{max}$ is the peak temperature reached while the pulse was on. The die will en cool until $t - t_p = \tau$ at which time the heating effect of the first pulse is ssipated. The second and all subsequent pulses repeat the process. To illustrate me possible variations, let $t_p$ = pulse length, $t_r$ = pulse repetition rate, and $\tau$ = ermal time constant. Depending on the relative length of these parameters, the ak temperature can vary considerably. Three possible cases are shown in Figure 9 (there are more but they are relatively unimportant):

Case 1. [Fig. 5.9(a)]: The pulse length is less than $\tau$, and the die never attains e CW temperature. Also, the interpulse interval $(t_r - t_p)$ is long compared to $\tau$ d, hence, the channel cools down to the heat sink temperature between pulses.

Case 2. [Fig. 5.9(b)]: The pulse length is longer than $\tau$, so the channel reaches e full CW temperature during the pulse but the interpulse interval $(t_r - t_p)$ is ghtly longer than $\tau$ so that the channel still cools down to the heat sink temperature tween pulses.

Case 3. [Fig. 5.9(c)]: The interpulse interval is less than $\tau$, so the channel starts heat from the second pulse before the heat from the first has dissipated com- etely. The time for the peak temperature to stabilize depends on the relative lues of $\tau$, $t_p$, and $t_r$.

e most important thing to remember is that when the pulse length exceeds $\tau$, lsed operation cannot increase the RF output power.

## 5 MEASUREMENT OF THERMAL RESISTANCE AND CHANNEL TEMPERATURE

urrently three principal methods exist for measuring the channel temperature d, consequently, the thermal resistance of a GaAs FET, but each involves some proximations. These three methods are now considered.

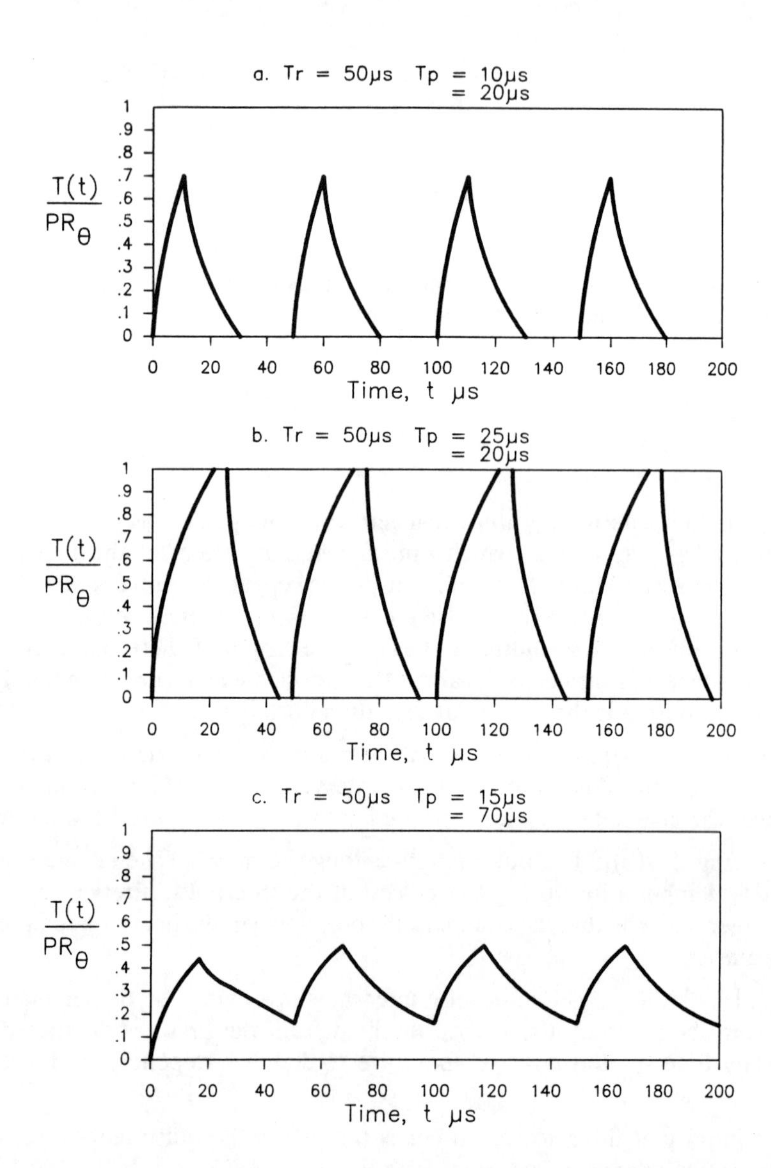


**Figure 5.9** Repetitive-pulse temperature rise for various pulse lengths, pulse repetition rates, a thermal time constants.

In the infrared method, the FET must be biased and mounted in an ope fixture (nonoscillating) and the surface scanned with an infrared microscope. Son very sophisticated equipment is available that can give an intensity versus positi ot. This method has three limitations. First, an initial calibration run must be ade with a material having a known emissivity. Secondly, the resolution of this ethod is about 5 $\mu$m, which may not be good enough for today's small geometries d, third, passivation can introduce errors. Figure 5.10 shows the temperature stribution of the surface of an FET measured by this method.

The liquid crystal method of measuring thermal resistance is based on the ct that above a certain critical temperature the molecules in the liquid crystal come randomly oriented. This change in the molecular orientation is detectable der a microscope using a polarized light source. In operation the GaAs FET is ounted in a test jig and a thin layer of the liquid crystal is placed on top of the rface of the device. Bias is applied to the FET and gradually increased until the ystal orientation changes at the hottest spot on the surface. The thermal resistance determined by dividing the difference between the heat sink temperature and e critical temperature by the dc power input. This technique has good spatial solution, so it can be used to detect hot spots, but it cannot provide a temperature ofile in the same way that the infrared method can. This technique also has two her limitations. First, the emissivity of the surface of the FET and, consequently, e indicated surface temperature are modified by the liquid crystal material. Secd, the test is destructive since the FET becomes contaminated. Further details this technique can be found in [13].

The third method for measuring thermal resistance is the forward diode volte method.[3] This method is based on the fact that the gate-source voltage of a rward-biased FET Schottky gate will decrease when the channel temperature creases if a constant forward current is made to flow between the gate and source. ver a limited temperature range this variation is linear:

$$\Delta V_{gsF} = K \Delta T_{ch} \tag{5.22}$$

ere $\Delta V_{gsF}$ is the change in gate-source voltage corresponding to a change $\Delta T_{ch}$ channel temperature and $K$, which is negative, is the thermal coefficient of the te-source voltage. Figure 5.11 shows the circuit used to measure FET thermal sistance. Naturally, the current in the voltage meter used to measure $V_{gs}$ must negligible in relation to the current $I_{gsF}$ between the gate and source. In the easurement sequence, the first step is to measure $K$. The device being measured placed in a constant-temperature container or on a heat block, $I_{gsF}$ is held constant, d the value of $V_{gsF}$ is recorded as the temperature of the hot block is changed. Figure 5.11 only switch $S_1$ is closed. The results are shown in Figure 5.12(a). e slope of the graph gives the value of $K$. Next, $\Delta V_{gsF}$ is measured as the switches

e material on this method of measuring thermal resistance was written by Y. Aoki.

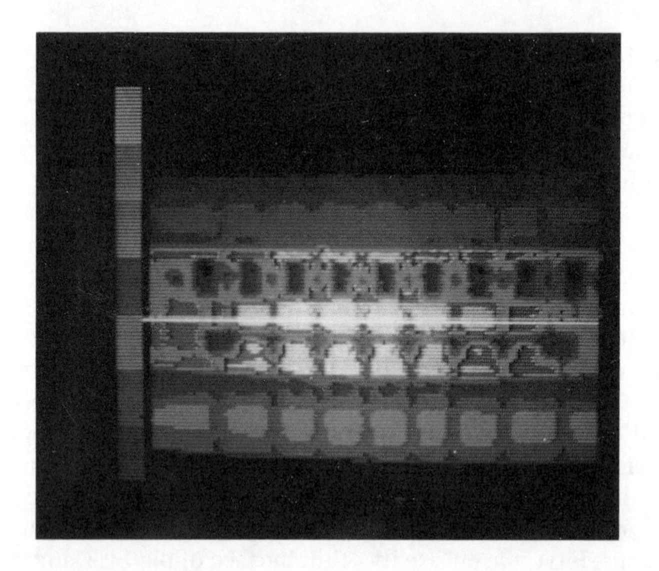

**Figure 5.10** Temperature distribution of an FET observed by infrared camera (Photo courtesy of Aoki).

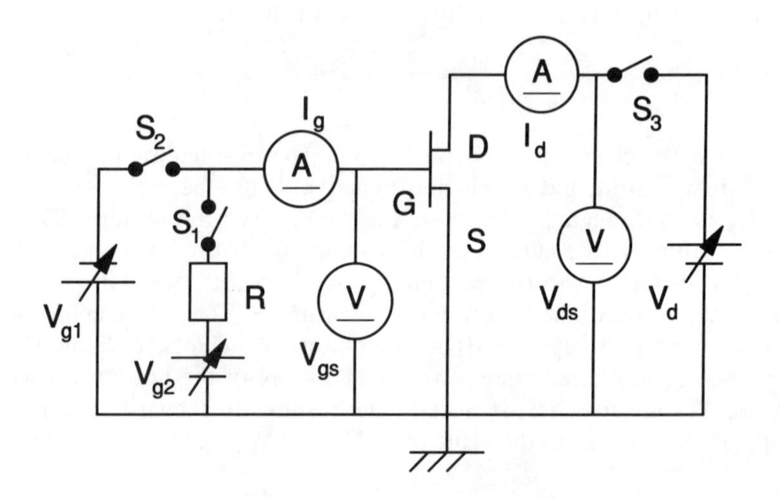


**Figure 5.11** Circuit diagram for measurement of thermal resistance of an FET using temperat dependence of forward gate current. (Diagram courtesy of Y. Aoki).

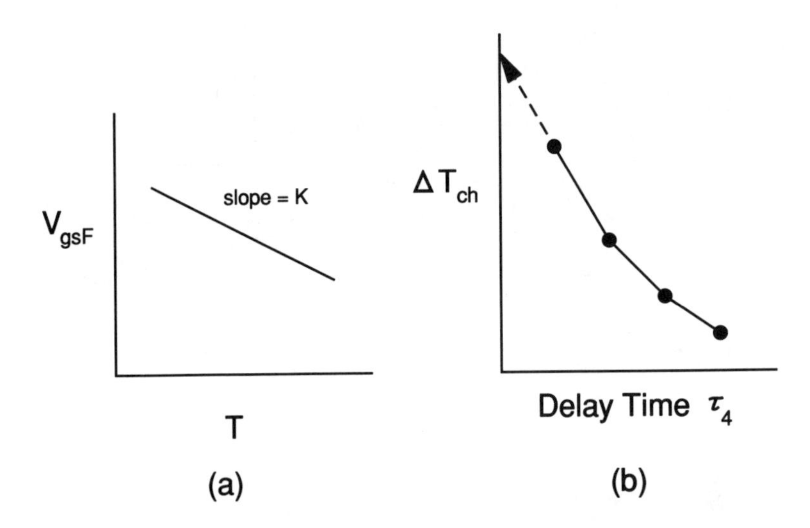


igure 5.12 (a) Temperature dependence of forward gate-source voltage for constant gate current and (b) sample plot of $\Delta T_{ch}$ versus $\tau_4$ to estimate channel temperature at $\tau_4 = 0$. (Courtesy of Y. Aoki).

re turned on and off with the timing shown in Figure 5.13. The concept in this ethod is to raise the channel temperature by applying a drain current to the evice, thereby causing a drop in $V_{gsF}$, which can be used in turn to calculate the annel temperature, and from this to compute the thermal resistance. With ref- rence to Figure 5.13, the drop in forward voltage $\Delta V_{gsF}$ can be calculated from

$$\Delta V_{gsF} = V_{gsF2} - V_{gsF1} \tag{5.23}$$

he change in temperature $\Delta T_{ch}$ can be calculated from (5.22). However, $\Delta T_{ch}$ is ot the actual increase in channel temperature during operation. This is because $T_{ch}$ is measured at time $\tau_4$ after switch $S_3$ is turned off. To find the actual temperature f the channel during operation, the delay time $\tau_4$ must be varied to develop a raph such as the one in Figure 5.12(b). By extrapolating this curve, it is possible derive a $\Delta T_{ch}$ for the case $\tau_4 = 0$, and to consider this to be the temperature crease in the channel during operation. The advantages of this method are that can be used to measure the thermal resistance of packaged devices or flip-chip ounted die, neither of which can be measured by either of the two preceding chniques, and it is nondestructive. Its disadvantage is that it determines an average alue for thermal resistance and cannot detect hot spots or uneven heating. Com- ercial equipment utilizing the forward $V_{gs}$ method of measuring temperature often ses a sample-and-hold circuit to acquire the value of $V_{gs}$ immediately following

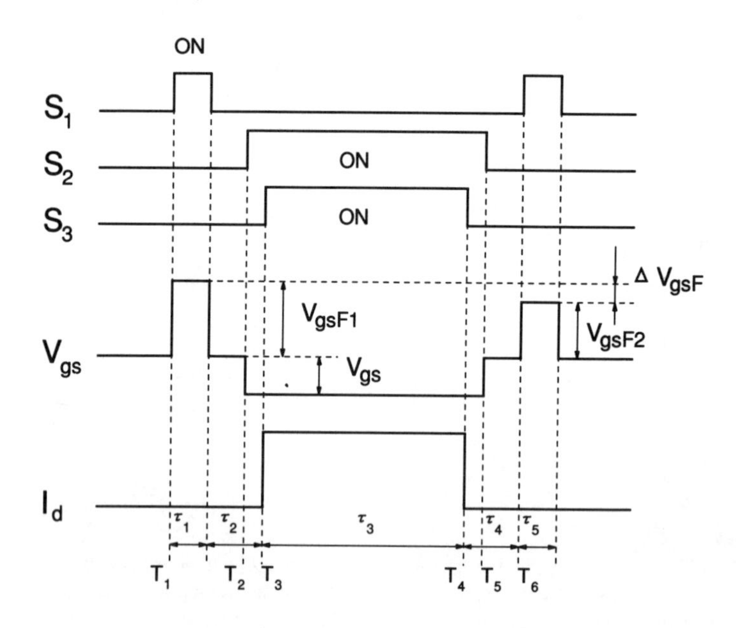


**Figure 5.13** Timing chart for apparatus to measure thermal resistance. (Courtesy of Y. Aoki).

the heat cycle. Such circuits now have such short acquisition times that there little error due to post-pulse cooling.

## 5.6 RELIABILITY

### 5.6.1 Failure Mechanisms

For any application, the user of a semiconductor device would like to be assure that the device will continue to function correctly over some time period and und a given set of conditions. An expendable system might have a useful life of 1 m or less, while transistors for use in a satellite must have a predicted life of the ord of 100 yr. Where a very long useful life is required, the user is often willing to pa a premium price for the device. This premium goes to pay for burn-in and fo extended accelerated life tests. However, there are certain limits to the predictic of FET life. In some situations an FET may fail for reasons that are beyond th control of the FET manufacturer, for example, failure of a cooling system, exposu to a very high RF input power, transients in a dc power supply. These problen are best handled by the systems engineer.

A second class of failure is a result of errors in manufacturing, such as weak onds, cracked die, interconnecting metal too thin, or contamination. This class f failure does not fit a predictable pattern but can be minimized by prescreening, hich might involve burn-in with bias (sometimes elevated), vibration, and tem-erature cycling. In military systems these procedures are often a requirement for high reliability device. They still do not guarantee a long life. In fact, if the burn-temperature is too elevated or too long, then the useful life of the device may e reduced.

Finally, certain types of failure may occur after many hours and can usually e predicted using statistical methods. Examples of this type of failure in power ETs are contact degradation, metal migration, and changes in the channel under e gate. All of these failures have one characteristic in common, they are tem-erature activated [14]. How a change is influenced by temperature is defined by e reaction rate, RR. The reaction rate RR can be defined by

$$\mathrm{RR} = A \exp(-E_a/kT) \tag{5.24}$$

here $A$ is a constant, $E_a$ is the activation energy in electron volts, $k$ is Boltzmann's onstant, and $T$ is the temperature in Kelvin. The activation energy is a measure how easily a reaction is influenced by temperature. A low activation energy ould then indicate potentially a short-life device where semiconductors are con-erned. Approximately 1 eV is the boundary between "short" and "long" life evices when considering high-reliability applications.

The activation energy for ohmic contact degradation is of the order of 1 eV. common form of contact degradation for AuGe/Ni contacts is gallium diffusion oward through the contact, causing the contact resistance to rise. This increases e source resistance and, consequently, reduces the FET gain. More complicated ontact systems have been reported that resist Ga migration but they do not seem be in general use as yet.

Aluminum gates will withstand temperatures up to 300°C, but they have some her problems such as greater susceptibility to electrostatic discharge and to elec-omigration with an activation energy of approximately 0.5 eV. Also, since the eposition thickness is limited by etching, it is not possible to deposit very thick etal and there is an interface problem where the gate metal meets the gold circuitry gate pad. Hence most power FETs use a Ti-Pt-Au or a similar refractory metal/ old system. For millimeterwave FETs the gate lengths are of the order of 0.25 m. To keep the gate resistance down, these gates are plated up using photoresist a mask. Aluminum gate FETs, which cannot (at present) be plated, tend to ave their performance dominated by gate resistance. Metal will migrate at points the FET where the current is highest, $10^5$ A/cm$^2$ being the current level where igration becomes a serious problem. The current to be considered is not only the component due to bias, but also that caused by large RF signal inputs. The

latter is not easy to predict, but its presence can be determined by tests, as will b
shown later.

In the case where the FET changes due to loss of carriers in the channe
several models have been proposed such as movement of the gate into the channe
movement of impurities from the substrate, and field-enhanced diffusion of impu
ities. Some of these are well documented, others are still a matter of contentio
The activation energy for this form of degradation is well over 1 eV and, in genera
this type of failure is not a problem. Tungsten silicide gates show some promi
of being less susceptible to interaction with the channel. The task of finding out
there is a degradation mechanism and what it is will be addressed in the ne
section.

Since temperature is the driving force behind the long-term degradation
FETs, this suggests that life tests carried out at elevated temperatures would ma
it possible to determine the life at lower, more realistic, temperatures. Before the
life tests can be used, some basic equations for the statistics must be defined.

### 5.6.2 Reliability Statistics

Suppose it is determined that a group of FETs is useful as long as the paramet
$I_{dss}$ does not decline by more than 20%. These FETs are then put on life test
an elevated temperature of, say, 250°C. Each FET is checked hourly to determi
if $I_{dss}$ has dropped below 80% of its initial value. When this occurs, it is declar
a failure and the time recorded. It will be noted very soon that the failures occ
over an extended period and that the failures peak at some time. The pattern
these failures is described by the distribution equation. The development of the
equations is found in [14] and [15]. Experiments have shown that for semiconducto
the log-normal distribution best describes the pattern of failures, and it is no
almost universally used by manufacturers. All of the equations that follow are f
this case. The log-normal probability density function describes how devices f
in time. This is given by

$$f(t) = \frac{1}{\sqrt{2\pi}\sigma t} \exp\left[-\frac{1}{2}\left(\frac{\ln t - \ln t_m}{\sigma}\right)^2\right] \quad (5.2$$

where $t$ is the elapsed time, $t_m$ is the median time to failure, and $\sigma$ is the dispersic
Figure 5.14(a) shows the probability density function for the log-normal distrib
tion. If the probability density function is integrated with respect to time, then t
failure density function will be obtained. It gives the fraction of devices $Q(t)$ th
have failed up to $t = t_1$:

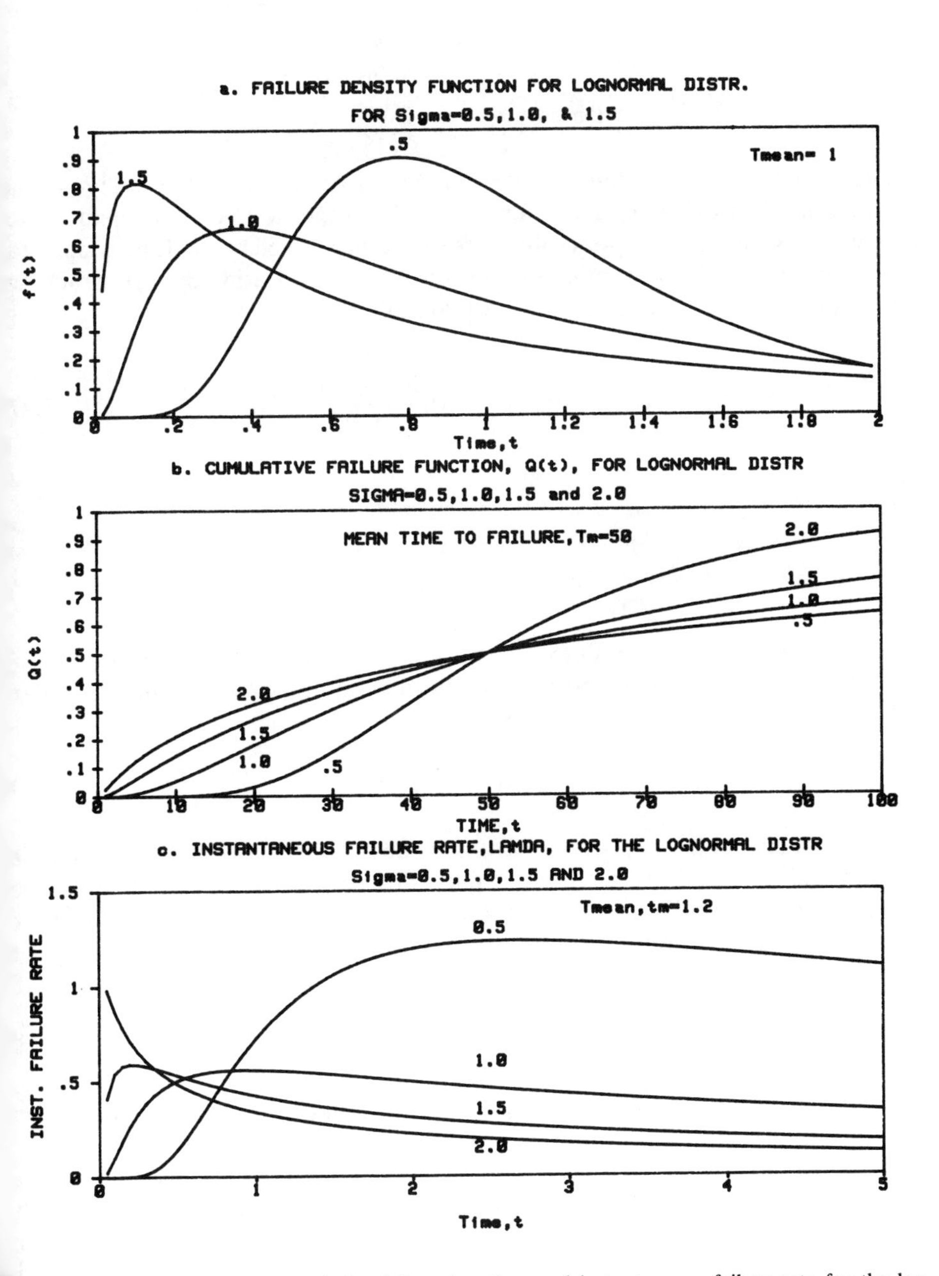


**Figure 5.14** Failure density, cumulative failure function, and instantaneous failure rate for the lognormal distribution.

$$Q(t_1) = \frac{1}{\sqrt{2\pi}\sigma} \int_0^{t_1} \frac{1}{t} \exp\left[-\frac{1}{2}\left(\frac{\ln t - \ln t_m}{\sigma}\right)^2\right] dt \tag{5.26}$$

Figure 5.14(b) shows the failure density function versus time for the log-normal distribution. It also shows the effect of $\sigma$, the dispersion, on the failure rate. Dispersion is determined primarily by the criteria by which a failure is defined. The term $Q(t)$ can be written in a more convenient form using the error function erf($z$) since rapidly converging series are available for erf($z$) [16]. Thus,

$$Q(t) = 0.5\left[1 + \text{erf}\left(\frac{\ln t - \ln t_m}{\sqrt{2}\sigma}\right)\right] \tag{5.27}$$

where

$$\text{erf}(z) = \frac{2}{\sqrt{\pi}} e^{-z^2} \sum_{n=0}^{\infty} \frac{2^n z^{2n+1}}{1.\ 2.\ 3.\ \ldots\ .\ 2n + 1} \tag{5.28}$$

It is immediately evident from (5.27) that $t_m$ is the median lifetime since $Q(t_m) = 0.5$. However, $t_m$ and $\sigma$ together uniquely determine the *mean time to failure* (MTTF), which is given by [14]

$$\text{MTTF} = t_m e^{\sigma^2/2} \tag{5.29}$$

It can immediately be seen from (5.29) that the MTTF is always greater than the median time to failure for any finite value of $\sigma$. Another parameter that is often required from reliability data is the instantaneous failure rate $\lambda(t)$. It is the failure rate divided by the number of units that have not yet failed:

$$\lambda(t) = \frac{f(t)}{1 - Q(t)} \tag{5.30}$$

Figure 5.14(c) is a plot of $\lambda(t)$ versus time for the log-normal distribution. Very often the failure rate is measured in FITs where 1 FIT is 1 failure in $10^9$ hr.

If a group of FETs is placed on life test at, say, 250°C and the fraction that have failed [i.e., $Q(t)$] is plotted versus time on log-normal graph paper, then the plot will usually be a straight line. Exceptions occur when there are two or more failure modes being triggered at the same time, or there are defective devices in the lot under test. For the first case, this is a difficult problem and requires special testing (see [14]). In the second case, random failures will occur early in the test that do not fit the line determined by later plots. These are called *infant failures* ecause they occur long before the patterned failures. Figure 5.15 is a sample of og-normal graph paper. After a plot is made and a line fitted to the plots, another ne may be drawn through the point labeled 0 shown in Figure 5.15 parallel to the lotted data line. This line intersects the sigma scale at the correct value of sigma or the data distribution. Alternatively, $\sigma$ can be determined from either $\sigma = \ln(t_m/t_{16})$ or $\sigma = \ln(t_{84}/t_m)$ where $t_{16}$ and $t_{84}$ are the times when 16% and 84% of the evices have failed. The line through the data points crosses the 50% failure line

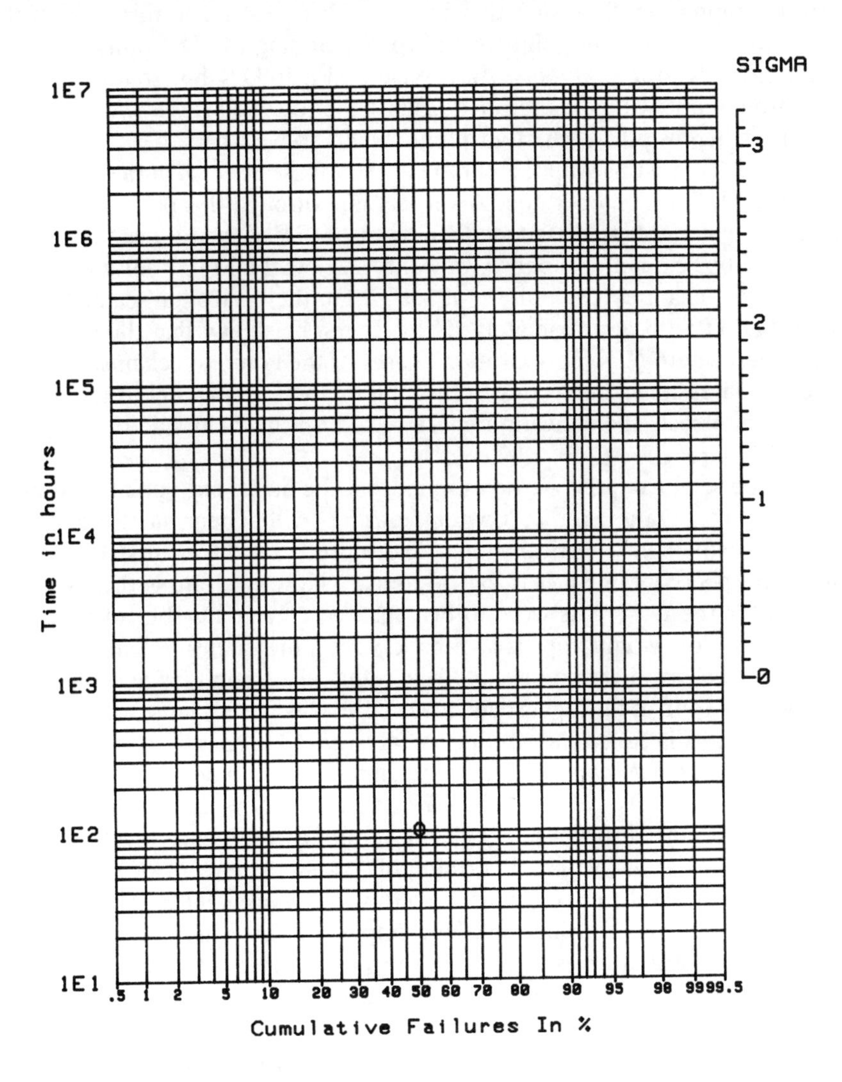


igure 5.15 Log-normal graph paper.

at time $t_m$, the median time to failure. This is not to be confused with MTBF, whic applies to vacuum tubes only. Unpublished, complete, reliability data are usuall treated as confidential by most companies, so the examples that follow have bee generated mathematically using a computer.

Figure 5.16 gives four plots of $Q(t)$ for a group of FETs stressed at 150°, 200° 225°, and 250°C. The median time to failure for these four cases is 25,000, 1300 400, and 130 hr, respectively. The slope of these lines is used with the right-han scale to determine the dispersion that has the value $\sigma = 1.1$ in this case. If thes four median time to failure points or the corresponding MTTF points are plotte on graph paper with a log-time vertical scale and a 1000/K horizontal scale, the another straight line will be established. This is the so-called *Arrhenius plot.* Figur 5.17 is an Arrhenius plot of the four points taken above with a best fit line draw through them. To use this graph, enter the horizontal scale (top) at the temperatur at which the FET will be operated. A vertical line through this point will intercep the data plot line at the predicted median life or MTTF of the FET at the operatin temperature. For example, suppose we want to know the MTTF for a power FE operated with a channel temperature of 100°C. On the horizontal scale, 1000/K = 2.68 and the MTTF is approximately $10^6$ hr. It can be shown that the slope of th plot is $5.04 \cdot E_a$ where $E_a$ is the activation energy of the failure mechanism. To mak this calculation easier, a line is drawn through the point 0 parallel to the data plc to where it intersects the activation energy scale on the right side of the plot. I the case of Figure 5.17, $E_a = 1.00$ eV.

It can be seen from the Arrhenius plot that the line is extrapolated to predic the projected life. Data taken at low temperatures will appear more credible, bu the test times will be unrealistically long. Therefore, after the estimated mean tim to failure has been obtained, it is logical to ask "How accurate is this estimate? If the tests for the $Q(t)$ plots were to be repeated a large number of times, the the Arrhenius plots would appear as a band with a high density near the middle The center of the band would be the "true" plot. If the confidence level require is very high, the ratio of the highest and lowest plot will be very large. The differenc between the limits is called the *confidence interval* (CI). A more useful number i the *confidence interval ratio* (CIR). Its meaning will be made more clear in th example that follows. To calculate the CIR one must know the number of device used in the tests, the dispersion $\sigma$, and the required confidence level. If in 9 ou of 10 tests the result lies within a certain plus-or-minus percent of nominal (i.e the confidence interval ratio), then the confidence level is 90%. The calculatio of CI for the log-normal distribution is not simple. To make the calculation, a tab of Student's $T$ distribution is required. The procedure is as follows: Given th required confidence level $\gamma$, the dispersion $\sigma$, and the sample size $n$, first calculat $F(z) = 0.5(1 + \gamma)$. Knowing $F(z)$ and $m = n - 1$, the number of degrees o freedom, determine $z$ from Student's $T$ distribution table. Table 5.3 is a shortene $T$ distribution table for several commonly used confidence levels and sample size

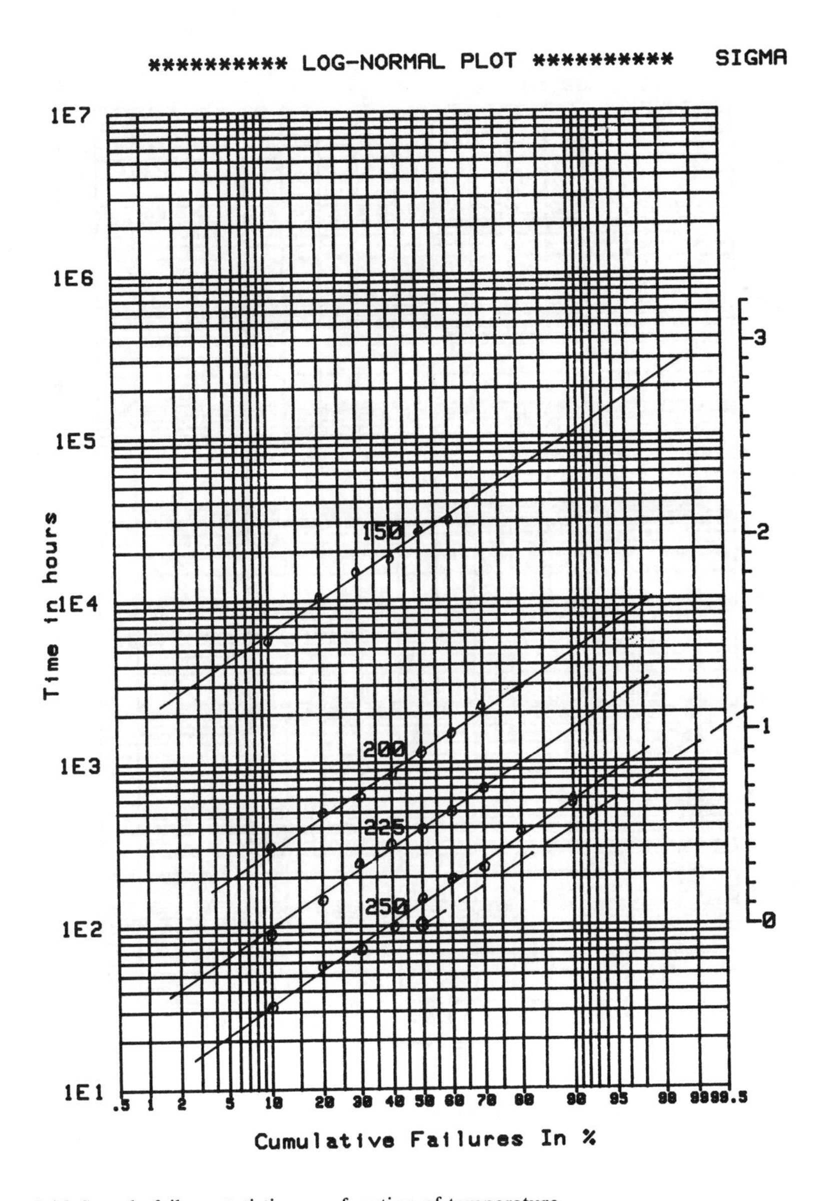


Figure 5.16 Sample failure statistics as a function of temperature.

Finally, calculate CIR $= \sigma \exp(z/\sqrt{n})$. As an example, suppose 10 FETs are placed on life test and an MTTF of $10^4$ hr is obtained with $\sigma = 1.0$. For a confidence level of 99%, what are the confidence limits of the estimated MTTF? We obtain

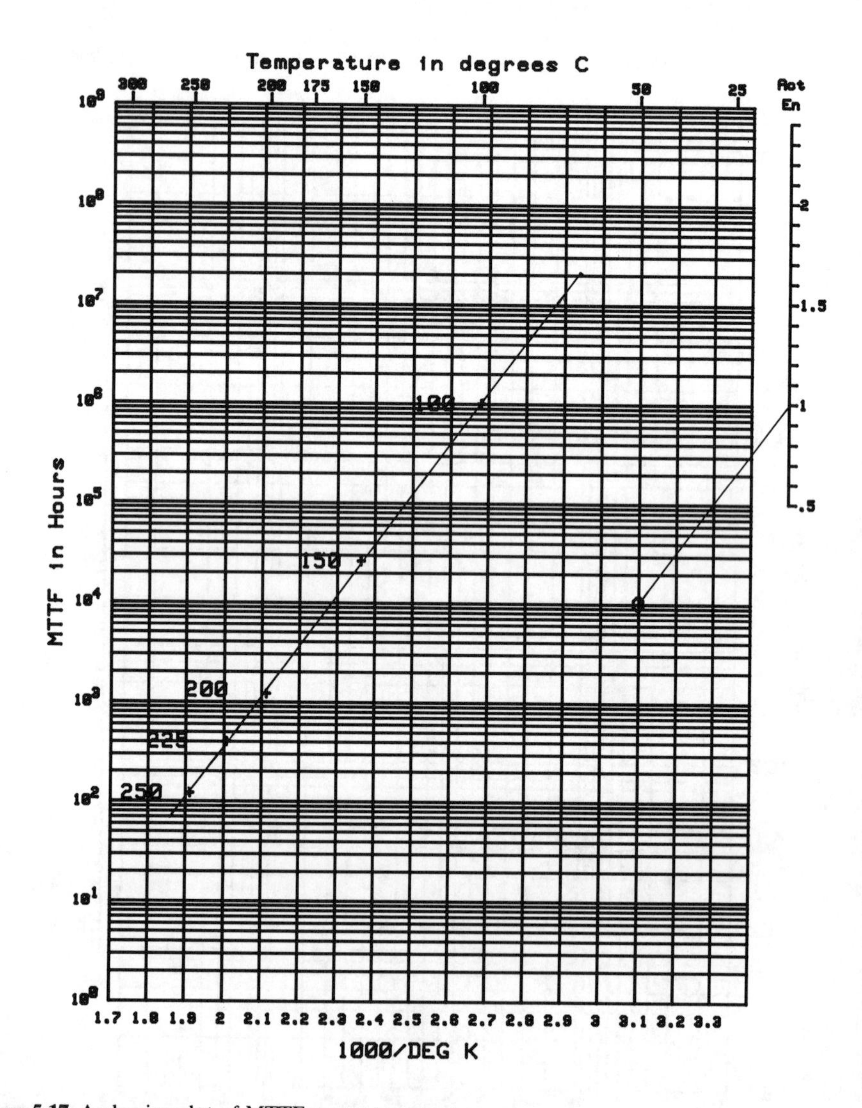


**Figure 5.17** Arrhenius plot of MTTF versus temperature.

$F(z) = 0.5(1 + 0.99) = 0.995$. From Student's $T$ distribution, $z = 3.25$ f
$F(z) = 0.995$ and $m = 10 - 1 = 9$ degrees of freedom. Hence CIR $= 1 \exp(3.2$
$\sqrt{10}) = 2.795$. The true value of the MTTF will lie between $10^4$/CIR and $10^4$
CIR or $3.59 \times 10^3 <$ MTTF $< 2.795 \times 10^4$ hr for a 99% confidence level. Finall
the CI is $2.795 \times 10^4 - 3.59 \times 10^3 = 2.436 \times 10^4$ hr. Reference [17] makes
possible to obtain the CIR graphically. It can be seen that the confidence level ca

**Table 5.3**
Student's $T$ Table

| $\gamma$ | $F(z)$ | $m$ 9 | 14 | 19 | 50 | 100 |
|---|---|---|---|---|---|---|
| .80 | 0.900 | 1.38 | 1.33 | 1.37 | 1.30 | 1.29 |
| .90 | 0.950 | 1.83 | 1.76 | 1.73 | 1.68 | 1.66 |
| .95 | 0.975 | 2.26 | 2.15 | 2.09 | 2.01 | 1.98 |
| .98 | 0.990 | 2.82 | 2.62 | 2.54 | 2.40 | 2.37 |
| .99 | 0.995 | 3.25 | 2.98 | 2.86 | 2.68 | 2.63 |

e increased by increasing the confidence interval or the size of the test sample. 'he latter is, of course, the only option if the given confidence interval is fixed.

## .6.3 Reliability Testing

'rescreening has already been mentioned as a method of eliminating certain types f failure. Tests to determine the potential life vary somewhat. The simplest of nese is the high-temperature storage test. In this test the FETs are simply placed n an oven at an elevated temperature and removed at intervals for testing. This continued until all, or most, of the units have failed. One of the reasons this ype of test is not a good one is that the act of removing the FETs from the oven ntroduces another and undesirable variable. In addition, units can be lost in the rocess of moving and testing them. A more sophisticated version of this test places ne FETs under monitored bias during the aging process. This adds another dimen-on to the failure process, but it does make it possible to detect failures by observing ne changes in the dc gate and drain currents. If RF drive is added to temperature nd bias as a stress factor, then the test becomes very complex and expensive. The nost practical answer to these problems is the computer-controlled life test.

Sometimes it is advantageous to use step-stress tests to determine the approx-nate temperature at which devices begin to fail. This is done by stressing the evices at some temperature, and then increasing the temperature periodically 'hile observing the failure rate. Sometimes these data can be used [14] to determine ne activation energy. The validity of this method depends on whether failed devices re replaced or not, and on the activation energy itself. Most semiconductor man-facturers prefer to use step-stressing as a preliminary to multiple tempera-ıre tests.

Automating a life test involves some relatively expensive equipment, but it still cost effective in that it eliminates most of the labor used to make the required eriodic tests. In addition, the computer does not object to working nights, weekends, and holidays. Automated testing can also be made very flexible by designin the software properly. The test setup to be described can do the following:

1. High temperature storage—bias applied periodically to determine when fail ures occur;
2. Stress with temperature and bias—dc bias monitored periodically;
3. Stress with temperature, bias, and RF drive—bias and RF gain monitore periodically.

In addition, the stress parameters of temperature, bias, and RF drive can be turne off individually as each device fails. This has an obvious benefit in that the faile FET is left in a condition suitable for diagnostic analysis. In nonautomated tests the stress continues after the device has failed and will often destroy it completel Figure 5.18 is the block diagram of such an automated test setup.[4] It can b configured for up to 10 FETs with two switch control units, and for more FET by adding more temperature controllers and switch control units. Stresses wit temperature, bias, and RF drive are available in any desired combination. Th number of FETs in the test, failure criteria (e.g., $I_{dss}$, RF gain, power output), te temperature, and test interval are all operator inputs. At each test time the device are measured and the result is compared with the initial test data. A decision i also made whether the device has failed or not. If it has, then all stress is turne off. At test time the result is printed, or stored, or both. One of the more importan benefits of this type of test is that these data are much more consistent than th taken with manual testing.

In devising a fixture for active life testing the first requirement is to ensur that the device will not oscillate during the test. In oven tests this is a major proble since ferrite beads lose their ability to suppress oscillations when they reach the Curie temperature. This is offset to some degree by the loss in mobility in th FET, but it does not solve the problem. Another difficulty arises from the change in the fixturing materials as the temperature is raised. Springs lose their tensio and plastics either polymerize or flow. Most of these problems can be eliminate by using an individually heated mount rather than an oven. Temperature contr can then be implemented on an individual basis and, in addition, RF testin becomes more practical. Figure 5.19 shows such a fixture. The heater block hea the FET only since it is thermally isolated from the rest of the fixture by the ceram posts and the polyimide circuit board. The device is mounted either on a carri or in a package. Due to the massive construction of the fixture the connectors ar kept cool by the cooling plate below. The polyimide circuit boards will surviv

[4]The automated life test setup and test fixture were developed in a program cosponsored by the Vari Solid State Microwave Division (now Litton Solid State) and the Naval Research Laboratory.

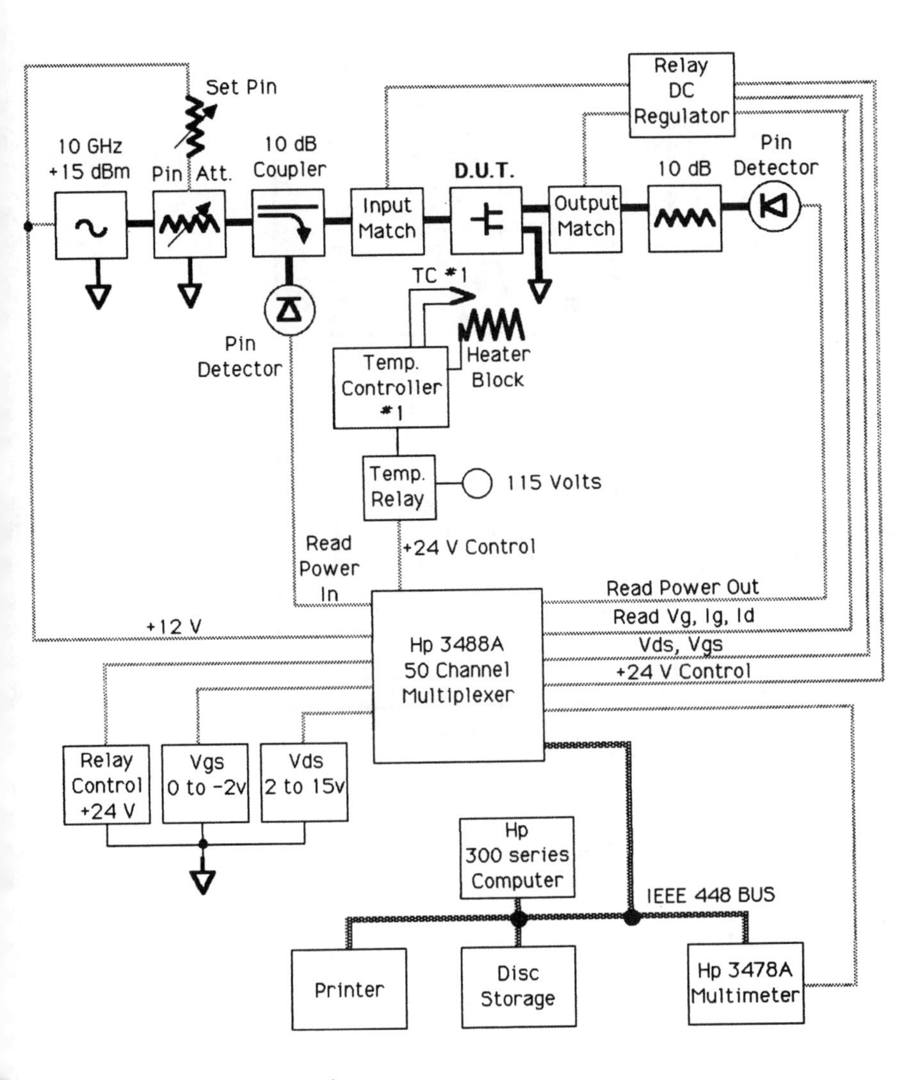


igure 5.18 Automated life test equipment.

00°C for several thousand hours although they change color somewhat. Since the ixture is kept at a constant low temperature it becomes practical to use detector iodes as power sensors. The diodes are matched for the same power-current haracteristic so that the same algorithm can be used for all 20 diodes to convert urrent to RF power output. The use of multiple power meters and sensors was onsidered impractical from the viewpoint of drift and excessive cost.

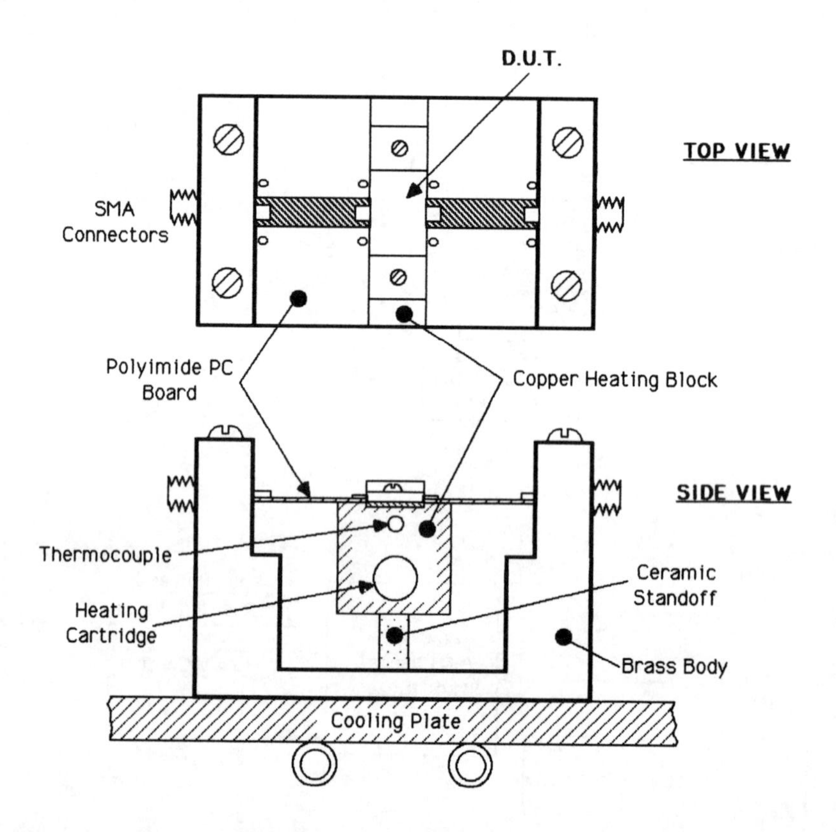


**Figure 5.19** Accelerated life test fixture for power FETs.

## 5.7 CONCLUSION

We can now calculate the thermal resistance of a given power FET design wit good enough accuracy to predict its channel operating temperature in a give ambient and dissipation. In addition, accelerated life tests can be carried out tha will enable the FET designer to determine if an FET will meet the useful lif specifications of the designer. The only negative aspect to the accelerated life test is that they are expensive and take time for accurate results. Nonetheless they ar now considered an essential part of modern semiconductor design. The impact o computer-directed life tests has made accelerated life tests a practical reality.

## REFERENCES

1. Carslaw, H. S., and J. L. Jaegar, *Conduction of Heat in Solids*, Oxford, UK: Clarendon Pres 1986, pp. 176–187.

2. Linsted, R. D., and R. J. Surty, "Steady-State Junction Temperatures of Semiconductor Chips," *IEEE Trans. on Electron Devices*, Vol. ED-19, January 1972, pp. 41–44.
3. Hewlett-Packard, "RF & Microwave AppCAD," Applications Programme Diskette, Part HAPP-0001.
4. Joyce, W. B., "Thermal Resistance of Heat Sinks With Temperature-Dependent Conductivity," *Solid-State Electronics*, Vol. 18, 1975, pp. 321–322.
5. Cooke, H. F., "Precise Technique Finds FET Thermal Resistance," *Microwaves & RF*, Vol. 25, August 1986, pp. 85–87.
6. Cohn, S. B., "Shielded Coupled-Strip Transmission Lines," *Proc. IRE*, Vol. MTT-3, October 1955, pp. 29–38.
7. Jones, Sidney, Consultant, Private Communication.
8. Higashisaka, A., Doctor Thesis, Tokyo Institute of Technology, 1981 (in Japanese).
9. Aono, Y., A Higashisaka, T. Ogawa, and F. Hasegawa, *Jpn. J. Appl. Phys.*, Vol. 17, Suppl. 17-1, 1978, p. 147.
0. Holway, L. H., and M. G. Adlerstein, "Approximate Formulas for the Thermal Resistance of IMPATT Diodes Compared with Computer Calculations," *IEEE Trans. on Electron Devices*, Vol. ED-24, February 1977, pp. 156–159.
1. Mortenson, K. E., "Transistor Junction Temperature As A Function of Time," *Proc. IRE*, Vol. 45, April 1957, pp. 505–513.
2. Pritchard, R. L., *Electrical Characteristics of Transistors*, New York: McGraw-Hill, 1967, pp. 614–634.
3. DiLorenzo, J. V., and D. D. Khandelwal, Eds., *GaAs FET Principles and Technology*, Dedham, MA: Artech House, 1982, pp. 313–347.
4. Reynolds, F. H., "Thermally Accelerated Aging of Semiconductor Components," *Proc. IEEE*, Vol. 62, February 1974, pp. 212–222.
5. *Accelerated Testing Handbook*, San Jose, CA: Technology Associates, 1978.
6. Abramowitz, M., and I. A. Stegun, *Handbook of Mathematical Functions*, New York: Dover Publications, 1972, p. 297.
7. DiLorenzo, J. V., and D. D. Khandelwal, Eds., *GaAs FET Principles and Technology*, Dedham, MA: Artech House, 1982, p. 362.

# Chapter 6
# Combining Techniques

J. L. B. Walker
*Thorn-EMI Electronics*

## .1 INTRODUCTION

lany applications require more output power than is available from a single tran-stor with the result that it is necessary to combine the outputs of two or more evices. The simplest method of doing this is to connect two or more transistor ips in parallel but this inevitably leads to lower impedance levels at the input d output, which makes impedance matching more difficult and reduces the band-idth. This technique also has another limitation, namely, that a phase difference ill exist between the input voltage applied to the outermost and central transistors the array, which leads to a loss of gain and power output. This problem becomes orse as the frequency and array size increase, but it can be overcome to some xtent by fabricating the array as a monolithic circuit because this enables the dividual transistor cells to be placed closer together. However, this creates other problem, heat dissipation (see Chap. 5, Sec. 5.3), and it does nothing to lve the impedance level problem. Cluster matching [1], in which the individual wer FET cells are partially matched before paralleling, is often employed as a eans of alleviating the impedance level problem, but this creates the possibility odd-mode oscillations [2]. Also, big chips with a very large gatewidth inevitably ve a low yield and are thus expensive. Eventually, therefore, true power com-ning must be considered rather than simple paralleling of unmatched or pre-atched transistors and this forms the subject of this chapter.

Successful power combining requires recognization of the need to maintain asonable impedance levels, to minimize thermal resistance, and to account for ut not necessarily eliminate) the phase difference between the transistors being mbined. All the power combining techniques to be described in this chapter fall

into two categories; those that prematch the transistor first (usually to $50\Omega$) prio
to combining, and those that do not. This latter category is dominated by th
*distributed amplifier* or *traveling-wave amplifier* as it is sometimes called. The forme
category, examples of which include tree, chain, and $N$-way power combining
constitutes the more traditional and most widely used method of power combining
particularly in the hybrid industry, while the latter category is most appropriat
for a monolithic implementation. The main performance advantage of the distrib
uted amplifier method of power combining compared with the other methods i
that it enables much wider bandwidth power amplifiers to be produced, for example
2- to 18-GHz, 0.5-W distributed amplifiers with input and output VSWRs < 2.5:
are commercially available from Texas Instruments. However, distributed ampl
fiers are restricted to less than 1-W power outputs in a 50-$\Omega$ system by curren
GaAs FET technology, and the overall efficiency is substantially lower than
possible using the other methods of power combining.

In this chapter the distributed amplifier method of power combining will b
described first followed by tree, chain, and $N$-way power combiners.

## 6.2 DISTRIBUTED AMPLIFIER POWER COMBINING

The concept of distributed amplification was first proposed and patented by Perciv
of EMI (later to become Thron-EMI) [3] in 1936. Originally the concept was applie
to vacuum tube amplifiers to achieve large gain-bandwidth products, but sinc
GaAs FETs can be considered to be the solid-state version of vacuum tubes,
was inevitable that sooner or later the concept would be applied to these device
also. The first to apply the concept to GaAs FETs was Ayasli et al. [4] of Raythec
in 1981 who achieved about a 7.5-dB gain with typically a 10-dB return loss ove
2 to 14 GHz. This amplifier was designed for small-signal applications, althoug
the output power at 10 GHz at 1-dB gain compression was reported to be +
dBm. Since then a number of papers have been published on distributed amplifie
specifically designed for high-power output [5–9].

### 6.2.1 Small-Signal Analysis

An ideal distributed amplifier is shown in Figure 6.1. Its operation can be unde
stood if each FET is replaced by its simplified equivalent circuit shown in Figu
6.2, where $R_{in}$ is the effective input resistance seen between the gate and sour
terminals and, as can be seen from Figure 1.10, is approximately the sum of $R_g$
$R_i$ + $R_s$. Similarly, $G_{out}$ is the effective output conductance. Assuming for t
moment that $R_{in} = G_{out} = 0$, then each gate-source capacitance $C_{gs}$ is symmetrica

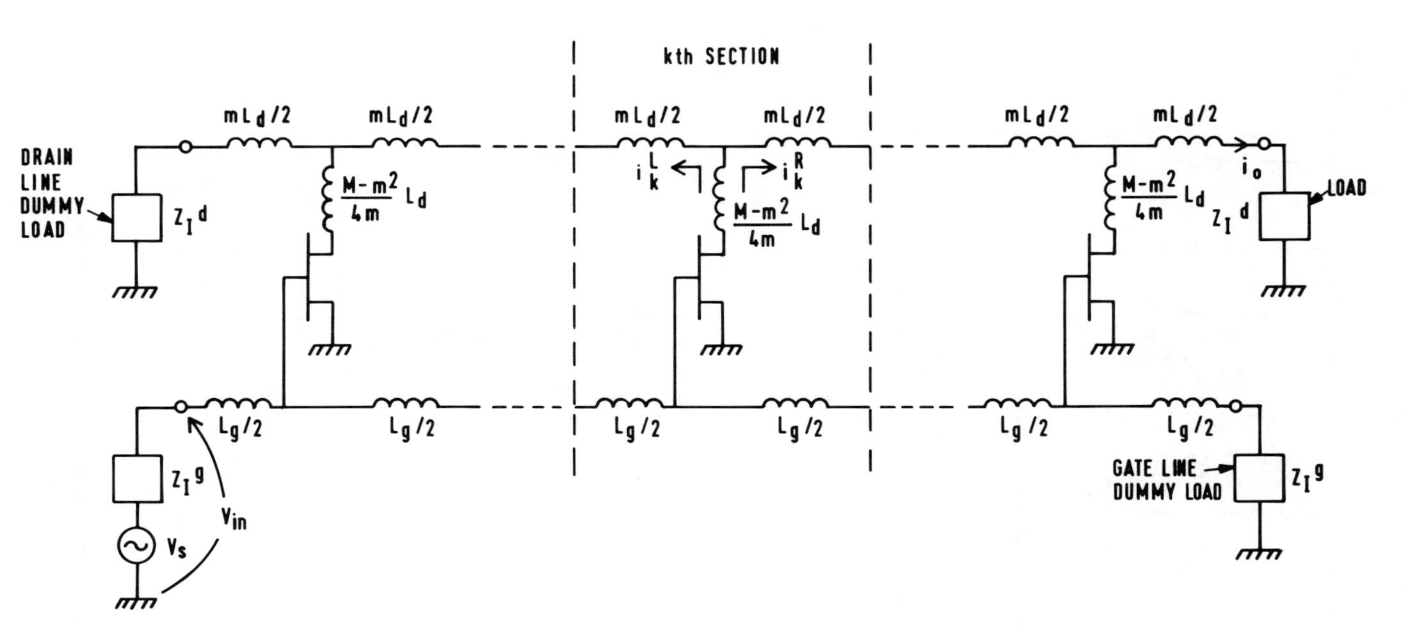


**Figure 6.1** Circuit diagram of an ideal distributed amplifier.

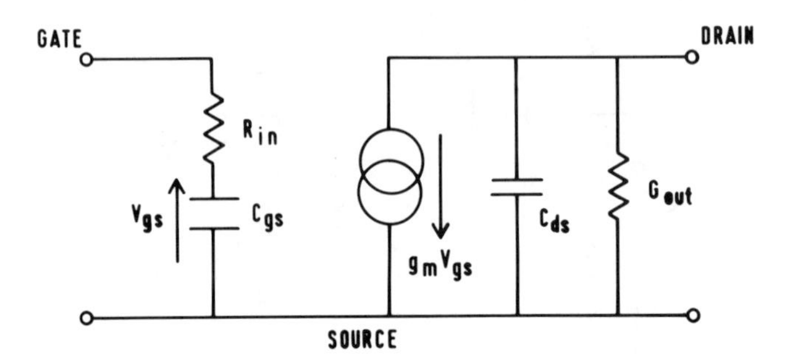


**Figure 6.2** Simplified equivalent circuit of an FET.

embedded between a pair of inductors and forms a constant $K$ filter section [1(
with image impedance

$$Z_I^g = Z_0^g \sqrt{1 - \left(\frac{\omega}{\omega_c^g}\right)^2} \quad (6.1$$

where

$$Z_0^g = \sqrt{\frac{L_g}{C_{gs}}} \quad (6.2$$

and

$$\omega_c^g = 2/\sqrt{L_g C_{gs}} \quad (6.$$

The cascade of all these constant $K$ filter sections forms an artificial transmissio
line of characteristic impedance $Z_I^g$, which is a purely real but frequency-depende
number below the cutoff frequency $\omega_c^g$. If the gate transmission line is terminat
in source and load impedances $Z_I^g$, then the line will be perfectly matched, that i
the distributed amplifier will have $S_{11} = 0$. Assuming for the moment that $M$
$m = 1$ in Figure 6.1, then $L_d$ and $C_{ds}$ form another constant $K$ artificial transmissio
line with characteristic impedance $Z_I^d$ and cutoff frequency $\omega_c^d$. If the drain line
also terminated in source and load impedances $Z_I^d$, then the distributed amplifi
with have $S_{22} = 0$ as well.

Since the gate transmission line is perfectly matched at both ends and lossle
the voltage source $v_s$ will cause a voltage wave to travel down the gate line su

at the magnitude of the voltage across each gate-source capacitance is constant, nd hence each FET will inject the same magnitude of RF drain current into the rain transmission line, half of which will travel to the left and half to the right. /ith proper design, the current components traveling to the right will add up in hase, while the current components traveling to the left tend to cancel at the drain ne dummy load. Thus, strictly speaking, the distributed amplifier is a current ombiner rather than a power combiner.

The power gain of the distributed amplifier, $|S_{21}|^2$, is by definition the ratio f the power delivered to the load to the power available from the source:

$$|S_{21}|^2 = \left| \frac{i_0^2 Z_I^d}{v_s^2/4Z_I^g} \right| \tag{6.4}$$

here $i_0$ is the current flowing through the load and is given by

$$\begin{aligned} i_0 = & \left[ \frac{1}{1 + \dfrac{j\omega C_{ds}}{2}(Z_I^d + j\omega L_d/2)} \right] \{ i_n^R + i_{n-1}^R \exp(-j\theta_d) \\ & + \ldots + i_1^R \exp[-j(n-1)\theta_d] \end{aligned} \tag{6.5}$$

here $\theta_d$ is the drain-line phase change per section, $i_k^R$ is the amount of RF drain urrent emanating from the $k$'th FET flowing to the right, and where the first term brackets accounts for the current division at the output between the load and e internal capacitance of the FET. Straightforward calculation shows that when $= m = 1$, $\theta_d$ is given by

$$\theta_d = \cos^{-1}\left[1 - 2\left(\frac{\omega}{\omega_c^d}\right)^2\right] \tag{6.6}$$

ow

$$i_k^R = (g_m v_{gs}^k)/2, \qquad \text{for } k = 1 \rightarrow n \tag{6.7}$$

d a further straightforward calculation shows that $v_{gs}^k$ is given by

$$v_{gs}^k = \frac{v_s}{2\sqrt{1 - \left(\dfrac{\omega}{\omega_c^g}\right)^2}} \exp[-j(k - \tfrac{1}{2})\theta_g] \tag{6.8}$$

Substituting (6.7) and (6.8) into (6.5) gives

$$i_0 = \frac{g_m v_s}{4\sqrt{1 - \left(\frac{\omega}{\omega_c^g}\right)^2}} \left( \frac{\exp\left[-j\left(n - \frac{1}{2}\right)\theta_g\right]}{1 + j\omega \frac{C_{ds}}{2}\left(Z_I^d + j\omega \frac{L_d}{2}\right)} \right) \{1 + \exp[-j(\theta_d - \theta_g)] + \ldots + \exp[-j(n-1)(\theta_d - \theta_g)]\} \quad (6.$$

Summing the series and substituting for $Z_I^d$ gives

$$|i_0| = \frac{1}{4} \frac{g_m v_s}{\sqrt{1 - \left(\frac{\omega}{\omega_c^g}\right)^2} \sqrt{1 - \left(\frac{\omega}{\omega_c^d}\right)^2}} \left| \frac{\sin \frac{n}{2}(\theta_d - \theta_g)}{\sin \frac{1}{2}(\theta_d - \theta_g)} \right| \quad (6.1$$

and substituting (6.10) into (6.4) results in the gain of the distributed amplifi being given by

$$|S_{21}|^2 = \frac{1}{4} \frac{g_m^2 Z_0^g Z_0^d}{\sqrt{1 - \left(\frac{\omega}{\omega_c^g}\right)^2} \sqrt{1 - \left(\frac{\omega}{\omega_c^d}\right)^2}} \left| \frac{\sin \frac{n}{2}(\theta_d - \theta_g)}{\sin \frac{1}{2}(\theta_d - \theta_g)} \right|^2 \quad (6.1$$

Normally in a distributed amplifier the following conditions are imposed: $Z_0^g$ $Z_0^d = Z_0$ and $\theta_g = \theta_d$. (The necessity for these conditions and their implication discussed in Appendix 6A.) Equation (6.6) shows that $\theta_g = \theta_d$ implies that $\omega_c^g$ $\omega_c^d = \omega_c$, in which case the gain simplifies to

$$|S_{21}|^2 = \frac{1}{4} \frac{(n g_m Z_0)^2}{1 - \left(\frac{\omega}{\omega_c}\right)^2} \quad (6.1$$

Thus the gain increases monotonically with increasing frequency, the gain bei within 1 dB of its dc value up to $\omega = 0.45\omega_c$.

If the same simple model for an FET shown in Figure 6.2 is used to calculate e gain of a conventional balanced amplifier [11], then at sufficiently low fre- iencies the balanced amplifier's gain is given by:

$$|S_{21}|^2 = 4g_m^2 Z_0^2 \tag{6.13}$$

Comparing (6.12) and (6.13) we see that $n$ must equal 4 if a distributed nplifier is to have the same gain (i.e., a distributed amplifier uses twice as many ETs and consumes twice as much dc power as a balanced amplifier for the same in).

We can see from (6.12) that the gain can be increased without limit by using ore FETs in the distributed amplifier. However, while this is true, there is an timum number of FETs, $n_{\text{opt}}$, that maximizes the gain per FET. Therefore, for gh gain levels it is best to cascade a number of distributed amplifiers rather than insert all the FETs into just one distributed amplifier. This optimum number n be determined from

$$\frac{d}{dn}\left(\frac{10 \log_{10}|S_{21}|^2}{n}\right) = 0$$

th $|S_{21}|^2$ given by (6.12). Performing the necessary differentiation results in $n_{\text{opt}}$ ing given by

$$n_{\text{opt}} = \frac{2e}{g_m Z_0}\left[1 - \left(\frac{\omega}{\omega_c}\right)^2\right] \tag{6.14}$$

ere $e = 2.7183\ldots$ For a typical small-signal FET with $g_m = 40$ mS, then $n_{\text{opt}}$ 2.7, that is, three FETs per distributed amplifier is the optimum resulting in a -dB gain at low frequencies. As an application of this result, suppose a 36-dB in amplifier is required, then this can be realized by cascading four distributed plifiers, each using 3 FETs or by using a single distributed amplifier with 63 Ts. Using the optimum configuration results in a saving of 51 FETs!

The reverse gain of the distributed amplifier, that is, the gain referred to the wer dissipated in the drain line dummy load, can be calculated in a similar manner that used to calculate the forward gain. The result is

$$|S_{21}^R| = \frac{1}{4}\frac{(g_m Z_0)^2}{1 - \left(\frac{\omega}{\omega_c}\right)^2}\left(\frac{\sin n\theta}{\sin \theta}\right)^2 \tag{6.15}$$

where it is assumed that the gate and drain transmission lines have equal prop gation constants, i.e., $\theta_d = \theta_g = \theta$. The forward and reverse gains are plotted Figure 6.3 in normalized form for a four-FET distributed amplifier. Except at ve low and very high frequencies, the reverse gain is always 10 dB below the forwa gain.

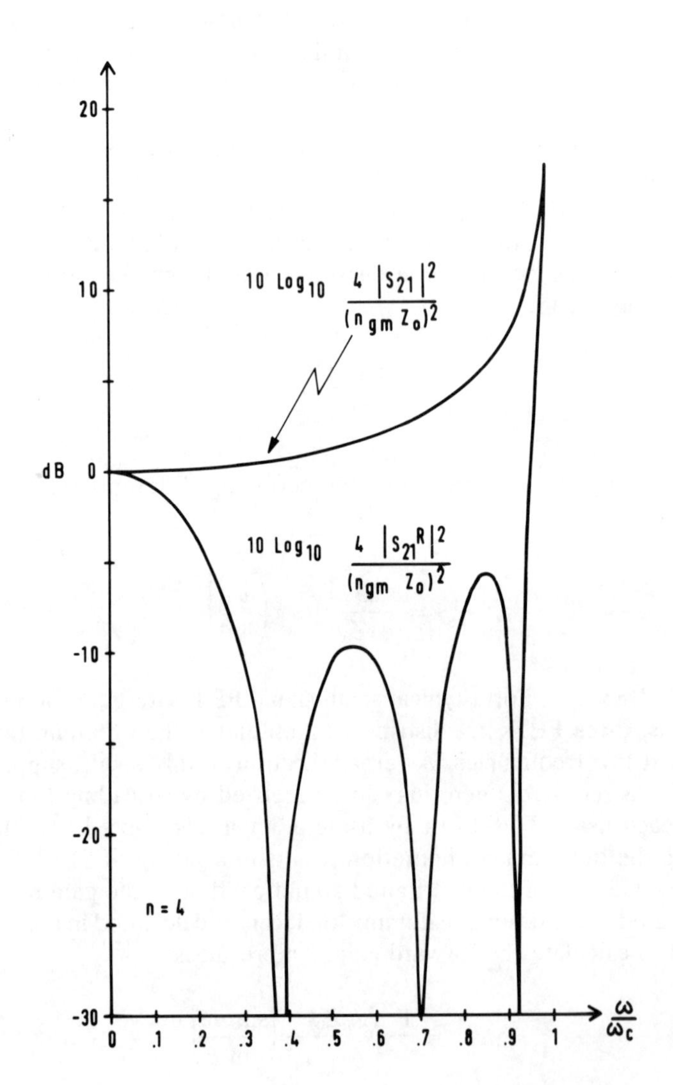


**Figure 6.3** Normalized forward and reverse gain of an ideal distributed amplifier with four FETs.

### .2.2 Effect of Resistive Terminations and Loss Within the FET on Small-Signal Analysis

Jp to this point, we have considered an idealized amplifier with zero internal loss erminated in image impedances. In the real world the gate and drain transmission nes will be terminated at both ends in pure resistances, and the effect of internal oss must also be accounted for. Terminating the drain and gate transmission lines a pure resistances instead of image impedances has two consequences. First, the istributed amplifier will now have nonzero input and output reflection coefficients t all frequencies other than dc; in fact, it can be shown that $|S_{ii}|$ is given by

$$|S_{ii}| = \frac{\left(\frac{\omega}{\omega_c}\right)^2 |\sin n\theta|}{\left\{4\left[1 - \left(\frac{\omega}{\omega_c}\right)^2\right] + \left(\frac{\omega}{\omega_c}\right)^4 \sin^2 n\theta\right\}^{1/2}}, \quad \text{for } i = 1 \text{ or } 2 \tag{6.16}$$

'his equation shows that the reflection coefficient increases steadily with an oscil- atory component superimposed as the frequency increases, and at $\omega = \omega_c$

$$|S_{ii}| = \frac{n}{\sqrt{1 + n^2}} \tag{6.17}$$

hat is, the more sections one has, the closer $|S_{ii}|$ approaches unity near cutoff. 'able 6.1 shows the effective upper frequency limit for an $n$ FET distributed mplifier obtained by numerical solution of (6.16) if a VSWR of 2:1 is taken as he upper acceptable limit. The second and less obvious consequence of terminating he transmission lines in a pure resistance is that the gate voltages are now no

**Table 6.1**
Frequency at Which a Distributed Amplifier's Input and Output VSWR = 2:1

| $n$ | $\omega/\omega_c$ |
|---|---|
| 1 | 0.707 |
| 2 | 0.815 |
| 3 | 0.705 |
| 4 | 0.775 |
| 5 | 0.845 |

longer equal; in fact, it can be shown that the voltage across the $k$'th gate-sourc capacitance in an $n$ FET distributed amplifier is given by

$$\left|v_{gs}^{k}\right| = \left|v_s\right| \frac{\sqrt{1 - \left(\frac{\omega}{\omega_c}\right)^2 \sin^2(n - k + 1/2)\theta}}{4 - 4\left(\frac{\omega}{\omega_c}\right)^2 + \left(\frac{\omega}{\omega_c}\right)^4 \sin^2 n\theta} \tag{6.18}$$

where $v_s$ is the source voltage as shown in Figure 6.1, and $\left|v_{gs}^{k}\right|$ increases wit frequency with a large oscillatory component superimposed unless $k \approx n$. Near $\omega = \omega_c$, very substantial differences in the individual gate-source voltages can occur. For example, $\left|v_{gs}^{4}\right| = 5\left|v_{gs}^{1}\right|$ for $n = 4$ at $\omega = \omega_c$. The unequal gate-source voltag excitation coupled with the mismatched drain transmission line results in consic erable gain ripple. These undesirable effects of gain ripple, bandwidth reductior and degraded port match can be minimized by inserting an impedance-matchin network between the terminating resistances and the artificial transmission line the most common matching network being an $m$-derived half-section [10] with $m = 0.6$. The insertion of such a network also results in all FETs having a near equal gate-source voltage, but this voltage shows a nearly monotonic increase wit frequency. An example of the application of this technique is given in Append 6A. It is an interesting observation that the rising gain characteristic of the ide distributed amplifier when terminated in image impedances as shown in Figure 6 is a consequence of the source and load impedances decreasing as frequenc increases, while the rising gain characteristic of the distributed amplifier whe terminated in pure resistances (with or without $m$-derived half-section matching is a consequence of the gate-source voltage rising with increasing frequency.

More serious in practical applications is the effect of nonzero values for th FET internal resistances, $R_{in}$ and $G_{out}$ in Figure 6.2. The $R_{in}$ causes attenuation of the input voltage wave as it travels down the gate transmission line such that th ac gate-source voltage of the right-hand end FETs in Figure 6.1 is smaller tha that of the left-hand ones with the effect being more pronounced as the frequenc increases, while $G_{out}$ causes a fraction of the output current to be dissipated with the FET rather than in the external load, even at very low frequencies. An analys of the effect of nonzero values of $R_{in}$ and $G_{out}$ can be found in [12, 13] and graphic interpretations of these analyses can be found in [13–15]. In practical applicatio gate-line loss is a more serious problem than drain-line loss, but steps need to b taken in the design of a distributed amplifier to minimize the detrimental effec caused by both loss mechanisms.

As far as the gate line is concerned the objective is to ensure that the gat source capacitance associated with each FET has the same magnitude of ac volta dropped across it despite the presence of gate-line loss. This can be achieved vo ways. The first method is to taper the characteristic impedance of the gate ansmission line. The principle behind this loss compensation mechanism can be :adily understood by considering an ideal lossless tapered transmission line con- ecting a source $Z_s$ to a load $Z_L$ with $Z_L > Z_s$. If $v_{in}$ is the voltage across the input f the transmission line and $v_{out}$ the voltage across the load, then, $v_{out}^2/Z_L = v_{in}^2/Z_s$ nce the line is lossless, and hence $v_{out} > v_{in}$ if $Z_L > Z_s$. With the correct choice f taper profile it is possible to achieve a balance at each FET between the voltage crease as a result of tapering and the voltage decrease due to attenuation and us ensure that the magnitude of the voltage across the gate-source capacitance each FET is approximately the same. Tapering of the gate line can be imple- ented in practice as shown in Figure 6.4(a) by placing an additional capacitance parallel with each FET and progressively reducing its value for each successive ET to raise the characteristic impedance of the gate transmission line. It is also ecessary to vary the value of inductance between each FET in order to achieve approximately equal phase change between each section of the gate line. The sadvantage of this method is that the cutoff frequency of the gate transmission e is reduced.

The second method is to insert a capacitor in series with each FET, as shown Figure 6.4(b), and to progressively increase its value for each successive FET , 7]. Of course, a high value resistor has to be connected in parallel with the ries capacitor in order to be able to apply dc bias to the gate. Each series capacitor d gate-source capacitance combination forms a potential divider; hence, pro- essively increasing the value of series capacitance enables a balance to be achieved each FET between the voltage increase as a result of series capacitance tapering d the voltage decrease due to attenuation. Once again, this ensures that the agnitude of the voltage across the gate-source capacitance of each FET is approx- ately the same but, as before, it is also necessary to taper the series inductors as to ensure that the phase change between each section of the gate line is proximately equal. This form of tapering results in the characteristic impedance the gate transmission line progressively decreasing toward the gate-line dummy ad. However, this technique has one major disadvantage, namely, that the poten- l divider action causes a substantial loss of gain, which can only be recovered in actice by increasing the gatewidth of each FET with a consequent increase in e amplifier's current consumption. Hence this technique is normally used only distributed amplifiers designed for high-power output where the increase in tewidth is very beneficial. This effect is discussed in further detail in Appendix . Note that with both techniques that exact equalization of the voltages can only achieved at a spot frequency.

Unlike gate-line loss, drain-line loss cannot be compensated for, only mini- zed. Minimization of the drain-line loss requires the use of a FET having a very w value of $G_{out}$ or using a multiple FET combination with a very low effective lue of $G_{out}$. The former option is difficult to implement in practice because it is

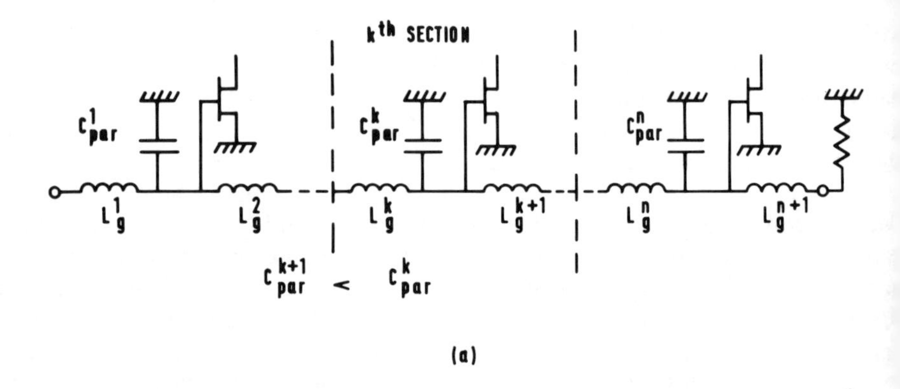


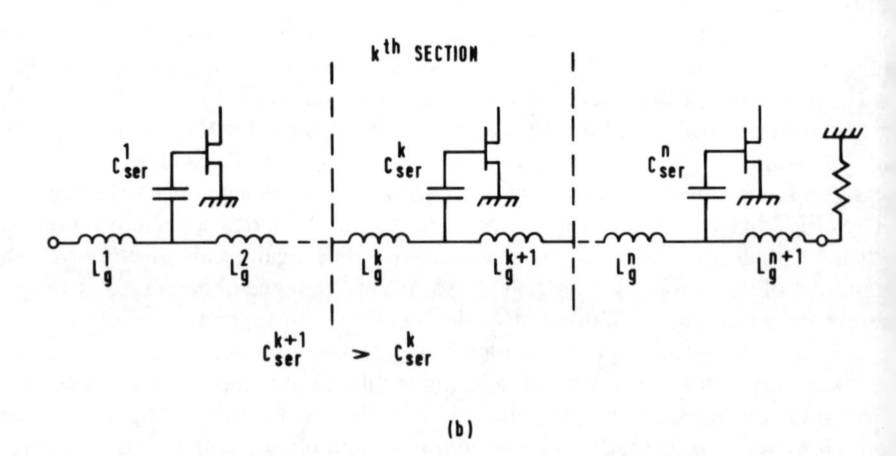


**Figure 6.4** Two methods of tapering the gate-line impedance: (a) insertion of an additional sh capacitance in parallel with the gate and (b) insertion of a series gate capacitor.

very difficult to control the value of $G_{out}$ during manufacture, hence the lat method is the preferred option. This can be implemented using either a casco FET combination [16, 17], that is, a common-source FET in cascode with a co mon-gate FET or a dual-gate FET [18] since a dual-gate FET is equivalent to cascode FET. The output conductance of a cascode FET combination is appro nately an order of magnitude lower [17] than that of a common-source FET and onsequently drain-line attenuation is substantially reduced and the gain of the istributed amplifier increased. The technique also provides three additional benfits, namely, the second gate can be used to provide *automatic gain control* (AGC); ne effective feedback impedance between the drain and gate transmission lines is ubstantially increased, which eases the design and results in lower gain ripple; and ne low-frequency output VSWR is improved. If a cascode FET combination is sed rather than a dual-gate FET then a series inductance can be incorporated etween the common-source and common-gate FETs to provide some beneficial igh-frequency gain peaking [16, 17].

### .2.3 Large-Signal Analysis

o consider the large-signal properties of the distributed amplifier, we must first alculate the drain-line voltage distribution [19]. For simplicity, assume that $M =$ $= 1$ in Figure 6.1 and that gate- and drain-line phase equalization is achieved y placing an additional capacitance to ground at each drain as discussed in Appenix 6A. Each FET injects a current of value $g_m v_{gs}^k$ into the drain line, half of which avels to the left and half to the right since the drain of each FET sees an impedance $Z_I^d$ in both directions. The voltage at the drain of each FET is readily calculated y replacing the left- and right-hand sections of the drain line by their Norton quivalents. Thus, the drain-source voltage across the $k$'th FET is given by

$$v_{ds}^k = \frac{Z_I^d}{2}\Big\{ i_1^R \exp[-j(k-1)\theta] + \ldots + i_{k+1}^R \exp(-j\theta) + i_k^R + i_k^L + i_{k+1}^L \exp(-j\theta) + \ldots + i_n \exp[-j(n-k)\theta]\Big\} \quad (6.19)$$

here $i_k^R = i_k^L = (g_m v_{gs}^k)/2$ with $v_{gs}^k$ given by (6.8). Thus,

$$v_{ds}^k = \frac{Z_I^d}{2} g_m v_{gs}^1 \sum_{i=1}^{n} \exp[-j(i - 1 + |k - i|\theta)] \quad (6.20)$$

his series can be summed in closed form as

$$v_{ds}^k = \frac{Z_I^d}{2} g_m v_{gs}^1 \exp[-j(k-1)\theta]\left\{k + \frac{\exp(-j2\theta) - \exp[j2(n-k+1)\theta]}{1 - \exp(-j2\theta)}\right\} \quad (6.21)$$

Straightforward but lengthy trigonometric manipulations show that $|v_{ds}^k|$ is given b

$$|v_{ds}^k| = \frac{Z_l^d}{2} g_m v_{gs}^1 \sqrt{k^2 + \frac{2k \cos(n - k + 1)\theta \sin(n - k)\theta}{\sin\theta} + \frac{\sin^2(n - k)\theta}{\sin^2\theta}} \quad (6.2$$

The voltage distribution along the drain line as given by (6.22) is plotted in no malized form for the case $n = 4$ in Figure 6.5, where we see that a very larg variation in the magnitude of the drain-source voltage with frequency occurs fc the early FETs. However, it can be shown from (6.22) that at all frequencies

$$|v_{ds}^k| \geq |v_{ds}^{k-1}|, \qquad \text{for } k = 2 \rightarrow n \quad (6.2$$

and hence at all frequencies the largest drain-source voltage appears across the la FET. The maximum linear power output from the distributed amplifier under Cla A conditions thus occurs when $|v_{ds}^n|$ assumes its greatest value, which is given b (1.3) as

$$|v_{ds}^n| = v_{dspeak} = \frac{V_{dgB} - |V_p| - V_k}{2} \quad (6.2$$

Under these conditions the maximum linear RF output power from the distribute amplifier is given by

$$P_{\text{RFmax}} = \frac{(V_{dgB} - |V_p| - V_k)^2}{8Z_l^d} \quad (6.2$$

Regardless of the number of FETs used, $P_{\text{RFmax}}$ is the maximum linear outp power from a distributed amplifier having the topology shown in Figure 6.1 and is determined principally by $V_{dgB}$ and $Z_l^d$. For a typical FET having $V_{dgB} = 20$V, $V_p = 2.5$V, and $V_k = 1$V, then $P_{\text{RFmax}} = 0.7$W if $Z_l^d = 50\Omega$. Thus, distribute amplifiers are restricted to less than 1-W power levels in a 50-$\Omega$ system [7]. Th only way to circumvent this limitation is to use a very low value for $Z_l^d$, whic results in a corresponding reduction in gain and then to cascade the distribute amplifier with a broadband impedance transformer, or to develop an FET structu with a very high breakdown voltage, but this requires a technological advance.

Equation (6.25) is the most fundamental large-signal limitation for a distri uted amplifier, but the power output may be limited to a lower level than th given by (6.25) with an improper design. For example, the distributed amplifi

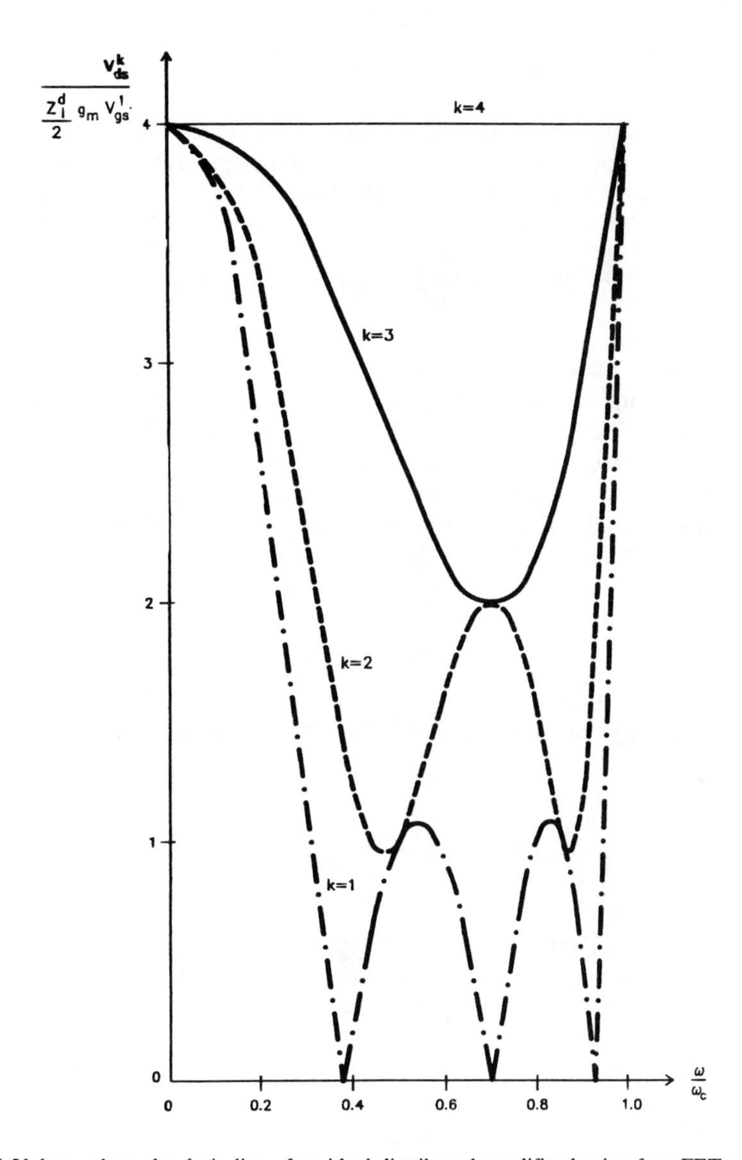


**gure 6.5** Voltage along the drain line of an ideal distributed amplifier having four FETs.

ust be capable of supplying sufficient RF output current into the load to support e voltage level given by (6.24). This requires each FET to have a minimum open-annel current of

$$I_F \geq 2\,(V_{dgB} - |V_p| - V_k)/nZ_I^d \tag{6.26}$$

for an $n$ FET design with maximum efficiency occurring when equality is achieve in (6.26). For the preceding example, $I_F = (0.66/n)$ amperes. Eliminating $Z_I^d$ fro (6.25) and (6.26) shows that the maximum linear RF power output is given by

$$P_{\text{RFmax}} = \frac{n}{16} I_F\,(V_{dgB} - |V_p| - V_k) \tag{6.27}$$

Comparing the value given by (6.27) with the corresponding value for a singl ended Class A amplifier as given by (1.4) shows that the maximum linear powe output and, hence, efficiency of the distributed amplifier is half that achievab from lossless power combining of the individual FETs.

To achieve the voltage and current modulation implied by (6.27), the gat source voltage of each FET must swing between $V_{gs} = -|V_p|$ and $V_{gs} = 0$, whic means that the RF input power to the amplifier must be

$$P_{\text{in}} = \frac{V_p^2}{8Z_I^g} \tag{6.28}$$

and hence the large-signal gain, which is obtained from the ratio of (6.25) ar (6.28), is given by

$$|S_{21}|^2 = \left(\frac{V_{dgB} - |V_p| - V_k}{V_p}\right)^2 \tag{6.29}$$

assuming $Z_I^g = Z_I^d$. That is, the gain of a distributed amplifier designed for ma imum power output and efficiency is determined only by the dc voltage paramete of the FET and is not an independently specifiable parameter. For a typical FE having $V_{dgB} = 20\text{V}$, $V_p = 2.5\text{V}$, and $V_k = 1\text{V}$, then the large-signal gain is 8 dB, although losses within the amplifier will result in a lower gain level. The larg signal gain will only be identical to the small-signal gain given by (6.12) if the FE has a linear $I_{ds}$ versus $V_{gs}$ characteristic.

It is of interest to examine the design of a distributed amplifier for maximu power and efficiency from a different viewpoint. Consider initially a distribut amplifier with just one FET. Then the FET will see a load impedance of $Z_I^d/2$, b for maximum power output and efficiency this load impedance must be identi to the optimum load impedance given by (1.10), and hence

$$Z_I^d/2 = (V_{dgB} - |V_p| - V_K)/I_F \tag{6.30}$$

For a typical FET having $V_{dgB} = 20\text{V}$, $V_p = 2.5\text{V}$, and $V_K = 1\text{V}$, then $I_F = 0.66\text{A}$ if $Z_I^d = 50\Omega$. From (1.4) such an FET has a maximum RF output power of 1.4W, but in the single-FET distributed amplifier this power divides equally between the external load and the drain-line dummy load. Hence, it is easily demonstrated that the single-FET distributed amplifier has only half the power output and efficiency capability of a single-ended Class A amplifier. Such a FET is likely to have a gate-source capacitance of about 2 pF, and hence from (6.2) and (6.3) the cutoff frequency of the gate and drain transmission lines will be about 3 GHz. The bandwidth of the amplifier can be increased $n$-fold by using an $n$ FET distributed amplifier with each FET having $1/n$'th the gatewidth of the single FET design. The seemingly simple step of replacing one large FET by $n$ smaller FETs has several interesting consequences. First, the power dissipated in the drain-line dummy load is now no longer identical in value to the power delivered to the external load because it depends on the vector addition of the current contribution from the various FETs. In fact, as shown in Section 6.2.1 the power dissipated in the dummy load is now substantially lower than that in the external load except at very low frequencies and near cutoff. Second, the maximum output power that can be delivered to the external load remains the same since the dc voltage parameters of an FET are not affected by gatewidth scaling. The FETs preceding the output device have to inject the correct magnitude and phase of current into the drain line to ensure that the drain-line voltage wave grows as shown in Figure 6.5, and thus ensure that the drain-source voltage of the output device swings between $V_K$ and $V_{dgB} - |V|_P$, otherwise the output power will be lower than that given by (6.25). Third, the FETs do not contribute equally to the output power. The power generated by each FET is given by

$$P_k = 1/2 \, \text{Re}[v_{ds}^k \, (i_k^L + i_k^R)] \tag{6.31}$$

Making the necessary substitutions from (6.7), (6.8), and (6.21), then

$$P_k = Z_I^d \, (g_m v_{gs}^1)^2 \left[ k + \frac{\cos\,(n - k + 1)\theta \, \sin\,(n - k)\theta}{\sin\,\theta} \right] \tag{6.32}$$

The power generated by each FET as given by the preceding expression is plotted in normalized form for the case $n = 4$ in Figure 6.6. Also plotted in normalized form is the sum of the power dissipated in the external and dummy loads given by

$$P_T = \frac{Z_I^d}{8} \, (g_m v_{gs}^1)^2 \left[ n^2 + \left( \frac{\sin\,n\theta}{\sin\,\theta} \right)^2 \right] \tag{6.33}$$

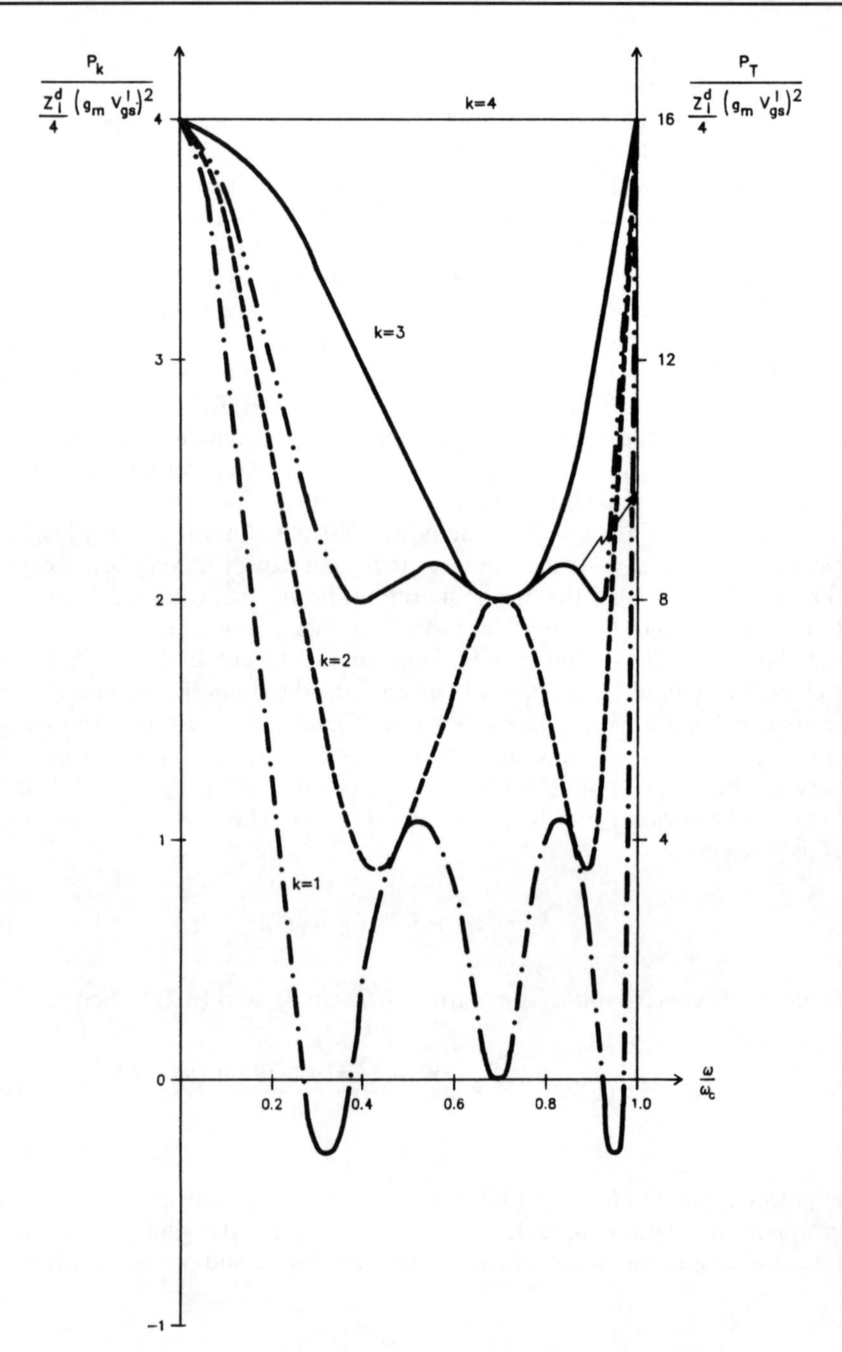


**Figure 6.6** Total power dissipated in the external loads and the power generated by each FET in an ideal distributed amplifier with four FETs.

d it is readily verified from Figure 6.6 that at any frequency

$$P_T = \sum_{k=1}^{4} P_k$$

gure 6.6 shows that over a small range of frequencies the first FET actually *sorbs* power rather than generating it. The last FET generates almost half the tput power over most of the frequency range, that is, $0.3 \leq \omega/\omega_c \leq 0.95$. The rd FET never contributes less than 25% of the output power over the same quency range, while the first FET contributes at most 12% and generally much s than this. Obviously, the second FET makes up the balance. This very non-iform power generation distribution is a direct consequence of the growing volt-e magnitude along the drain line as a result of each FET injecting a current into line. An alternative but entirely equivalent way of interpreting this analysis is consider the load line seen by each FET [20]. Each FET has the same current ing but very different magnitudes and phases of voltage swing, which are fre-ency dependent, with the result that each FET operates along a different load e. It should be noted that the load line the FET operates along is the electronic d line [20], not the circuit element value of $Z_I^d/2$. In this respect the distributed plifier method of power combining is fundamentally different from that in con-ntional power combining because in the latter there is no interaction between FETs, at least in principle. Therefore, the load line seen by any FET is deter-ned only by circuit element values, whereas in the distributed amplifier the load e seen by an FET is determined not only by the circuit but also by the operating rformance of all the other FETs.

Finally, tapering of the drain line has been used [21, 22] to force more of the ected current to travel to the right than to the left. Tapering of the drain line not increase the output power from a distributed amplifier, assuming that the tput impedance remains unchanged, but it can increase the efficiency by requiring ver FETs and, therefore, less dc current to achieve a given output power. This hnique requires a step discontinuity in the drain-line characteristic impedance occur at each FET, and hence this technique results in degraded output VSWR er wide bandwidths.

## PASSIVE POWER COMBINER/DIVIDER NETWORKS

the other methods of power combining to be described in this chapter depend the properties of passive power combining networks, which are briefly reviewed e. A detailed derivation of the properties can be found in the cited references. ce these networks are reciprocal, the same network can be used either as a ver combiner or a power divider, and henceforth the network will be described either as a combiner or a divider depending on the context. Power divider/combin networks can be realized in all of the main microwave transmission media, such microstrip, stripline, fin-line, waveguide, but microstrip and stripline are the norm media when the device is used for combining high-power GaAs FET amplifiers.

The simplest type of power combiner/divider is the two-way splitter in whi a fraction of the incident signal appears out of one port with the remainder of t incident signal appearing out of another port. Ideally, the splitter should be lossle reciprocal, perfectly isolated, and frequency independent. However, it is a we known fact that it is impossible to construct a lossless, reciprocal, perfectly match three-port network [23]. Hence, the two-way splitter is realized either as a lo three-port or a lossless four-port device. In the ideal case, no power emerges fr the fourth port so it is termed the *isolated port*. The lossless four-port network c be considered to be a lossy three-port network if the load connected to the isolat port is regarded as an internal resistance rather than an external load. Note th the term *lossy* refers to the presence of resistive elements in the network, but th may not necessarily be any power dissipated in the resistive elements.

For the purposes of this chapter, two-way power splitters can be categoriz according to the phase relationship between the two output signals as being eit in-phase or quadrature-phase splitters. One can also produce power splitters t have the two output signals in antiphase, but while such power splitters find wi spread use in balanced mixers and push-pull amplifiers (see Sec. 1.2.5) they seldom if ever used in power combining. Two-way in-phase power dividers will considered first followed by quadrature splitters and finally $N$-way splitters.

### 6.3.1 Two-Way In-Phase Power Combiner/Divider Networks

The Wilkinson splitter [24] is the simplest and most widely used two-way in-ph power divider/combiner network and is shown in Figure 6.7. It is easily designed a fabricated and it has a wide bandwidth capability. Furthermore, due to the inher symmetry of the structure, zero amplitude and phase difference exist between the t outputs. The $S$ parameters of the Wilkinson splitter are readily calculated by p forming an even-odd mode analysis and can be shown to be given by [25]

$$
\begin{aligned}
|S_{11}|^2 &= \frac{\cos^2\theta}{8 + \cos^2\theta} \\
|S_{22}|^2 = |S_{33}|^2 &= \frac{\cos^4\theta}{(8 + \cos^2\theta)\,(1 + 7\sin^2\theta)} \\
|S_{12}|^2 = |S_{13}|^2 &= \frac{4}{8 + \cos^2\theta} \\
|S_{23}|^2 &= \frac{4(1 + \sin^2\theta)\cos^2\theta}{(8 + \cos^2\theta)\,(1 + 7\sin^2\theta)}
\end{aligned}
\qquad (6.
$$

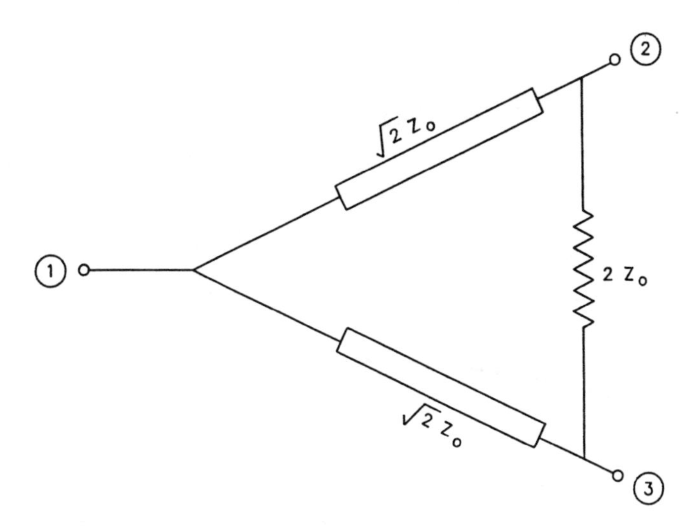


gure 6.7 Two-way Wilkinson power splitter.

here $\theta$ is the electrical length of the transmission lines shown in Figure 6.7. quation (6.34) assumes that the characteristic impedance of the transmission lines $\sqrt{2}Z_0$ and that the resistor connecting ports 2 and 3 has a resistance of $2Z_0$ ohms. he above expressions are shown plotted in Figure 6.8 from which it can be seen at the bandwidth of the device is determined by the minimum acceptable isolation nd port return loss rather than by the coupling, which shows very little variation ith frequency. A wider bandwidth can be achieved using a multisection design 25].

In practical applications ports 2 and 3 will be terminated in imperfect loads. -phase power combiner/divider networks suffer from one major problem under ese circumstances—the resulting input VSWR seen at port 1 will, in general, be orse than the VSWR of the load. Thus the amplifiers to be combined using an -phase power combiner/divider must themselves have an excellent input and utput VSWR if the resulting power-combined amplifier is to have a good VSWR.

In some applications one may want to implement the power divider/combiner s part of a GaAs MMIC, in which case the use of distributed circuit elements to alize the Wilkinson splitter is precluded except at very high frequencies by chip ze considerations. Therefore, in these circumstances the Wilkinson splitter is often alized using lumped elements [26]. However, monolithic lumped-element Wilkson splitters have substantially higher excess insertion loss than their nonmonothic distributed counterparts.

The rat-race coupler [27] is another type of in-phase power splitter, but it is substantially narrower bandwidth device in its standard form than the Wilkinson litter with no compensating advantages. The bandwidth of the rat-race coupler

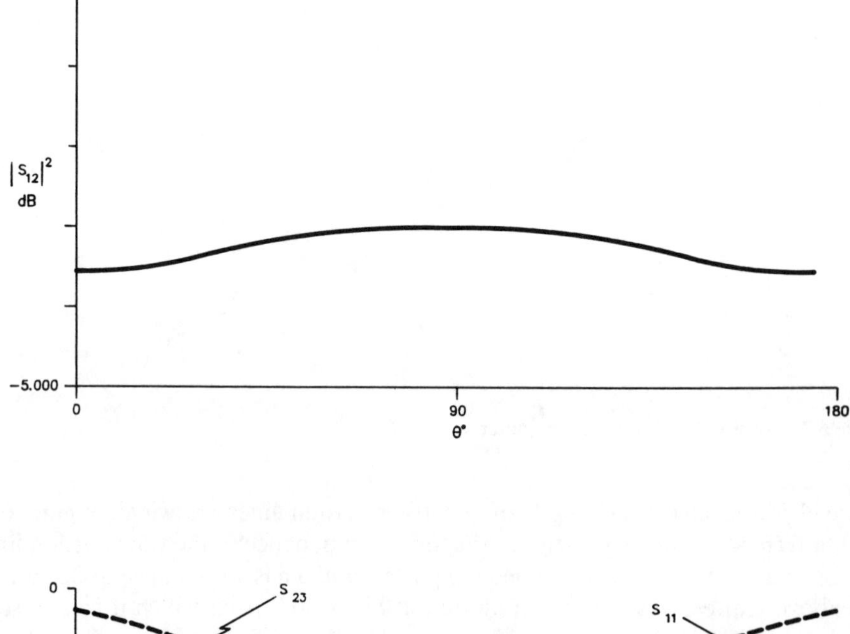


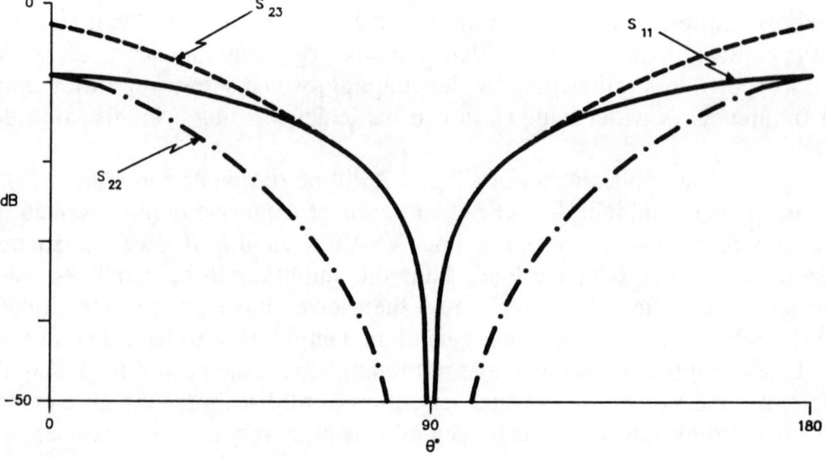


**Figure 6.8** Performance of an ideal Wilkinson power splitter.

an be increased to about an octave [28] but then it is substantially more difficult ) fabricate because of the requirement for a pair of tightly coupled lines.

## .3.2 Two-Way Quadrature-Phase Power Combiner/Divider Networks

'he simplest type of quadrature power divider to fabricate is the branch-arm coupler ?9] shown in Figure 6.9. The element values shown are for a 3-dB splitter, but an nequal power split can be achieved with different element values. Unlike the Vilkinson splitter case, an even-odd mode analysis of this seemingly simple structure results in very complex expressions for the scattering parameters and hence hese will not be given. However, the performance of the branch-arm coupler is eadily determined from any standard microwave circuit analysis program and the esults are given in Figure 6.10 for the 3-dB equal power split case. It should be oted that a 90-deg phase difference between the two outputs occurs only at the enter frequency. As can be seen from Figure 6.10, this device has a usable total andwidth of about 20%, but the bandwidth can be significantly enhanced by using nultiple branches [29].

The parallel coupled-line directional coupler shown in Figure 6.11 is a more ompact and broader bandwidth type of quadrature power divider than the branch-rm coupler. It also has the advantage that, theoretically, the phase difference etween the two outputs is exactly 90 deg at all frequencies and, theoretically, it as infinite isolation and perfect match at all ports at all frequencies. The disad-

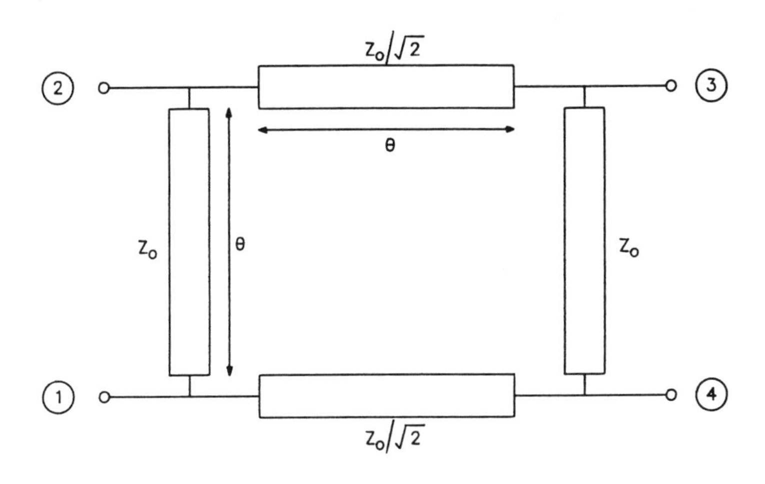


gure 6.9 Branch-arm coupler.

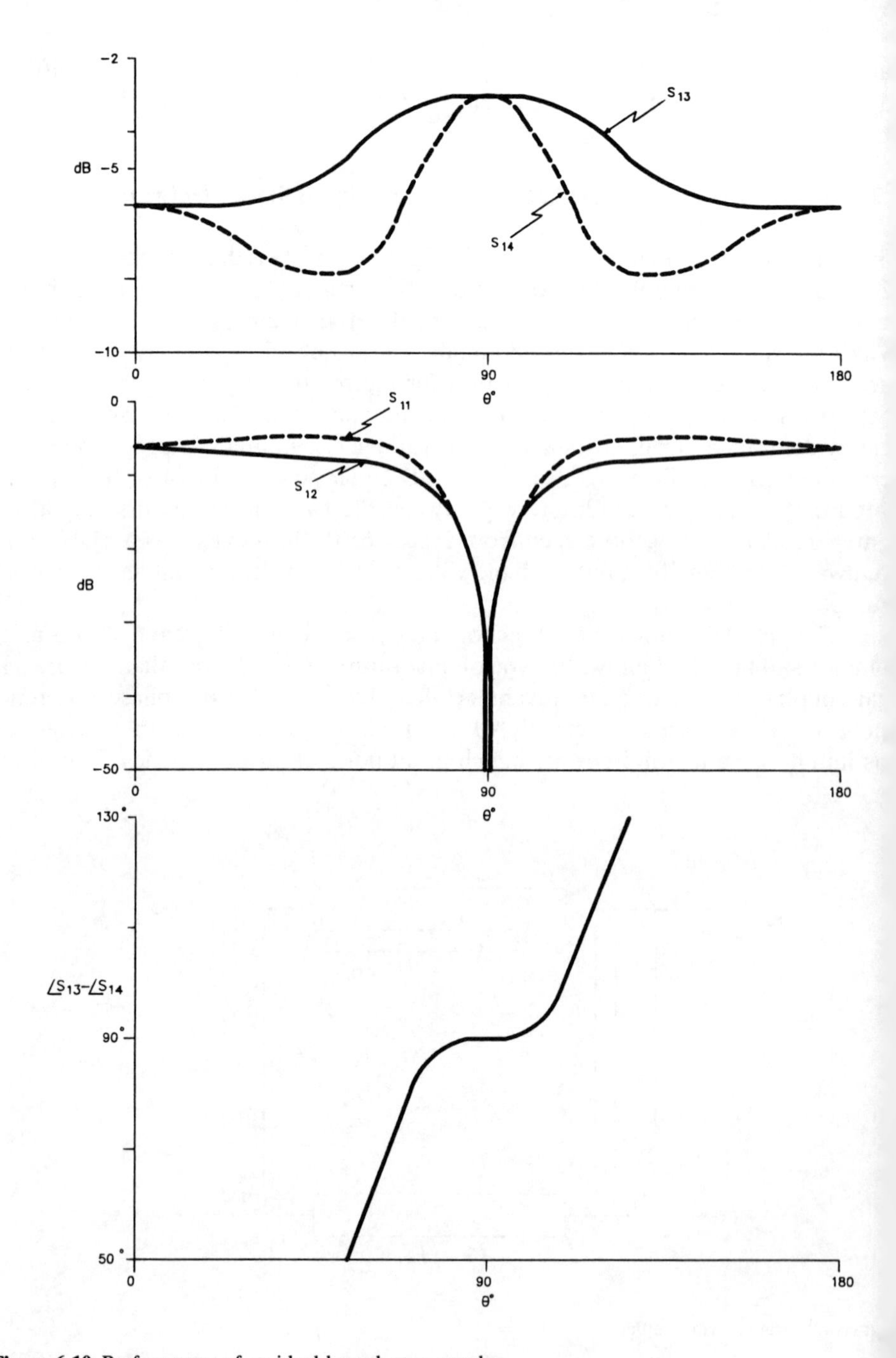


**Figure 6.10** Performance of an ideal branch-arm coupler.

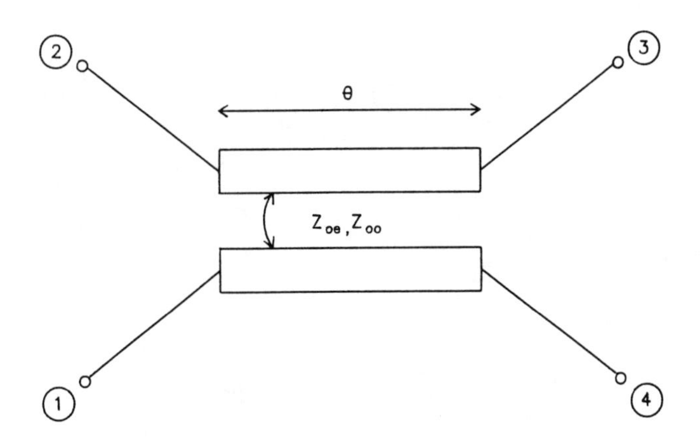


ure 6.11 Parallel coupled-line directional coupler.

ntage is that it is more difficult to fabricate when an equal or nearly equal power lit is required if realized in stripline or microstrip.

The $S$ parameters of the parallel coupled-line directional coupler are readily lculated by performing an even-odd mode analysis [30]. Provided that the even- d odd-mode characteristic impedances, $Z_{0e}$ and $Z_{00}$, are such that $Z_0 = Z_{0e}Z_{00}$, en the $S$ parameters are given by

$$
\begin{aligned}
S_{ii} &= 0,\ i = 1 \rightarrow 4 \\
S_{12} = S_{34} &= \frac{jk \sin \theta}{\sqrt{1 - k^2} \cos \theta + j \sin \theta} \\
S_{13} = S_{24} &= 0 \\
S_{14} = S_{23} &= \frac{\sqrt{1 - k^2}}{\sqrt{1 - k^2} \cos \theta + j \sin \theta}
\end{aligned}
\tag{6.35}
$$

here $k = (Z_{0e} - Z_{00})/(Z_{0e} + Z_{00})$. Thus the coupling at band center is given by $_{12}|^2 = k^2$. The variation in coupling with frequency is shown in Figure 6.12 from hich it can be seen that the device has a usable total bandwidth of about an octave the coupling is required to be within 3 ± 1 dB. The bandwidth can be increased 3:1 if the midband coupling is set at 2 dB instead of 3 dB [31].

For an equal power split (i.e., $k^2 = 0.5$), $Z_{0e} = 121\Omega$ and $Z_{00} = 21\Omega$ in a -$\Omega$ system, which requires a large capacitance per unit length between the two upled lines but a small capacitance per unit length to ground, that is, a pair of rrow lines that are extremely close together, the separation being a more sig-

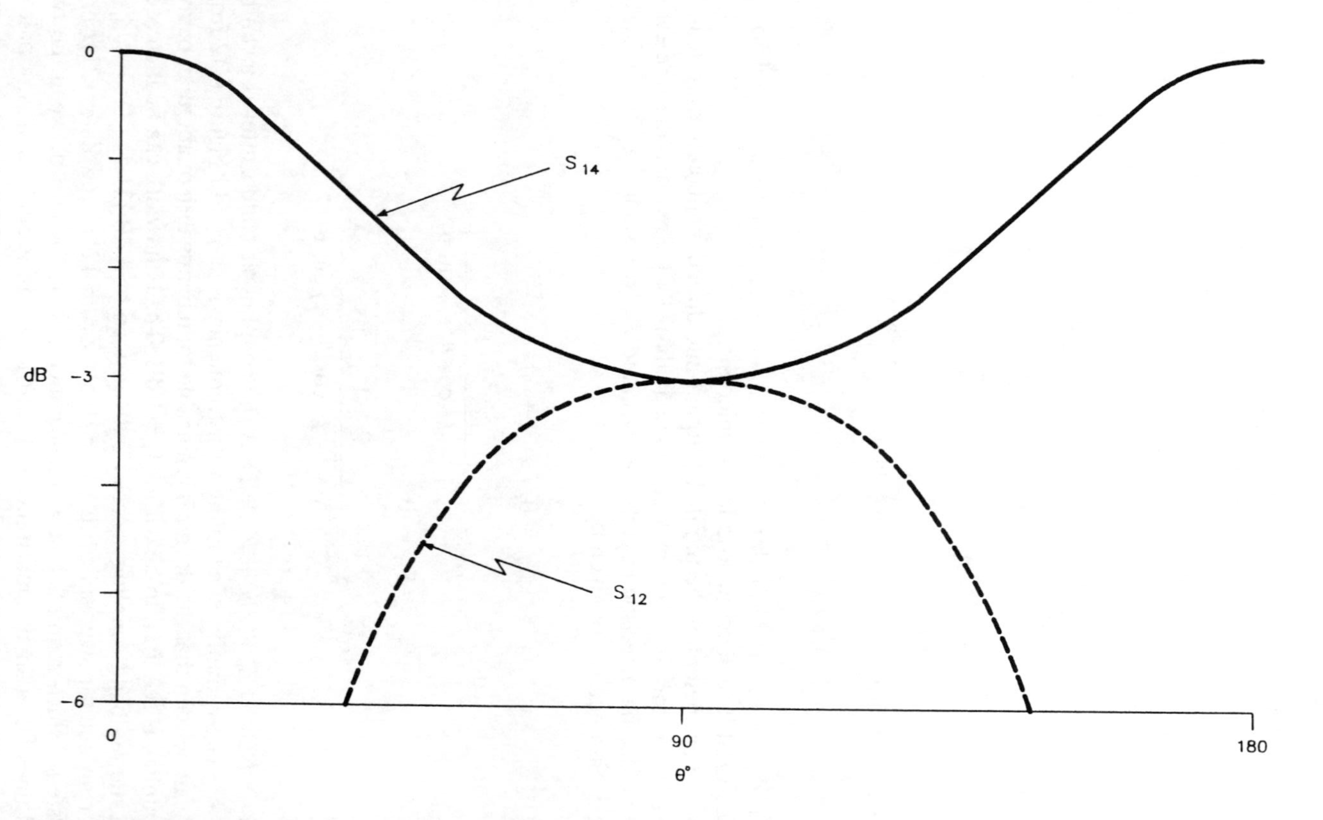


**Figure 6.12** Performance of an ideal parallel coupled-line directional coupler.

nificant problem than the width of the tracks. For a given interline capacitance, he separation between the conductors can be doubled by using the arrangement shown in Figure 6.13(a) if the two outside conductors are connected in parallel. Obviously the outer lines have to be narrower than the inner one in order to naintain the same capacitance to ground. The separation between the conductors can be increased still further, with a further narrowing of the conductors, by connecting another line in parallel as shown in Figure 6.13(b). This process of connecting more lines in parallel to increase the separation can be continued [32], but t results in the widths of the conductors being too narrow rather than the separation oo small. Two pairs of lines connected in parallel as shown in Figure 6.13(b) is generally considered to be the optimum. In many practical applications, one would ike the two output ports to appear on the same side of the coupler and so a crossover is often introduced halfway along the coupler as shown in Figure 6.13(c). The configuration shown in Figure 6.13(c) is known as a *Lange coupler* [33]. Finally, umped-element realizations need to be considered if the coupler must be realized nonolithically due to chip size considerations.

### 6.3.3 *N*-Way Power Combiner/Divider Networks

The two-way in-phase Wilkinson splitter described in Section 6.3.1 is in fact a particular case of the *N*-way Wilkinson power divider [24] shown in Figure 6.14. Figure 6.14 demonstrates that all the output ports are indistinguishable from one another, that is, the device is completely symmetric, so absolute phase and amplitude balance exists between all the outputs, but very high impedance transmission ines are required for large values of *N* unless an external quarter-wave impedance transformer is incorporated [34]. It can be shown [35] that the bandwidth over which acceptable isolation occurs increases as the value of *N* increases, but the nput VSWR is degraded. For example, the two-way Wilkinson has 20-dB isolation over a total bandwidth of 36% with the input VSWR $<$ 2:1 over more than two octaves, while an eight-way Wilkinson achieves the same isolation over a 100% otal bandwidth but the same input VSWR is only achieved over a 30% bandwidth.

A modification that can be made to the structure in Figure 6.14 is to replace the star network of resistors with a ring structure [36]. The microwave performance of the resulting structure is unaltered for $N = 3$ provided the resistors have the correct value, but for $N > 3$ the performance is degraded and the structure is no onger symmetric with the result that the isolation depends on which pair of ports t is being measured between, the worst case isolation occurring between adjacent output ports. The radial combiner [36, 37] is a practical realization of this configuration.

The main problem with the *N*-way Wilkinson and radial combiner/divider networks is that they cannot be realized with a planar construction. One obvious

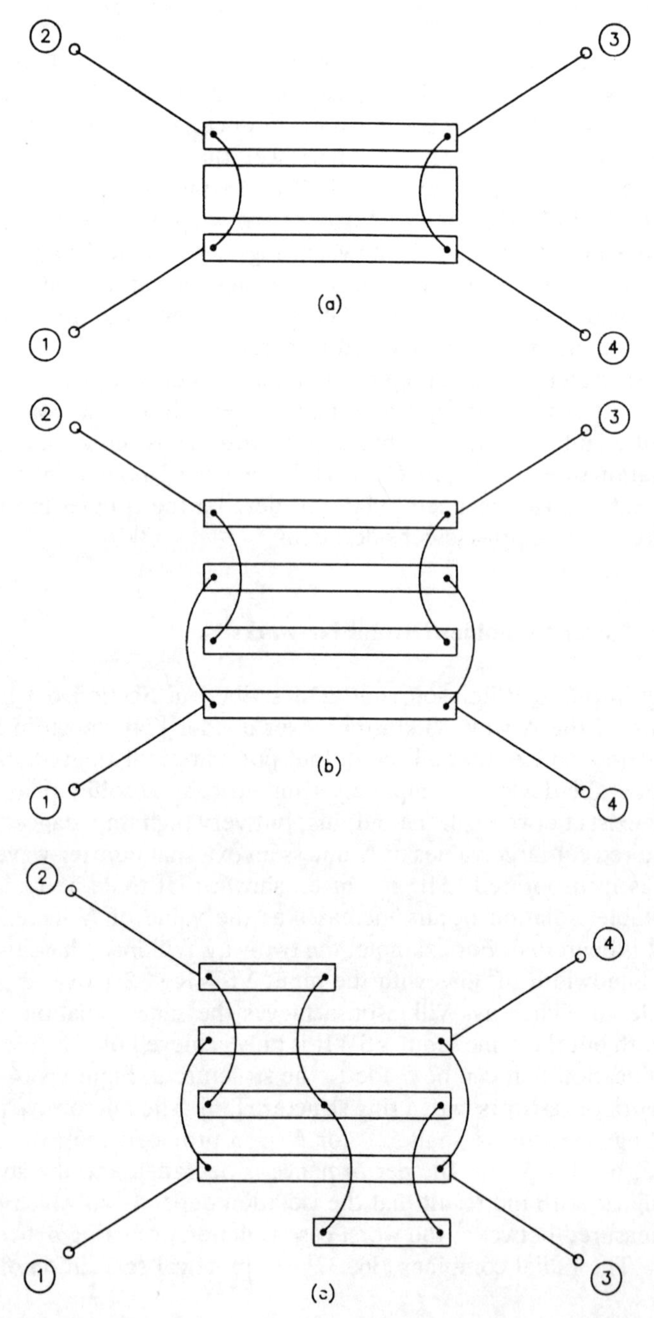


**Figure 6.13** Methods of easing the fabrication of the parallel coupled-line directional coupler in m crostrip: (a) connecting an additional line in parallel, (b) connecting two additional lin in parallel, and (c) connecting two additional lines in parallel with a crossover, the Lang coupler. [33]

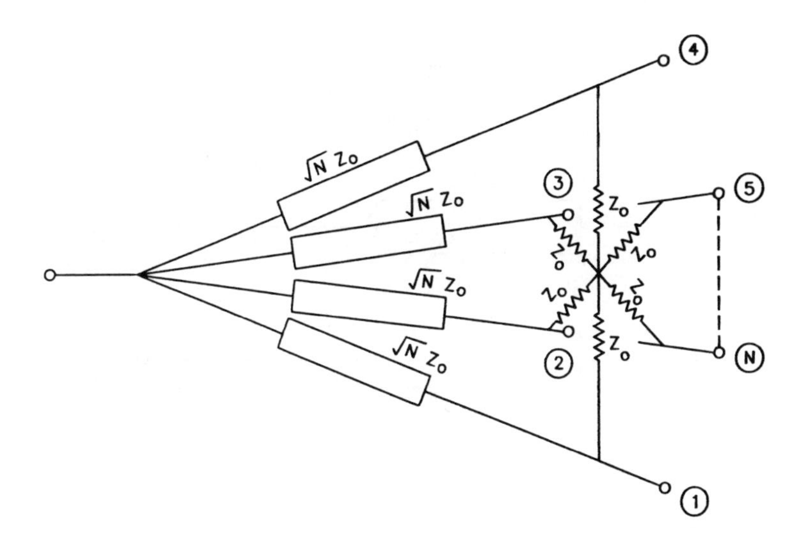


igure 6.14 An $N$-way Wilkinson power splitter.

olution to this problem is to cascade a number of two-way Wilkinson splitters to roduce an $N$-way combiner/divider. This approach is only practical for moderate alues of $N$ due to size and insertion loss considerations. The worst case isolation ccurs between adjacent ports and is similar in value to that of an individual two-vay Wilkinson, while the input VSWR has been shown [38] to be better than that f the equivalent $N$-way Wilkinson at all frequencies but worse than that of a two-vay Wilkinson for total bandwidths in excess of 60%. This method of power division bviously results in the number of output ports being a power of two, but a nodification [39] enables dividers with an odd number of output ports to be realzed. Also, the in-phase power dividers can be replaced by quadrature power lividers if desired so that the outputs of pairs of ports are in phase quadrature.

The other method of producing a planar divider is to generalize the two-way Vilkinson to that shown in Figure 6.15. This type of divider is sometimes referred o as a *fork divider* [36]. Galani and Temple [40] numerically optimized the transnission line characteristic impedance and resistor values to achieve the best comromise between isolation and VSWR. They achieved 20-dB minimum isolation nd 2:1 maximum VSWR over a 25% total bandwidth from a seven-way combiner. )ther authors [41, 42] have achieved broader bandwidths by using multiple-section ersions of the structure shown in Figure 6.15.

Finally, lumped-element realizations [43] need to be considered if we want o realize the coupler monolithically due to chip size considerations. However, uch realizations have substantially higher excess insertion loss than their non-

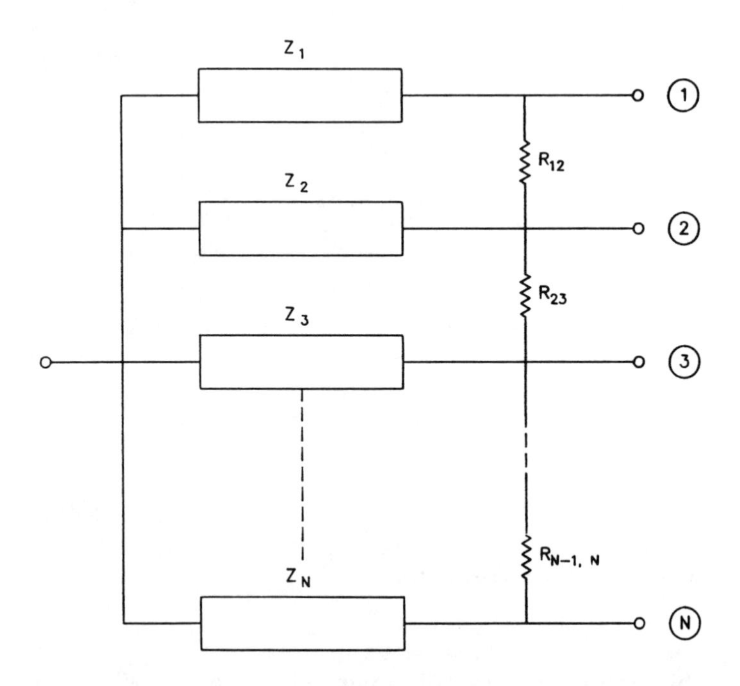


**Figure 6.15** An *N*-way fork divider.

monolithic counterparts; for example, Staudinger [43] reported 1-dB excess inse tion loss per path for a three-way lumped-element monolithic C- and X-band powe splitter.

## 6.4 POWER COMBINING METHODS

### 6.4.1 Corporate Power Combining

Corporate power combining is illustrated in Figure 6.16, which shows why th technique is sometimes referred to as the *tree method* of power combining. Th limiting case of Figure 6.16 in which $N = 2$ is of course the balanced amplifier quadrature couplers are used [31]. In the general case, the number of devic combined is restricted to a power of two (i.e., $N = 2^k$, where $k$ is an integer unless the technique described by Goldfarb [39] is used, and for this reason th technique is also known as the *binary power combining method*. In the ideal ca in which all $N$ amplifier modules are identical and perfectly matched and in whic the power divider/combiner networks have zero loss, the total output power

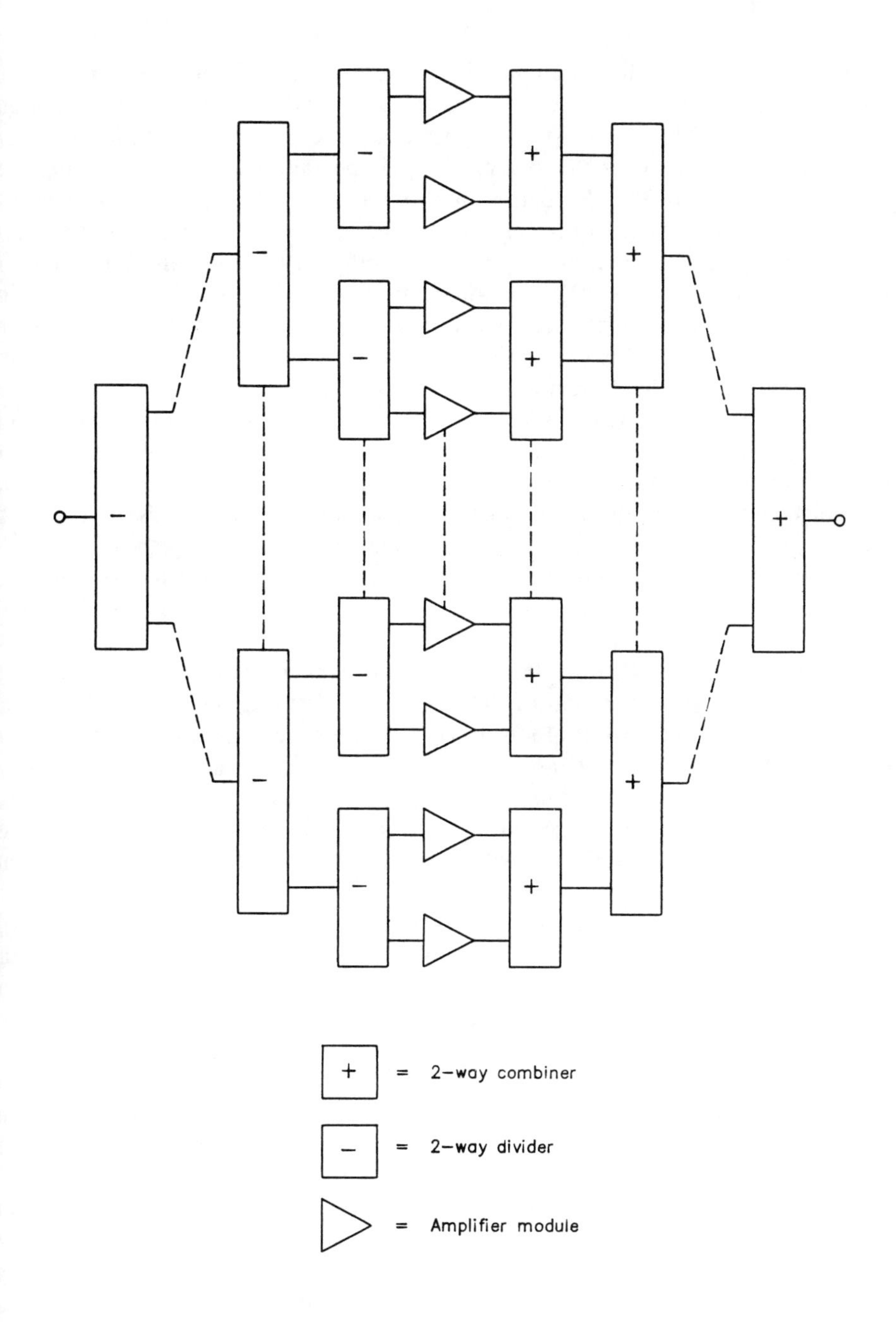


**igure 6.16** Corporate or "tree" method of power combining.

band center is $N$ times that of an individual amplifier module and 100% combinin efficiency is achieved, but the total gain is the same as that of an individual module

The individual two-way power splitters can be either the in-phase or quad rature variety but whatever choice is made it is important that the combiner networ present a very good VSWR to the amplifier modules because, as explained i Section 4.4.3, a mismatch at the combiner input reduces the power output by mor than the resulting mismatch insertion loss—in effect, the power amplifier module become detuned. If the amplifier modules have poor VSWRs, then high isolatio in the divider/combiner networks is required to prevent interaction between th amplifier modules. In this case it is important to consider not only the in-ban performance of the divider/combiner networks but also the out-of-band perfo mance as well, because GaAs FETs have substantial gain at low frequencies an hence high out-of-band isolation is a desirable attribute to help prevent out-o band oscillations. In this respect parallel coupled-line splitters have an inheren advantage over Wilkinson and branch-arm couplers because the isolation of th latter degrades away from band center, while for the former it is theoreticall infinite at all frequencies but in practice it is finite and improves as the frequenc is reduced. Of course, the isolation resistor must have an adequate power handlin capability.

One must also examine the effect of the type of power divider/combine network used on the operation of the power-combined amplifier in the presenc of either an internal or external mismatch. Consider the situation shown in Figur 6.17 in which a pair of power amplifier modules is combined using quadratur couplers. As explained in Section 1.5, the input VSWR of a power amplifier modu will be large if each amplifier module employs a single GaAs FET and is require to operate over a large bandwidth with flat gain because of the need to compensat for the inherent 6 dB/octave gain roll-off of the GaAs FET using the input matchin network. At band center the coupler has $|S_{12}|^2 = 0.5$; hence, as can be seen fro Figure 6.17, all of the reflected energy is dissipated in the dummy load if quadratur couplers are used and none appears at the input port of the coupler. Away fro band center the majority of the reflected energy is still dissipated in the dumm load and so the VSWR at the input port of the power combiner is still good despit the poor input VSWR of the individual power amplifier modules. A similar situatio exists at the output because, as explained in Section 1.5, power amplifiers inherent have a larger VSWR than their small-signal counterparts because the load require for maximum output power is not the same as that required for a conjugate matc Once again, the VSWR at the output port of the power combiner is still goo despite the poor output VSWR of the individual power amplifier modules. If, o the other hand, in-phase power divider/combiner networks are used, then th VSWR at the input and output ports of the power combiner will at best be identic to and, in general, worse than that of the power amplifier modules.

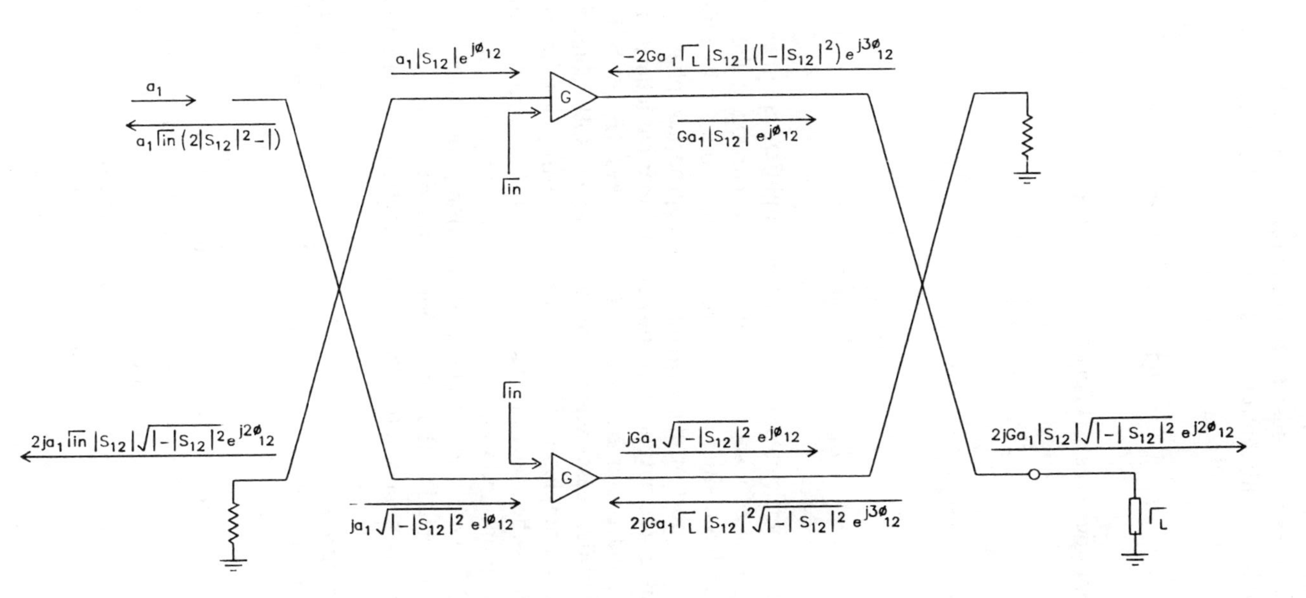


**Figure 6.17** Analysis of a balanced amplifier in the presence of an external mismatch.

The effect of an external mismatch on the operation of the power-combine amplifier can be deduced with reference to Figure 6.17. The upper amplifier modul sees a reflection coefficient of

$$\Gamma_{\text{upper}} = -2(1 - |S_{12}|^2)\Gamma_L e^{j2\phi_{12}} \tag{6.36}$$

while the lower amplifier sees a reflection coefficient of

$$\Gamma_{\text{lower}} = 2|S_{12}|^2 \Gamma_L e^{j\phi_{12}} \tag{6.37}$$

Thus at band center when $|S_{12}|^2 = 0.5$,

$$\begin{aligned} \Gamma_{\text{upper}} &= -\Gamma_{\text{lower}} \\ |\Gamma_{\text{upper}}| &= |\Gamma_{\text{lower}}| = |\Gamma_L| \end{aligned} \tag{6.38}$$

Hence, the reflection coefficients seen by the two amplifiers are in exact antiphas and in the limiting case of a short or open circuit occurring at the output of th power combiner, each amplifier module will operate into a totally reflecting loa which may itself be a short or open circuit if the appropriate phasing occurs. A open or short circuit at the output of the power combiner has a far more significa effect away from band center because, as (6.36) shows, $|\Gamma_{\text{upper}}| > 1$ in this cas (i.e., the upper amplifier sees a negative resistance and absorbs some of the pow generated by the lower amplifier). This can lead to catastrophic failure of the upp amplifier. Of course, the roles of the upper and lower amplifier modules a interchanged if the input and dummy ports of the input power divider are inte changed. This problem of excessive energy dissipation in one transistor if a total reflecting load is connected to the output port of the combiner is eliminated if i phase power divider/combiner networks are used due to the inherent symmetry such amplifiers.

As the preceding analysis shows, in-phase and quadrature couplers have co plementary strengths and weaknesses when used for power combining. Bert a Kaminsky [44] introduced the traveling-wave divider/combiner method to combi the best features of both schemes. This technique is now widely used for bo monolithic and hybrid applications [45, 46] and further examples of its applicati can be found in Chapter 7. Figure 6.18 shows one type of traveling-wave divide combiner and its application to power combining. It is seen to be constructed on of two-way Wilkinson splitters and lengths of transmission line that are λ/4 lo at band center. It is thus easily fabricated. At band center any reflected ener from the amplifier modules is dissipated in the dummy load of the Wilkins splitters, so a good VSWR is seen at the input and output ports of the pow combiner. While some degradation of VSWR away from band center occurs co

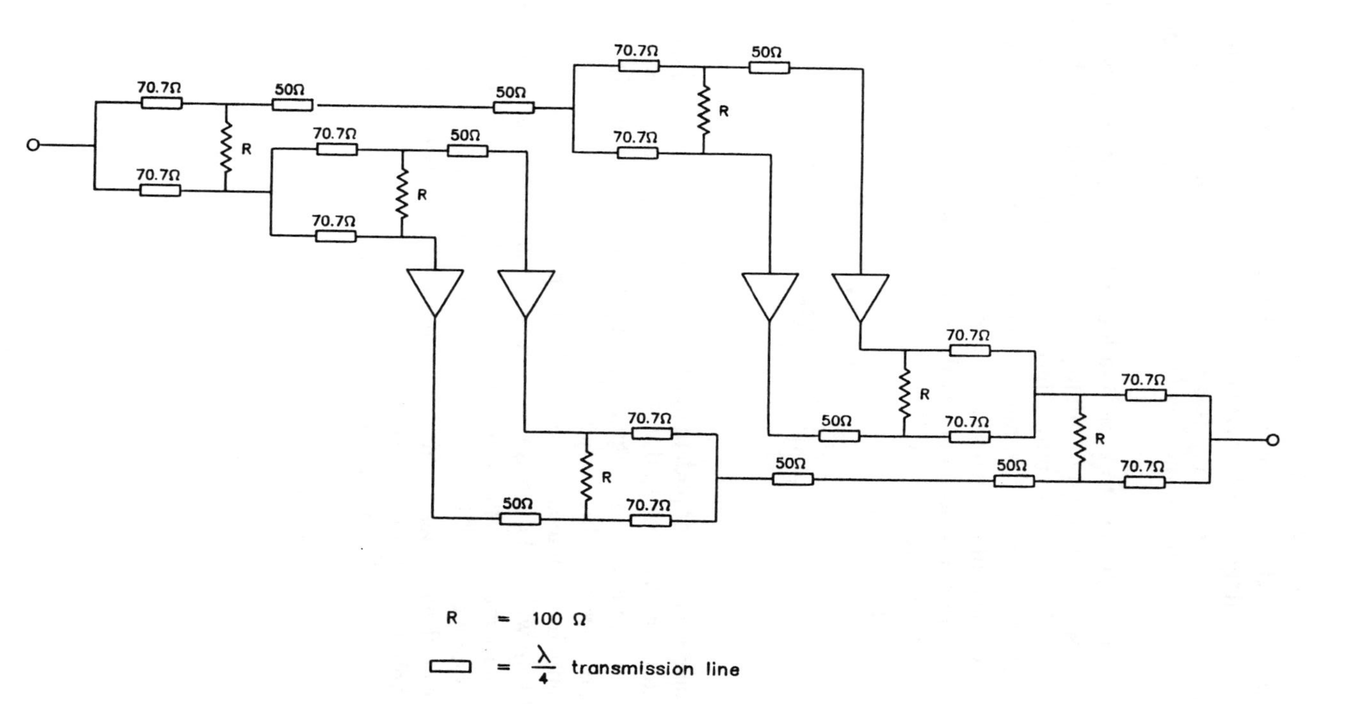


**Figure 6.18** Traveling-wave method of power combining.

pared with true quadrature-combined amplifiers, the amount of degradation is acceptable for up to an octave bandwidth [44]. Also, if a similar analysis to that given in Figure 6.17 is undertaken to examine the effect of an external mismatch we will find that the reflected energy is dissipated in the amplifiers equally at all frequencies. This method of power combining offers broad bandwidth and low loss.

Finally, it is very important to achieve a high combining efficiency in order to minimize the size and the amount of prime power needed and hence the cooling requirements, all of which translate into reduced cost. The combining efficiency is degraded in practice by resistive loss within the combiner network and by variations in the amplitude and phase of the signals to be combined. If one assumes that each part of the individual two-way splitters has the same resistive loss and that they are all identical, then the effect of the loss on combiner efficiency is readily calculated, the result being

$$\eta_c = 10^{\left(\frac{-|L|}{10}\right)\log_2 N} \tag{6.39}$$

where $|L|$ is the insertion loss per path in decibels and $N$ is the number of amplifiers combined. Equation (6.39) is plotted in Figure 6.19 for various values of $N$ and $L$. Resistive loss has an even more significant effect on the overall gain since the loss occurs at both the input and output of the amplifier. For example, suppose we combine 32 amplifier modules with a 0.3-dB loss/path in each two-way splitter, then the overall gain will be 3 dB less than that of the ampilifer module—a figure that is probably unacceptable. The more amplifiers one tries to combine, the lower the loss of the combiner must be in order to achieve an acceptable combining efficiency and overall gain, hence this combining technique is only suitable for combining relatively few amplifier modules.

The effect of amplitude and phase nonuniformities on combining efficiency, whether caused by variations within the amplifiers or by port to port variations in the combiner, can be readily calculated and has been considered by a number of authors [47–50]. Consider an $N$-way combiner with port $N + 1$ as the output port. Then the scattering matrix for the combiner is given by

$$\begin{pmatrix} b_1 \\ \vdots \\ b_{N+1} \end{pmatrix} = \begin{pmatrix} 0 & S_{12} & \cdots & S_{1,N+1} \\ S_{12} & 0 & \cdots & \\ \vdots & & & \vdots \\ S_{1,N+1} & \cdots & & 0 \end{pmatrix} \begin{pmatrix} a_1 \\ \vdots \\ a_{N+1} \end{pmatrix} \tag{6.40}$$

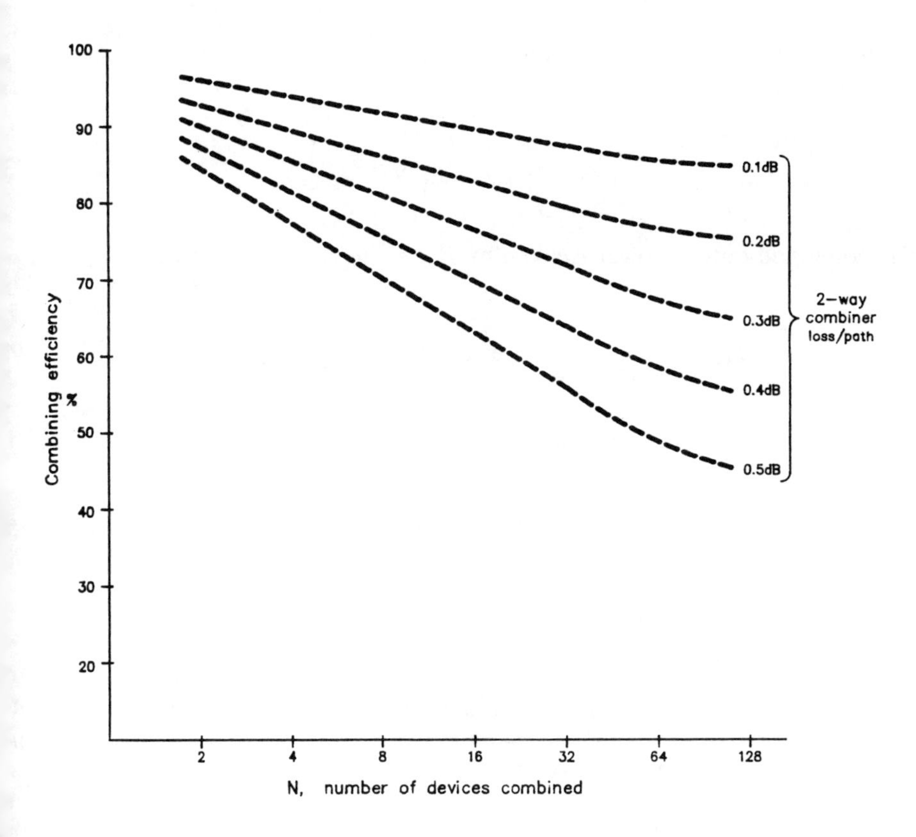


**Figure 6.19** Effect of loss within the two-way splitters on combining efficiency of a corporately combined amplifier.

Hence the output signal is given by

$$b_{N+1} = S_{1,N+1}\, a_1 + \ldots + S_{N,N+1}\, a_N \tag{6.41}$$

Since we are considering the product of $S_{i,N+1}$ and $a_i$ in each case, it is immaterial as far as the analysis is concerned whether the amplitude and phase nonuniformities occur in the amplifiers (i.e., the $a_i$ inputs), or in the combiner (i.e., $S_{i,N+1}$). For the purpose of this analysis, we assume that no mismatch exists and that all of the nonuniformities occur in the $a_i$ inputs and thus $S_{i,N+1} = 1/\sqrt{N}$. Assume that the ports are numbered in such a way that the largest input power $P$ is applied to port 1 of the combiner, then

$$a_i = \sqrt{x_i P}\, e^{j\theta i} \tag{6.42}$$

where $0 \leq x_i \leq 1$ and $\theta_i$ is the phase angle of the $i$'th input with respect to th input to port 1. Thus,

$$b_{N+1} = \frac{1}{\sqrt{N}}\left(\sum_{i=1}^{N} \sqrt{x_i P e^{j\phi}}_i\right) \quad (6.43)$$

The combined output power is given by $P_{out} = |b_{N+1}|^2$; thus,

$$P_{out} = \frac{P}{N}\left[\left(1 + \sum_{i=2}^{N} \sqrt{x_i}\cos\theta_i\right)^2 + \left(\sum_{i=2}^{N} \sqrt{x_i}\sin\theta_i\right)^2\right] \quad (6.44)$$

The total input power to the combiner is of course given by

$$P_{in} = P\sum_{i=1}^{N} x_i \quad (6.45)$$

and hence the combining efficiency $\eta_c$ is given by

$$\eta_c = \frac{P_{out}}{P_{in}} = \frac{\left(1 + \sum_{i=2}^{N} \sqrt{x_i}\cos\theta_i\right)^2 + \left(\sum_{i=2}^{N} \sqrt{x_i}\sin\theta_i\right)^2}{N\sum_{i=1}^{N} x_i} \quad (6.46)$$

A parameter related to combining efficiency is *excess insertion loss* $IL_{ex}$, which i defined as

$$IL_{ex} = 10\log_{10}\eta_c \quad (6.47)$$

Maximum power output occurs when $\theta_i = 0$ and $x_i = 1$ and is given by

$$P_{outmax} = NP \quad (6.48)$$

and hence the output power degradation factor is

$$\frac{P_{out}}{P_{outmax}} = \frac{\left(1 + \sum_{i=2}^{N} \sqrt{x_i}\cos\theta_i\right)^2 + \left(\sum_{i=2}^{N} \sqrt{x_i}\sin\theta_i\right)^2}{N^2} \quad (6.49)$$

Equations (6.46) and (6.49) enable the effect of amplitude and phase errors pon power output and combining efficiency to be readily examined. For example, $N = 2$ (i.e., a balanced amplifier), then the excess insertion loss will be less than .1 dB ($\eta_c = 97.7\%$) provided the simultaneous amplitude and phase imbalances re less than 1.3 dB and 15 deg, respectively, while the total output power will be nly 0.7 dB below the level that would occur if no amplitude or phase imbalance xisted. This example demonstrates that relatively little of the 0.7-dB reduction in utput power is caused by loss of combining efficiency as a result of amplitude and hase errors and most of the reduction in output power occurs because of the eduction in input power. Expressed another way, combining efficiency is not a ery sensitive function of amplitude and phase errors. For arbitrary values of $N$, iupta [51] has determined a lower bound for combining efficiency resulting from mplitude and phase variations in the amplifier outputs. For example, if the amplinde and phase errors are within $\pm 1.3$ dB and $\pm 15$ deg, then the combining fficiency will never be less than 91% (excess insertion loss of 0.4 dB). To ensure hat the excess insertion loss is never less than 0.1 dB requires that the amplitude nd phase errors are within $\pm 0.9$ dB and $\pm 5$ deg, respectively.

An extreme case of amplitude imbalance is amplifier failure when $x_i = 0$, nd its effect on power output is readily evaluated by means of (6.49). If $M$ out f $N$ amplifiers in Figure 6.16 fail, then the output power becomes

$$P_{\text{out}} = \left(1 - \frac{M}{N}\right)^2 P_{\text{outmax}} \tag{6.50}$$

Thus failure of one or more amplifier modules does not result in total failure of he combined power amplifier—a property known as *graceful degradation*. Equaion (6.50) shows that when half the amplifiers fail the output power drops by 6 B rather than 3 dB. This aspect is considered further in Section 6.4.3 where a nethod of achieving a higher output power in the presence of amplifier failures han that given by (6.50) is discussed.

The preceding analysis is based on certain idealistic assumptions such as a erfect match existing at all internal and external ports. Galani et al. [52] have xamined the effect of amplifier failures in the presence of imperfect matches. While the analysis was originally undertaken for $N$-way combiners, it is also appliable for corporate combining. The main conclusion is that the performance degadation resulting from a single amplifier module failure can be reduced by minmizing the VSWR that the combiner presents to the amplifier modules but it is ndependent of the combiner's interport isolation. If, however, more than one mplifier module should fail, then the performance degradation depends on both he input VSWR and isolation of the combiner.

### 6.4.2 Serial Power Combining

The serial or chain method of power combining [53, 54] is illustrated in Figur 6.20. As in the corporate power-combining method, the function of the divide network is to ensure that each amplifier module receives the same input level. Thu if $N$ amplifier modules are to be combined then $1/N$'th of the input power mus be supplied to each amplifier module, that is, the first coupler in Figure 6.20 mus have a coupling coefficient of $10 \log_{10} N$ dB. The power injected into the secon coupler will be $P_{in} (N - 1)/N$ watts and hence the second coupler needs a couplin coefficient of $10 \log_{10}(N - 1)$ dB and, in general, the $i$'th coupler needs a couplin coefficient of $10 \log_{10} (N + 1 - i)$ dB. In principle, either in-phase or quadratur couplers can be used, but the most appropriate type to use is the parallel couplec line type described in Section 6.3.2 because of the need to achieve a wide sprea in coupling values. The coupling coefficients required for the couplers in the com biner can be deduced in a similar manner and are identical to those in the divide as shown in Figure 6.20. Because a phase difference exists between the signal applied to each amplifier module, it is important that the combiner network hav the same phase properties as the divider and this is most readily achieved by makin the combiner identical to the divider. As in the corporate method of power con bining, it is important that each coupler in the combiner have a very good inp VSWR in order to avoid detuning the amplifier modules, and the couplers ne the output of the combiner also need to have a high isolation. As shown in Figur 6.20 some finite length of transmission line $l$ must be inserted between each couplec line section to make the structure physically realizable, and a corresponding lengt has to be added at each coupler output to preserve the phase relationships.

The main advantages of the serial combiner over the corporate method ( power combining are that an extra amplifier module can be added very readily t the chain to increase the output power if necessary, and one can combine an number of amplifiers rather than just a power of two, which has economic as we as size and power consumption advantages. However, the serial combiner also h a number of disadvantages. First, the signal entering any coupler in the divide network is attenuated by the loss in all the preceding couplers, so the couplin coefficient of each coupler must be modified from the value shown in Figure 6.2 to compensate for the preceding losses. A similar situation exists for the combine but in this case the effect of these losses is to reduce the combining efficiency. Th effect can be readily quantified if one assumes that all the couplers have the san loss and that it is divided equally between the two signal paths in each coupler, i which case combining efficiency is given by

$$\eta_c = \frac{1}{N}\left[10^{-(N-1)|L|/10} + \sum_{k=1}^{N-1} 10^{-k|L|/10}\right] \quad (6.5$$

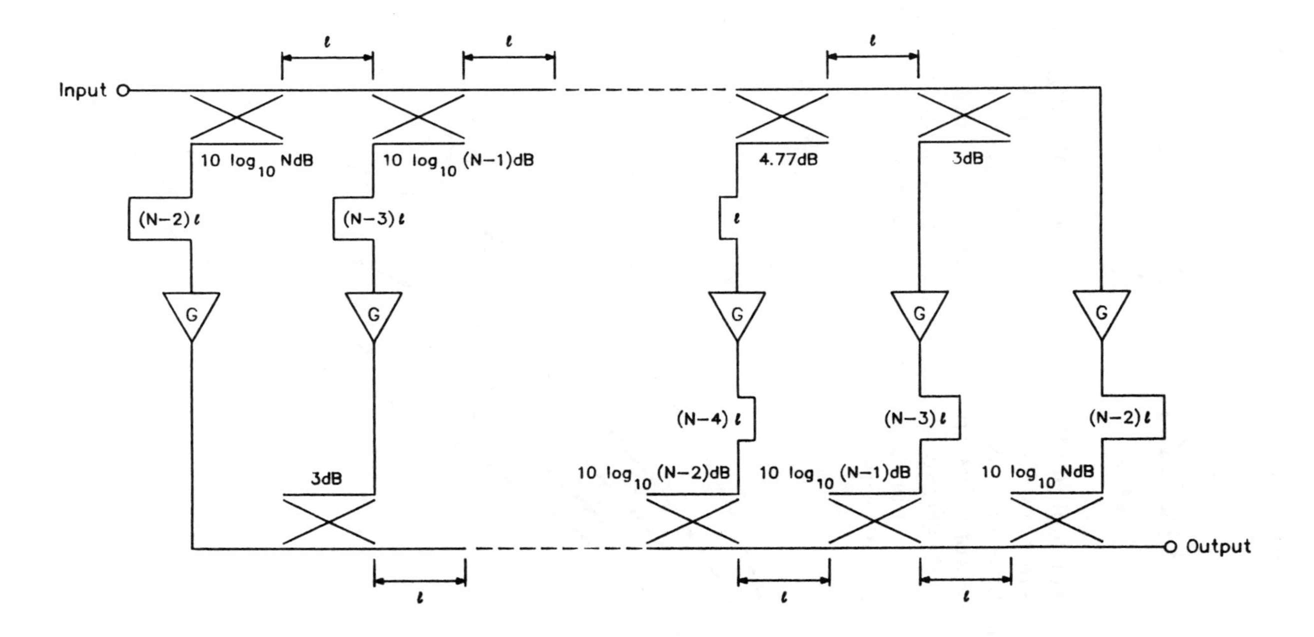


**Figure 6.20** Serial or "chain" method of power combining.

where $L$ is the insertion loss per path in decibels. Equation (6.51) can be expressed in closed form as

$$\eta_c = \frac{1}{N}\left[\frac{2\cdot 10^{-N|L|/10} - 10^{-(N-1)|L|/10} - 10^{-|L|/10}}{10^{-|L|/10} - 1}\right] \tag{6.52}$$

and Figure 6.21 is a graph of (6.52) for various values of $N$ and $L$. Comparison with Figure 6.19 shows that losses have a more serious effect on combining efficiency for serial combining than for corporate combining and thus serial combiners are restricted to lower values of $N$ than are corporate combiners. However, in practice the weaker couplers are likely to have a lower loss than the tight couplers and hence (6.52) and Figure 6.21 are a pessimistic assessment of the effect of loss on the combining efficiency of a serial combiner. The maximum value of $N$ for serial

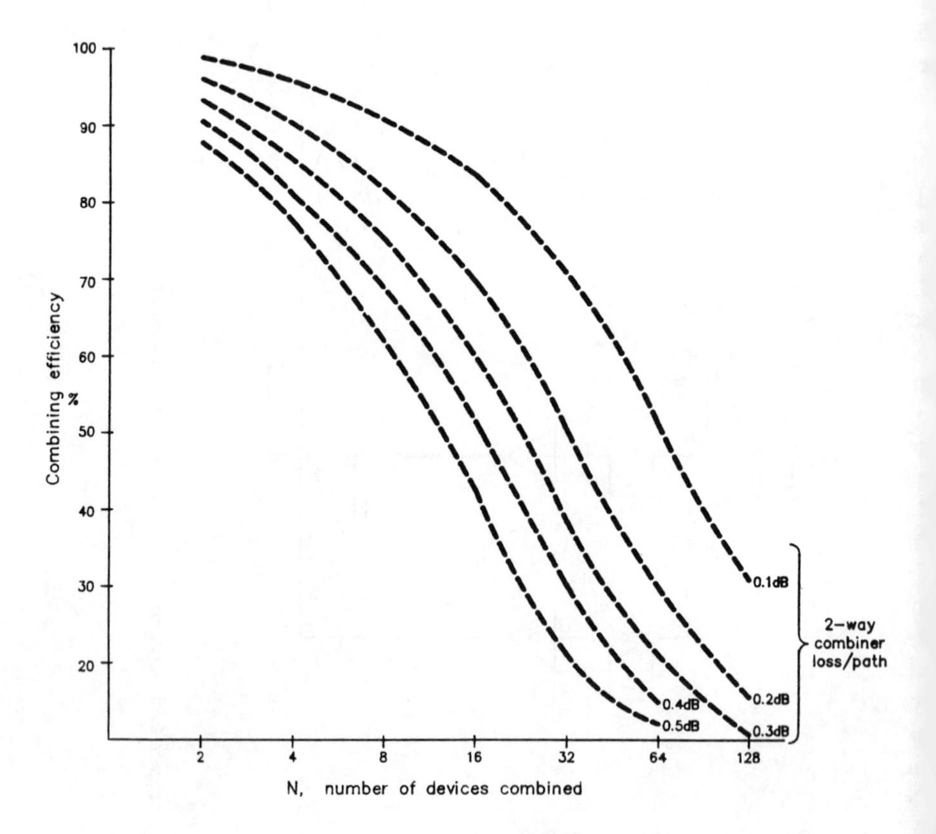


**Figure 6.21** Effect of loss within the couplers upon combining efficiency of a serially combined amplifier

mbining is also restricted in practice by the problem of achieving the very fine crements in coupling that are required. One of the biggest disadvantages of serial mbining is the limited bandwidth over which high combining efficiency and hence wer output can be achieved. This problem arises because although the relative ase difference between the two outputs of a quadrature coupler is exactly 90 deg all frequencies, the absolute phase shift through the coupler depends not only frequency but also on the value of coupling. At band center all the outputs to combined are in phase, but away from band center the outputs from the couplers e no longer in phase and this leads to a lower combining efficiency for a given lue of $N$ and bandwidth for serial combining than for corporate combining [54].

### .3 $N$-Way Power Combining

th corporate and serial combined power amplifiers suffer from poor combining iciency when large numbers of amplifiers have to be combined, which is a nsequence of the microwave signals having to travel along multiple sections of arter-wavelength transmission lines. To achieve a high combining efficiency for ge values of $N$, the path lengths must be minimized and this is achieved using way power divider/combiner networks as described in Section 6.3.3. Some very pressive results have been achieved using this technique, for example, Schellen-rg and Cohn [55] have reported a 12-way radial power combiner at X-band ving a 1-dB bandwidth of 31%, a minimum adjacent channel isolation of 13 dB th ±0.25 dB and ±5 deg amplitude and phase imbalances with a midband ertion loss of 0.25 dB implying a combining efficiency of 94.4%, excluding the ects of amplitude and phase variation from the amplifiers and the combiner. A mbining efficiency of 87.4% was achieved in practice when 12 GaAs FET ampli-r modules having amplitude and phase variations of ±1.2 dB and ±13 deg, spectively, were combined. Stones [56] has reported a 16-way radial power com-ner operating over 760 to 920 MHz with less than 0.3 dB of insertion loss, ±0.1 of amplitude imbalance, and ±1 deg of phase imbalance, a minimum adjacent rt isolation of 10 dB, and a maximum output port VSWR of 1.5:1. A saturated tput power of 1.3 kW was achieved when 16 nominally 60-W amplifiers were mbined. Finally, Sanders [57] has reported a 110-way radial combiner operating er 1.2 to 1.4 GHz with 21-dB minimum adjacent channel isolation, 1.2:1 input d output VSWR, a 0.23-dB insertion loss, and with a ±0.15-dB amplitude balance. A peak output power of 110 kW was achieved from this structure when ) nominally 1.1-kW amplifier modules were combined implying a combining iciency of 91%.

All of these power combiners will exhibit graceful degradation, which has en verified in practice by Schellenberg and Cohn [55] who found that the mea-ed output power in the presence of simulated amplifier failures was very close

to that predicted by (6.50). Under certain circumstances, better performance c
be achieved when an amplifier fails by using a radial or fork combiner without a
isolation resistors [58, 59]. For this to work properly, all of the amplifier modul
must have a low-output VSWR and be as near identical as possible so that negligib
interaction effects occur due to the absence of the isolation resistors. The meth
is based on the assumption that the mechanism of amplifier failure is predictab
for example, that the FETs will always fail short circuit, in which case the inte
vening network between each FET and the combiner input is arranged such th
a failed FET will present a short circuit to the input of the combiner. This w
ensure that an open circuit appears at the combiner output, which minimizes t
effect on the remaining working amplifier modules. The $N$-way combiner m
present a good input VSWR to the amplifier modules in order to minimize t
performance degradation resulting from a single amplifier module failure [52], b
the required good input VSWR is achieved in an interesting way for a resistorl
radial or fork combiner. Since this type of combiner is lossless, application of t
unitary and symmetry conditions to the scattering matrix enables one to show th

$$|S_{ii}| = 1 - 1/N \text{ and } |S_{ij}| = 1/N \qquad \text{for } i, j = 1 \rightarrow N, i \neq j \qquad (6.5$$

assuming that the output port is perfectly matched, that is, it appears that ea
amplifier is required to operate into an almost totally reflecting load. Howev
each amplifier sees not only its own reflected signal but also a contribution via
from all the other $N - 1$ amplifiers and the vector sum of all these signals equa
to zero, that is, an active match condition is established such that each amplif
appears to operate into a matched load, but the input VSWR in the presence
amplifier failures gets steadily worse in proportion to the number of failed amplif
modules. Saleh [58] has shown that this scheme results in the output power falli
by only 3.5 dB when half the amplifiers fail even though each working amplif
sees an effective input return loss of 9.5 dB. The danger with this scheme is th
if for any reason the failed amplifier module presents an open circuit to the inp
of the combiner, then all the remaining working amplifier modules see a reflecti
coefficient of unity magnitude, which could cause catastrophic failure of all t
amplifier modules unless they are all capable of withstanding such a mismatch.

# APPENDIX 6A

## THE NECESSITY OF HAVING $Z_0^g = Z_0^d = Z_0$ AND $\theta_g = \theta_d$ AND THE IMPLICATIONS

amination of (6.2) and (6.3) shows that the requirement to have $Z_0^g = Z_0^d$ and $= \omega_c^d$ means that one must have $L_g = L_d$ and $C_{gs} = C_0$. Unfortunately, all FETs ve $C_{gs} > C_0$. Thus we can either achieve a perfect impedance match (i.e., $= Z_0^d$) and a mismatched phase response (i.e., $\theta_g \neq \theta_d$), or vice versa. Examition of (6.11) shows that the gain is a very sensitive function of $|\theta_d - \theta_g|$; in fact, $- \theta_g|$ must be less than 25 deg for the gain ripple to be within 1 dB for $n = 4$ d thus the perfect impedance match and mismatched phase response solution is t a viable option. If the alternative option is selected of a perfect phase match, en for a typical FET a good output match ($S_{22} = 0$) requires that the load pedance be $Z_0^d = 3Z_0^g$, but in many applications it would be unacceptable to ve unequal source and load impedances. If the amplifier had $Z_0^d = 3Z_0^g$ but was erated into a load impedance of $Z_0 = Z_0^g$ then the output match would be very or (3:1 VSWR) and this would also cause substantial gain ripple.

Thus, in general, we must attempt to equalize both the phase response and e characteristic impedances of the gate and drain transmission lines. Thus it is cessary either to add some additional capacitance to ground, $C_{add}$, at each drain ch that $C_0 + C_{add} = C_{gs}$, or to insert a capacitor in series with each gate of value $C_{gs}/(C_{gs} - C_0)$. The former scheme suffers from two drawbacks, namely, the trinsically higher cutoff frequency of the drain line is degraded to that of the gate e, and the very low value of $C_{add}$ (around 0.2 pF for a typical small-signal FET) difficult to realize without significant series parasitics. As a result this approach rarely used. The series gate capacitor approach results in the cutoff frequency the amplifier being $\omega_c^d$ rather than $\omega_c^g$, but it still requires a very low value of pacitance. Obviously, each series gate capacitor has to be shunted with a high lue resistor in order to provide dc gate bias to the FET. However, the series te capacitor approach has another limitation as well, namely, that this capacitor d $C_{gs}$ form a potential divider resulting in an effective $g_m$ of $(C_0/C_{gs})g_m$, which sults in about a 10-dB gain reduction if typical FETs are used.

In most designs this level of gain reduction could not be tolerated, and thus appears at first sight that the series capacitor approach must be discounted as ing impractical. However, if the gatewidth of each FET is multiplied by the ratio ${}_s/C_0$ then the FET's $g_m$ will increase in the same ratio and the gain of the disbuted amplifier will be restored to its original value. The input and output pacitances of the FET are also increased in the same ratio with the result that e cutoff frequency of the distributed amplifier is reduced by the factor $C_0/C_{gs}$, at is, the cutoff frequency is now $\omega_c^g$ of the unscaled amplifier. An added benefit using the larger FETs is that the value of the additional gate series capacitance needed is increased in the factor $C_{gs}/C_0$ as well, which eases the realizability prol lem. However, for small-signal, low-power applications the series gate capacitc technique is still unattractive because it results in a very high dc power consumptic for a given gain level. The series gate capacitor technique does, however, com into its own for power amplifiers as discussed in Section 6.2.2.

A common solution to the problem of achieving $\theta_g = \theta_d$ and $Z_0^g = Z_0^d$ is have nonunity values for $m$ and $M$ in Figure 6.1, in which case the drain lin becomes a cascade of modified constant $m$-derived filter sections [10, 60]. For th modified constant $m$-derived filter we let $C_0 = mC_{gs}/M$, in which case

$$Z_I^d = Z_0^d \sqrt{1 - \left(\frac{\omega}{\omega_c^d}\right)^2} \quad \text{(6A.1}$$

where

$$Z_0^d = \sqrt{\frac{ML_d}{C_{gs}}}, \qquad \omega_c^d = 2/\sqrt{L_d C_{gs}} \quad \text{(6A.2}$$

and

$$\theta_d = \cos^{-1}\left[\frac{1 - \dfrac{M + m^2}{M}\left(\dfrac{\omega}{\omega_c^d}\right)^2}{1 - \dfrac{M - m^2}{M}\left(\dfrac{\omega}{\omega_c^d}\right)^2}\right] \quad \text{(6A.3}$$

If $M = 1$, then one has a conventional $m$-derived filter section and the requireme that $Z_0^d = Z_0^g$ means that $L_d = L_g$ and thus $\omega_c^d = \omega_c^g$. Unfortunately, however, differs considerably from $\theta_g$ if a typical value of $m$ is used, such as $m = 1/3$, whic results in a very poor gain response. Thus a nonunity value must be used for also, which implies an imperfect output match as shown by (6A.2), but $\theta_d$ is the much closer to $\theta_g$. Varying the value of $M$ enables a trade-off to be made betwee output VSWR and gain flatness. Figure 6A.1 shows the circuit of a four-stag distributed amplifier designed using a constant $K$ gate line and a modified $m$-derive drain line with $m$-derived half-sections inserted at each end of the gate and dra transmission lines to improve the match as discussed in Section 6.2.1. The amplifi is designed to use a FET with $C_{gs} = 0.3$ pF, $C_0 = 0.1$ pF, and $g_m = 40$ mS ar to operate into 50-$\Omega$ source and load impedances. The cutoff frequency is th

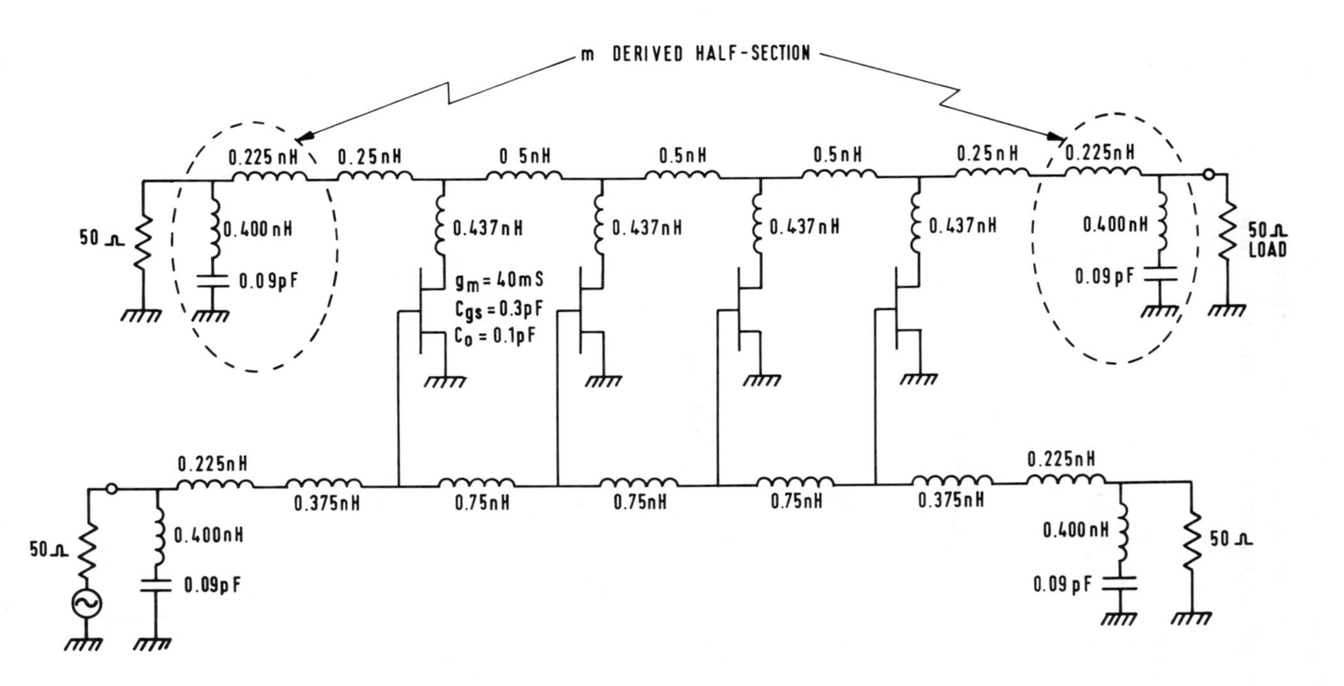


**Figure 6A.1** Circuit diagram of a distributed amplifier with constant-$K$ gate line and a modified $m$-derived drain line.

21.2 GHz and $m/M$ = 1/3. In this design $M$ has been chosen to have the value $M$ = 2, and Figure 6A.2 shows the computed performance from which it can be seen that the input and output return losses are better than 10 dB and the gain is flat

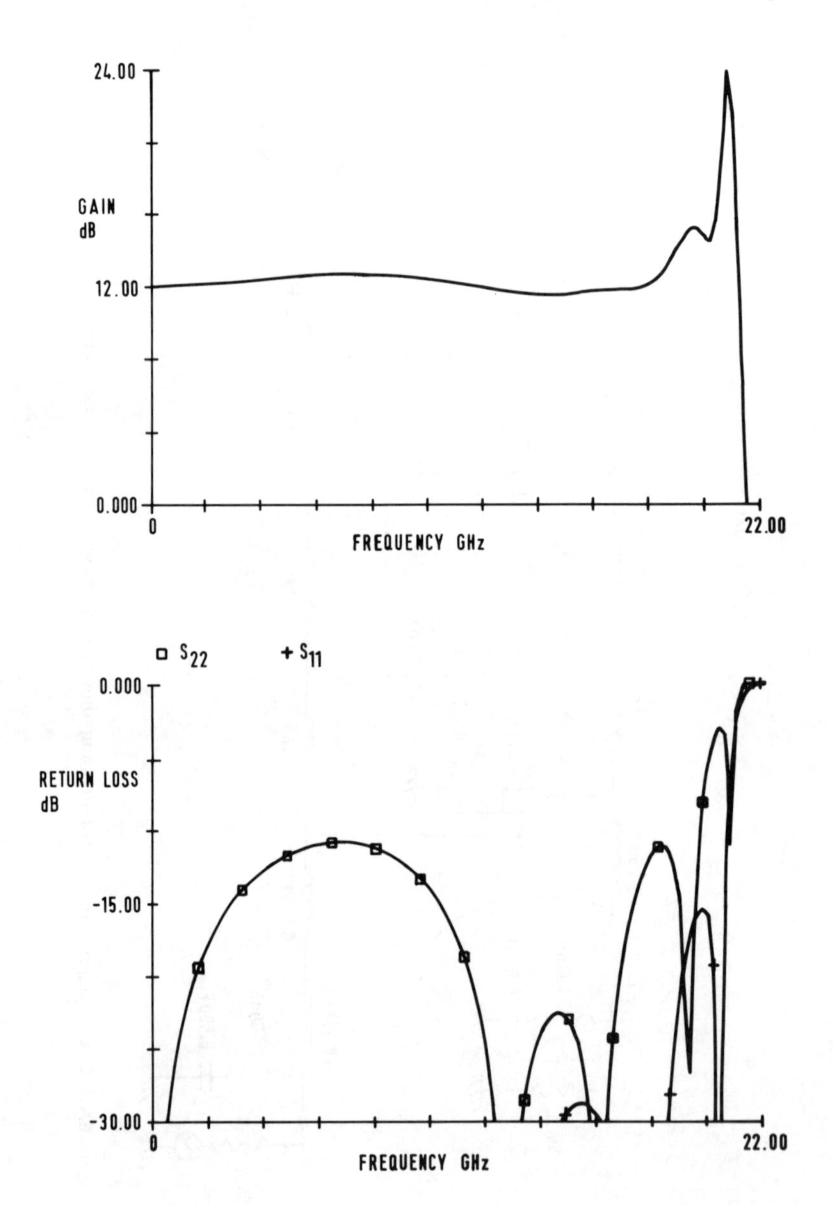


**Figure 6A.2** Computed performance of the distributed amplifier shown in Figure 6A.1.

within ± 0.6 dB up to 18 GHz. If losses within the amplifier were included, then with this value of $M$ the gain would exhibit a high-frequency roll-off, but increasing the value of $M$ can compensate for this effect.

## REFERENCES


1. Freitag, R. G., J. E. Degenford, D. C. Boire, M. C. Driver, R. A. Wickstrom, and C. D. Chang, "Wideband 3W Amplifier Employing Cluster Matching," *IEEE 1983 Microwave and Millimetre-Wave Monolithic Circuits Symp. Digest*, Boston, MA, May 31–June 1, 1983, pp. 62–66.
2. Freitag, R. G., S. H. Lee, D. M. Krafcsik, D. E. Dawson, and J. E. Degenford, "Stability and Improved Circuit Modeling Considerations for High Power MMIC Amplifiers," *IEEE 1988 Microwave and Millimetre-Wave Monolithic Circuits Symp. Digest*, New York, NY, May 24–25, 1988, pp. 125–128.
3. Percival, W. S., British patent Specification No. 450562, applied for July 24, 1936.
4. Ayasli, Y., J. L. Vorhaus, R. Mozzi, and L. Reynolds, "Monolithic GaAs Travelling Wave Amplifier," *Electronics Letters*, Vol. 17, No. 12, June 1981, pp. 413–414.
5. Ayasli, Y., L. D. Reynolds, R. L. Mozzi, J. L. Vorhaus, and L. K. Hanes, "2-20 GHz GaAs Travelling-Wave Power Amplifier," *IEEE 1988 Microwave and Millimetre-Wave Monolithic Circuits Symp. Digest*, Boston, MA, May 31–June 1, 1983, pp. 67–70.
6. Kim, B., and H. Q. Tserng, "0.5W 2-21GHz Monolithic GaAs Distributed Amplifier," *Electronics Letters*, Vol. 20, No. 7, March 1984, pp. 288–289.
7. Ayasli, Y., S. W. Miller, R. Mozzi, and L. K. Hanes, "Capacitively Coupled Travelling-Wave Power Amplifier," *IEEE Trans. on Microwave Theory and Techniques*, Vol. MTT-32, December 1984, pp. 1704–1711.
8. Chase, E. M., and W. Kennan, "A Power Distributed Amplifier using Constant-R Networks," *IEEE 1986 Microwave and Millimetre-Wave Monolithic Circuits Symp. Digest*, Baltimore, MD, June 4–5, 1986, pp. 13–17.
9. Schindler, M. J., J. P. Wendler, M. P. Zaitlin, M. E. Miller, and J. R. Dormail, "A K/Ka-Band Distributed Power Amplifier with Capacitive Drain Coupling," *IEEE 1988 Microwave and Millimetre-Wave Monolithic Circuits Symp. Digest*, New York, NY, May 24–25, 1988, pp. 5–8.
10. Matthaei, G. L., L. Young, and E. M. T. Jones, *Microwave Filters, Impedance-Making Networks, and Coupling Structures*, Norwood, MA: Artech House, 1980.
11. Walker, J. L. B., "Improving Operation of Classic Broadband, Balanced Amplifiers," *Microwaves & RF*, Vol. 26, September 1987, pp. 175–182.
12. McKay, T., J. Eisenberg, and R. E. Williams, "A High-Performance 2-18.5 GHz Distributed Amplifier—Theory and Experiment," *IEEE Trans. on Microwave Theory and Techniques*, Vol. MTT-34, December 1986, pp. 1559–1568.
13. Beyer, J. B., S. N. Prasad, R. C. Becker, J. E. Nordman, and G. K. Hokenwarter, "MESFET Distributed Amplifier Guidelines," *IEEE Trans. on Microwave Theory and Techniques*, Vol. MTT-32, March 1984, pp. 268–276.
14. Becker, R. C., and J. B. Beyer, "On Gain-Bandwidth Product for Distributed Amplifiers," *IEEE Trans. on Microwave Theory and Techniques*, Vol. MTT-34, June 1986, pp. 736–738.
15. Prasad, S. N., J. B. Beyer, and I. S. Chang, "Power-Bandwidth Considerations in the Design of MESFET Distributed Amplifiers," *IEEE Trans. on Microwave Theory and Techniques*, Vol. MTT-36, July 1988, pp. 1117–1123.
16. Dawson, D. E., M. J. Salib, and L. E. Dickens, "Distributed Cascade Amplifier and Noise Figure Modeling of an Arbitrary Amplifier Configuration," *Proc. of IEEE Int. Solid-State Circuits Conf.*, 1984, pp. 78–79.

17. LaRue, R. A., S. G. Bandy, and G. A. Zdasiuk, "A 12dB High Gain, Monolithic Distributed Amplifier," *IEEE Trans. on Microwave Theory and Techniques*, Vol. MTT-34, December 1986, pp. 1542–1547.
18. Kennan, W., T. Andrade, and C. C. Huang, "A 2-18GHz Monolithic Distributed Amplifier Using Dual-Gate GaAs FETs," *IEEE Trans. on Electron Devices*, Vol. ED-31, December 1984, pp. 1926–1930.
19. Walker, J. L. B., "Some Observations on the Design and Performance of Distributed Amplifiers," *IEEE Trans. on Microwave Theory and Techniques*, Vol. MTT-40, January 1992, pp. 164–168.
20. Salib, M. L., D. E. Dawson, and H. K. Hahn, "Load-Line Analysis in the Frequency Domain with Distributed Amplifier Design Examples," *IEEE MTT-S Int. Microwave Symp. Digest*, Las Vegas, NV, June 9–11, 1987, pp. 575–578.
21. Ginzton, E. L., W. R. Hewlett, J. H. Jasberg, and J. D. Noe, "Distributed Amplification," *Proc. IRE*, Vol. 36, August 1948, pp. 956–969.
22. Jones, K. E., G. S. Barta, and G. C. Herrick, "A 1 to 10 GHz Tapered Distributed Amplifier in a Hermetic Surface Mount Package," *Proc. of IEEE GaAs I.C. Symp.*, Monterey, CA, November 12–14, 1985, pp. 137–140.
23. Collin, R. E., *Foundations for Microwave Engineering*, McGraw-Hill, 1966, p. 308.
24. Wilkinson, E. J., "An N-Way Hybrid Power Divider," *IRE Trans. on Microwave Theory and Techniques*, Vol. MTT-8, January 1960, pp. 116–118.
25. Cohn, S. B., "A Class of Broadband Three-part TEM-Mode Hybrids," *IEEE Trans. on Microwave Theory and Techniques*, Vol. MTT-16, February 1968, pp. 110–116.
26. Beckwith, W., and W. Staudinger, "Wide Bandwidth Monolithic Power Dividers," *Microwave J.* Vol. 32, Feburary 1989, pp. 155–160.
27. Malherbe, J. A. G., *Microwave Transmission Line Couplers*, Norwood, MA: Artech House, 1988, pp. 31–42.
28. March, S., "A Wideband Stripline Hybrid Ring," *IEEE Trans. on Microwave Theory and Techniques*, Vol. MTT-16, June 1968, p. 361.
29. Levy, R., and L. F. Lind, "Synthesis of Symmetrical Branch-Guide Directional Couplers," *IEEE Trans. on Microwave Theory and Techniques*, Vol. MTT-16, February 1968, pp. 80–89.
30. Matthaei, G. L., L. Young, and E. M. T. Jones, *Microwave Fitters, Impedance-Matching Networks, and Coupling Structures*, Dedham, MA: Artech House, 1980, pp. 802–805.
31. Walker, J. L. B., "Improving Operation of Classic Broadband Balanced Amplifiers," *Microwaves & RF*, Vol. 26, September 1987, pp. 175–182.
32. Ou, W. P., "Design Equations for an Interdigitated Directional Coupler," *IEEE Trans. on Microwave Theory and Techniques*, Vol. MTT-23, February 1975, pp. 253–255.
33. Lange, J., "Interdigitated Stripline Quadrature Hybrid," *IEEE Trans. on Microwave Theory and Techniques*, Vol. MTT-17, December 1969, pp. 1150–1151.
34. Saleh, A. A. M., "Computation of the Frequency Response of a Class of Symmetric N-Way Power Dividers," *Bell System Technical J.*, Vol. 59, October 1980, pp. 1493–1512.
35. Taub, J. J., and B. Fitzgerald, "A Note on N-Way Hybrid Power Dividers," *IEEE Trans. On Microwave Theory and Techniques*, Vol. MTT-12, March 1964, pp. 260–261.
36. Saleh, A. A., "Planar Electrically Symmetric N-Way Hybrid Power Dividers/Combiners," *IEEE Trans. on Microwave Theory and Techniques*, Vol. MTT-28, June 1980, pp. 555–563.
37. Bearse, S. V., Ed., "Compact Radial Power Combiner Teams up a Dozen Power GaAs FETs," *Microwaves*, Vol. 16, October 1977, p. 9.
38. Hindin, H. J., "Standing-Wave Ratio of Binary TEM Power Dividers," *IEEE Trans. on Microwave Theory and Techniques*, Vol. MTT-16, February 1968, pp. 123–125.
39. Goldfarb, M. E., "A Recombinant In-Phase Power Divider," *IEEE Trans. on Microwave Theory and Techniques*, Vol. MTT-39, August 1991, pp. 1438–1440.

Galani, Z., and S. J. Temple, "A Broadband Planar N-Way Combiner/Divider," *IEEE MTT-S Int. Microwave Symp. Digest*, 1977, pp. 499–501.

Nagai, N., E. Maekawa, and K. Ono, "New N-Way Hybrid Power Dividers," *IEEE Trans. on Microwave Theory and Techniques*, Vol. MTT-25, December 1977, pp. 1008–1012.

Yau, W., J. M. Schellenberg, "A N-Way Broadband Planar Power Combiner/Divider," *Microwave J.*, Vol. 29, November 1986, pp. 147–151.

Staudinger, J., "Wide Bandwidth MMIC Power Dividers: Implementation and a Practical Design Technique," *Microwave J.*, Vol. 33, February 1990, pp. 73–90.

Bert, A. G., and D. Kaminsky, "The Travelling-Wave Divider/Combiner," *IEEE Trans. on Microwave Theory and Techniques*, Vol. MTT-28, December 1980, pp. 1468–1473.

Tserng, H. Q., and P. Saunier, "10-30 GHz Monolithic GaAs Travelling-Wave Divider/Combiner," *Electronics Letters*, Vol. 21, 1985, pp. 950–951.

Camilleri, N., B. Kim, H. Q. Tserng, and H. D. Shih, "Ka-Band Monolithic GaAs FET Power Amplifier Modules," *IEEE 1988 Microwave and Millimetre Wave Monolithic Circuits Symp. Digest*, New York, NY, May 24–25, 1988, pp. 129–132.

Franke, E. A., "Excess Insertion Loss at the Input Ports of a Combiner Hybrid," *RF Design*, November 1985, pp. 43–48.

Blocksome, R. K., "Predicting RF Output Power From Combined Power Amplifier Modules," *RF Design*, February 1988, pp. 40–49.

Ernst, R. L., R. L. Camisa, and A. Presser, "Graceful Degradation Properties of Matched N-Port Power Amplifier Combiners," *IEEE Int. Microwave Symp. Digest*, 1977, pp. 174–177.

Lee, K. J., "A 25KW Solid State Transmitter for L-Band Radars," *IEEE Int. Microwave Symp. Digest*, 1979, pp. 298–302.

Gupta, M. S., "Degradation of Power Combining Efficiency Due to Variability Among Signal Sources," *IEEE Trans. on Microwave Theory and Techniques*, Vol. MTT-40, May 1992, pp. 1031–1034.

Galani, Z., J. L. Lampen, and S. J. Temple, "Single-Frequency Analysis of Radial and Planar Amplifier Combiner Circuits," *IEEE Trans. Microwave Theory and Techniques*, Vol. MTT-29, July 1981, pp. 642–654.

Russell, K. J., "Microwave Power Combining Techniques," *IEEE Trans. on Microwave Theory and Techniques*, Vol. MTT-27, May 1979, pp. 472–478.

Cappucci, J., "Combining Amplifiers? Try Serial—Feed Arrays," *Microwaves*, Vol. 15, October 1976, pp. 36–41.

Schellenberg, J. M., and M. Cohn, "A Wideband Radial Power Combiner for FET Amplifiers," *IEEE Int. Solid State Circuits Conf.*, 1978, pp. 164–165.

Stones, D. I., "A UHF 16-Way Power Combiner Designed by Synthetic Techniques," *Microwave J.*, Vol. 32, June 1989, pp. 117–120.

Sanders, B. J., "110-Way Parallel-Plate RF Divider/Combiner Network and Solid-State Module," *1980 Military Microwaves Conf. Proc.*, London, UK, October 22–24, 1980, pp. 185–198.

Saleh, A. A. M., "Improving the Graceful-Degradation Peformance of Combined Power Amplifiers," *IEEE Trans. on Microwave Theory and Techniques*, Vol. MTT-28, October 1980, pp. 1068–1070.

Foti, S. J., R. P. Flam, and W. J. Scharpf, "60-Way Radial Combiner Uses No Isolators," *Microwaves and RF*, Vol. 23, July 1984, pp. 96–118.

Chen, Y., and J. B. Beyer, "A 11GHz Paraphase Amplifier," *IEEE Int. Solid-State Circuits Conf.*, 1986, pp. 236–237.

# Chapter 7

# Systems Applications of GaAs FET Power Amplifiers

**G. Railton and J. L. B. Walker**

*Pascall Microwave and Thorn-EMI Electronics*

## INTRODUCTION

*lid-state power amplifiers* (SSPAs) are essential elements in a broad spectrum of crowave systems applications including spaceborne, airborne, and ground-based ‹ed and mobile) satellite communications, terrestrial broadcast and telecom- ınications, radar and *electronic warfare* (EW), and tactical weapons, together :h a wide variety of medical, scientific, and industrial instrumentation.

The need for improved channel capacity, transmission quality, and perfor- ınce agility coupled with the low cost of maintenance required in modern tactical d communications systems is leading to the development of GaAs FET power ıplifiers that are subsystems in their own right. In this respect, the inherently ›dular solid-state approach allows high levels of functional integration and control :ess, not feasible in traditional *traveling-wave tube amplifier* (TWTA) technology, ›viding expanded system flexibility. Additional benefits include the elimination high-voltage power supplies, low mass and size, enhanced linearity and noise :formance at high dc and RF efficiency, instantaneous switch-on, long mainte- ıce-free storage life, and improved ruggedness and reliability especially for ›und mobile, airborne, and satellite applications.

This chapter provides an overview of microwave GaAs FET power amplifier ›abilities, operational features, and performance aspects, illustrated through the : of selected applications examples.

## 7.2 SATELLITE APPLICATIONS

In recent years, significant advances have been made in GaAs FET power devic through the development of internally matched multicell topologies and therma efficient structures [1]. Consequently, the use of SSPAs in space has increas dramatically, primarily due to their advantages such as reliability, low mass a size, linearity and higher system flexibility. The latest device developments (r yet space qualified) report 25W of output power at −1-dB gain compression, 1 dB linear gain, and 40% power-added efficiency in a highly linear Class A mo at C-band frequencies [2]. Each major increase in device power provides n opportunities for system enhancement. The potential benefits from SSPA linear in future space programs include a 30% to 50% increase in SCPC/FDMA traf handling capability [3].

### 7.2.1 Reliability

The design life of new satellites continues to increase (from 7 to 20 yr) and her reliability is a critical factor. Solid-state power devices are still a relatively you technology, so the majority of FET reliability data is based on accelerated agi tests at elevated temperatures (250° to 280°C) together with stepped stress me surements to derive a predicted *mean time to failure* (MTTF) within six months so, rather than tens of years (see Sec. 5.6 for a detailed discussion on reliabilit A typical MTTF of between $10^7$ and $10^8$ hr is obtained when semiconductor juncti temperatures are limited to a maximum of 110°C. Environmental stress screeni and "burn-in" of amplifiers and subsystems is carried out to eliminate infant m tality as shown in Figure 7.1. GaAs FETs are static sensitive, and suitable ca must be taken in production to eliminate total failures or weakening of devic

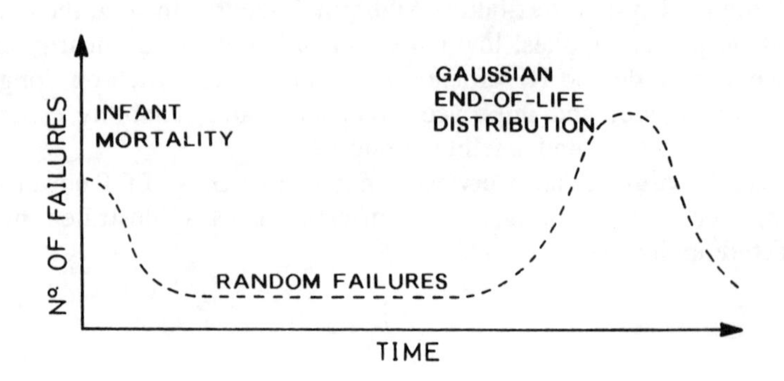


**Figure 7.1** Typical failure distribution.

at can lead to infant mortality. People and furnishings are capable of storing
etween 6 and 15 kV of static charge.

Because there are almost as many SSPA designs as there are applications, it
impossible to be specific with respect to SSPA reliability, but in general com-
ercial earth station application MTTFs of 50,000 to 100,000 hr are typical (10,000
50,000 hr for TWTAs), and with spaceborne SSPAs the MTTF can exceed
)0,000 hr (100,000 hr for TWTAs) [3]. The TWTA is still an art, and small changes
TWT design require a new qualification program, which is not usually the case
ith SSPAs. Many thousands of hours of spaceborne operational experience have
en accumulated with discrete FET devices that provide up to 10W of output
wer, and the gathering of statistical data will continue in years to come, but it
significant that no in-orbit failures of SSPAs have been reported to date.

**2.2 Active Phased Arrays**

he *very small aperture terminal* (VSAT) is a low-cost miniature earth station,
oviding data and video conference communications between remote locations
d central facilities [4]. This requirement, in common with mobile satellite services,
lls for the implementation of spaceborne reconfigurable antennas, such as the active
ased array, providing multibeam frequency reuse technology to maximize spectrum
ility. Array transmission elements are typically small-power (2- to 3-W) SSPAs with
od modular gain and phase tracking and high linearity and power-added efficiency
]. Integration of miniature GaAs FET hybrid power modules [6] with attenuators
d phase shifters in *monolithic microwave integrated circuit* (MMIC) form for
am synthesis, together with thick/thin film dc and control circuits, provides sub-
antial savings in size and mass with high levels of efficiency. These distributed
stems avoid the multipaction/discharge problems associated with traditional, sin-
e high-power TWTA phased arrays.

The *high-voltage FET amplifier* (HVFA) technique provides enhanced effi-
ency for active array modules [7]. This design consists of multiple FETs dc-
nnected in series via linking inductors, as shown in Figure 7.2, with source
pacitors providing the necessary RF ground. Increasing the nominal 10-V single
vice drain bias voltage to, say, 40V using a four-device HVFA reduces the dc/
conversion losses from the satellite bus (24 to 48V) and ohmic loss in the supply
stribution network. For example, an array of 1000 elements, each requiring
10-V, 1-A supply from a distribution cable of 0.001-$\Omega$ resistance, results in a
00-W ohmic loss. A four-device HVFA strategy reduces this loss to 62.5W—a
xteenfold improvement.

In the area of earth observation, *synthetic aperture radars* (SARs) require
tive phased arrays utilizing hundreds to thousands of solid-state transmitting and
ceiving (T/R) modules operating in pulsed conditions. Prime power and cooling

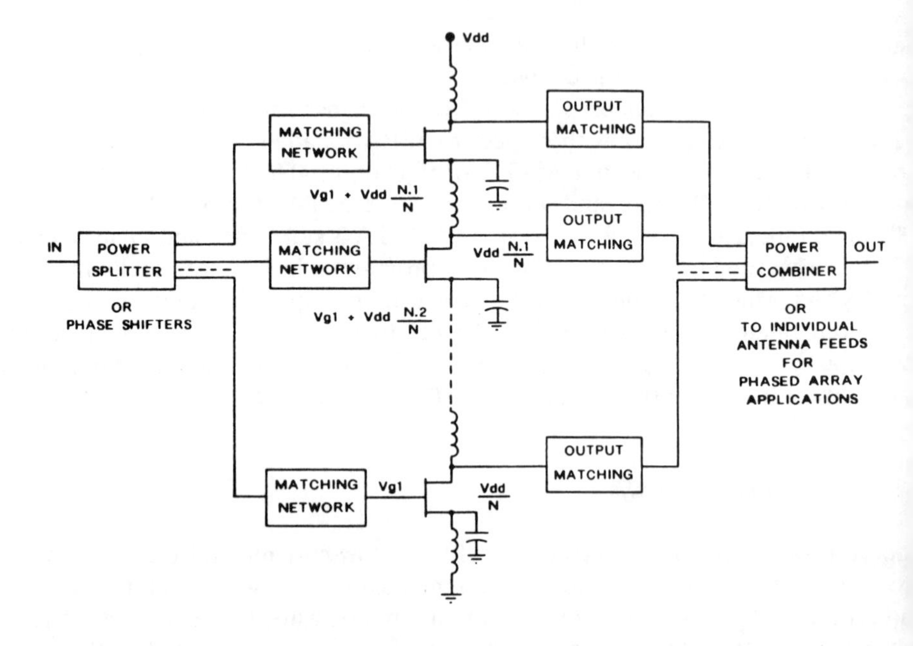


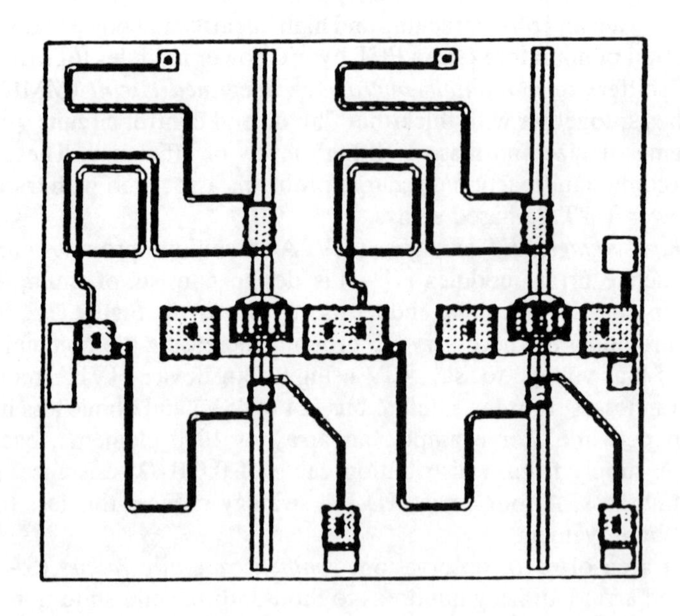

**Figure 7.2** High-voltage FET amplifier. (Reproduced with permission from [7].)

ɔnstraints are eased by using a high-efficiency Class B operation as discussed in ection 1.2. Class B power MMIC amplifiers have been developed [8] providing 0% power-added efficiency, an 8-dB gain, and 1.7W of output power over 5 to GHz, and a 20-W amplifier has been developed at 2.5 GHz with 51% power-dded efficiency, a 13.6-dB gain and two-tone intermodulation products at −22 Bc for 3 dB of output power back-off from saturation [9].

### .2.3 Power Combined Amplifiers

'o realize higher output power than is available from discrete internally matched evices or modules, parallel combining in planar or radial form is an essential nabling technology as discussed in Chapter 6. An added advantage with combined mplifiers is the capacity for graceful degradation [10], illustrated in Figure 7.3, nd the possible suppression of harmonic outputs [11].

The gain limitation of high-power transistors restricts the allowable fan-out etween a single driver and the output stage to 3 or 4 ways in the X-band and less t higher frequencies. Extension to a greater number of ways is feasible where arious types of combiner are linked in a corporate manner, although efficiency is egraded due to increased insertion losses. Also, care must be taken to ensure that hase and amplitude tracking remain within tolerable levels. A realistic upper limit ɔr planar schemes is about 16 ways, since above this a radial (cylindrical) approach less cumbersome and more efficient, although constraints on thermal manage-ent are more severe. This is illustrated in Figure 7.4, where the potential power vels (at −1 dB of gain compression) from 6-GHz power combined SSPAs are ɪown utilizing nominal 8- and 15-W GaAs FETs after allowance has been made ɔr typical combining losses [12].

The block diagram of a C-band SSPA designed for use in a satellite tran-onder application is shown in Figure 7.5, with performance details shown in igure 7.6 [13]. The amplifier comprises six FET driver stages with two PIN diode tenuators, followed by five high-power FETs arranged with one driving four in arallel. More than 11W of output power is provided at a dc to RF efficiency of ver 28% within the 3.7- to 4.2-GHz downlink band. Although this level of power currently available from a single FET device, the combined output stage allows se of more efficient, lower power devices with better thermal management and ence higher reliability. The design life of this SSPA exceeds 10 yr.

The overall amplifier gain is adjustable by telecommand signal to be between 4 and 66 dB via an attenuator control circuit. The gain is also compensated over is control range and over a −10° to +60°C temperature range by an integral emory circuit controlling the second PIN attenuator, and providing less than 0.3-B peak-to-peak output power variation. Overdrive protection over a 24-dB range provided by sensing the RF power at the driver output via a coupler and diode

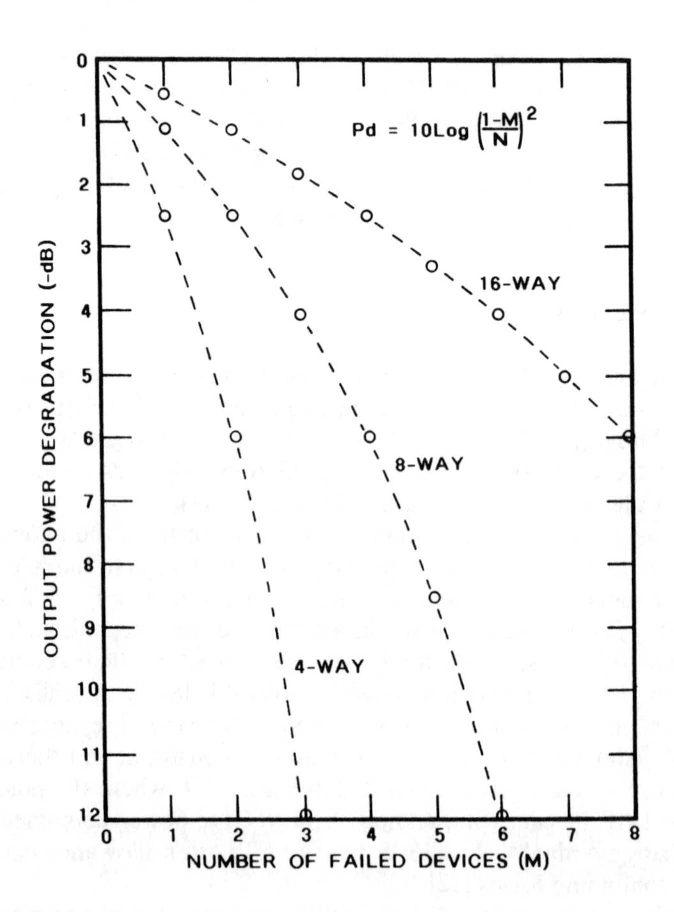


**Figure 7.3** Graceful degradation.

detector and attenuating the RF input drive when a preset limiting level is detecte
Two-tone intermodulation levels are −15 dBc maximum, and second harmoni
are maintained below −50 dBm using an output bandpass filter. The *electror
power conditioner* (EPC) converts the satellite bus power (26 to 43V) to 7.5-
+5-V, and −5-V secondary voltages with an efficiency of 85%.

A 50-W, 12.5-GHz transponder SSPA [14] is realized using an eight-w
radial waveguide cavity combiner as shown in Figure 7.7 with an insertion loss
under 0.3 dB. Eight high-efficiency, thermally and electrically independent pow
modules are used, comprising a driver FET and a balanced stage. Hence sixte
4-W output FETs are effectively combined in parallel with a dc to RF efficien
of 26%. This structure provides high isolation between devices, which is importa
when graceful degradation is required.

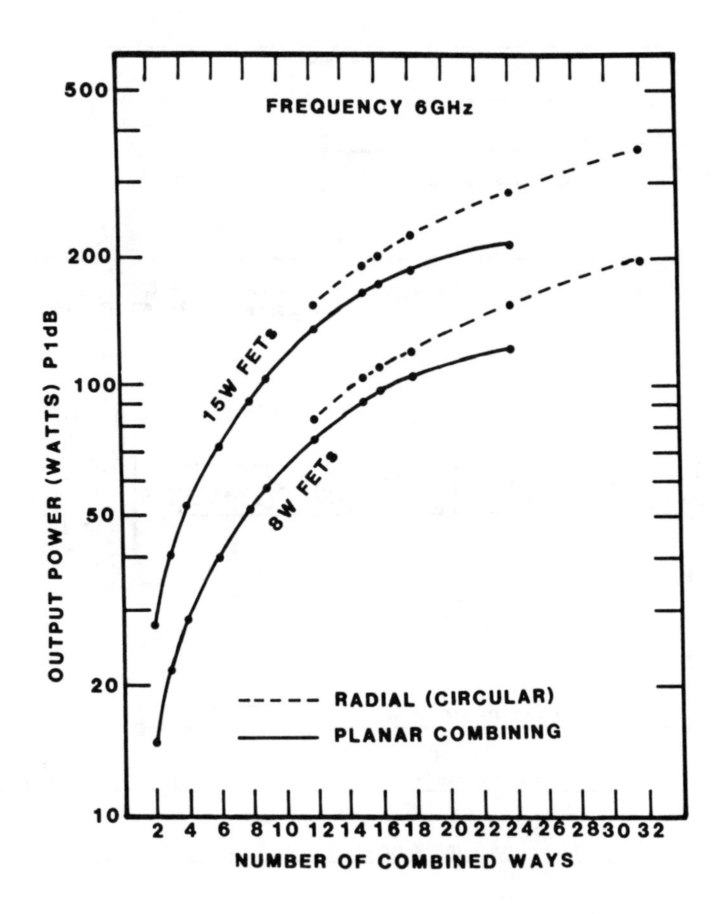


**gure 7.4** Power combining potential.

At higher frequencies, combining techniques have provided output powers f 8W at 18 GHz and 1W at 28 GHz, and an elegant six-way traveling wave scheme ee Sec. 6.4.1) in MMIC form, shown in Figure 7.8, produces in excess of 0.5W t 33 GHz [15].

## .2.4 Mobile Tactical

attlefield mobile satellite communications demand low mass, rugged, reliable, nd high-efficiency SSPAs. Transmission is usually of short duration (burst mode) avoid detection and conserve power, and hence the instantaneous switch-on pability of SSPAs is of great benefit. With portable systems, enhanced battery

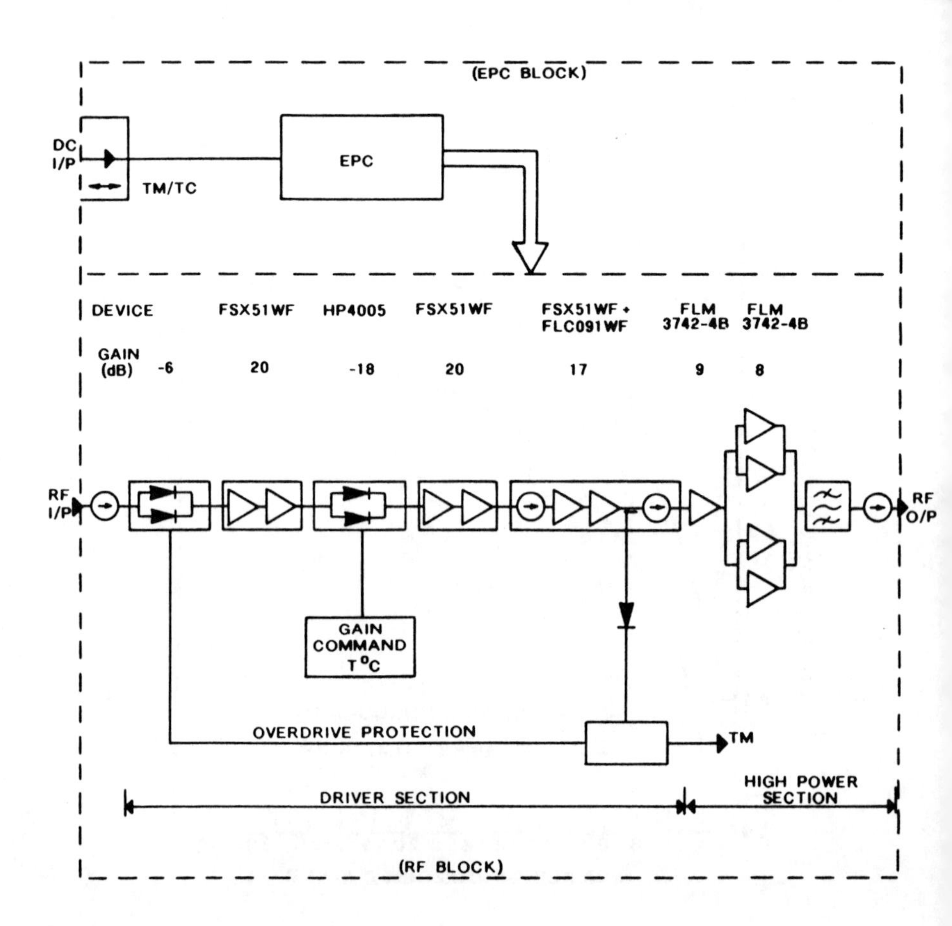


**Figure 7.5** Transponder SSPA block diagram.

conservation is available by utilizing adjustable power mode amplifiers where combination of switched FET drain voltage and constant current bias reduces consumption from, for example, 21 to 2.5W for output RF powers between 2 and 0.1W.

Small physical size is provided through the use of miniature hybrid circui on ceramic substrates; however, low mass is limited mainly by the choice of therm management strategy. An effective technique for short-duration communicatio (and airborne missile systems) is to make use of latent heat energy exchange [16 A *phase change material* (PCM) involving a solid-to-liquid transition is most co venient since a small volume change is involved, allowing the material to be seal inside a heat sink chamber and used repeatedly without replenishment. The pe formance of a PCM heat sink is illustrated in Figure 7.9, where the absorption

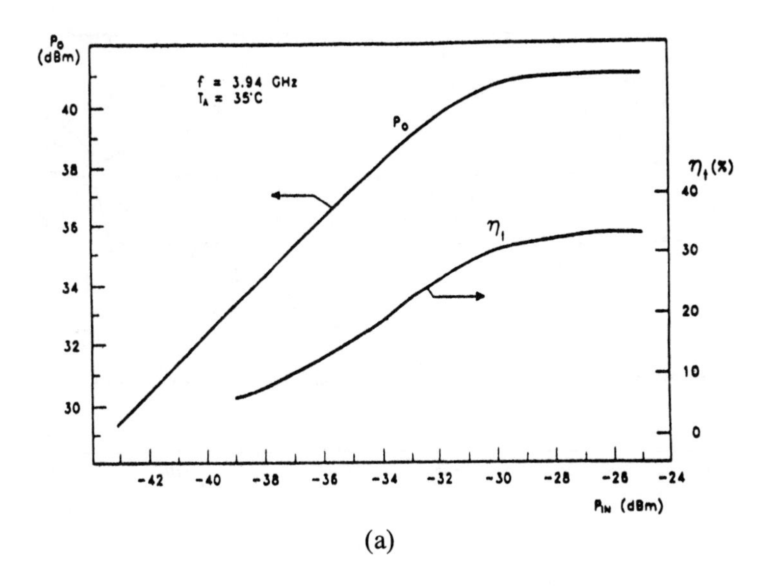


(a)

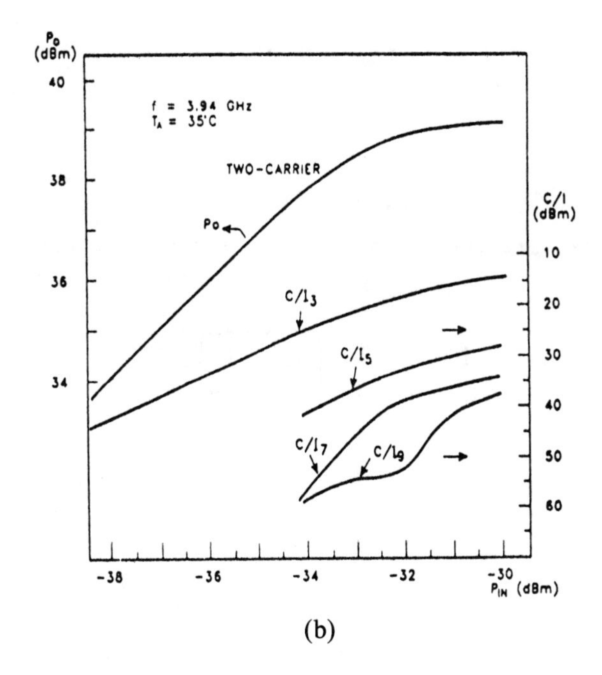


(b)

**Figure 7.6** Transponder SSPA performance: (a) AM/AM transfer function and efficiency, (b) intermodulation products, and (c) AM/PM transfer function.

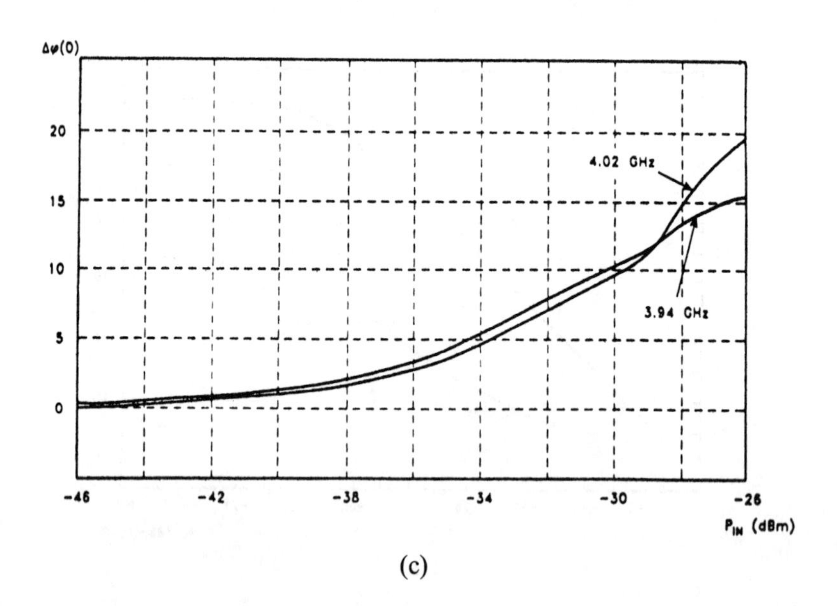


(c)

**Figure 7.6** (Contd.)

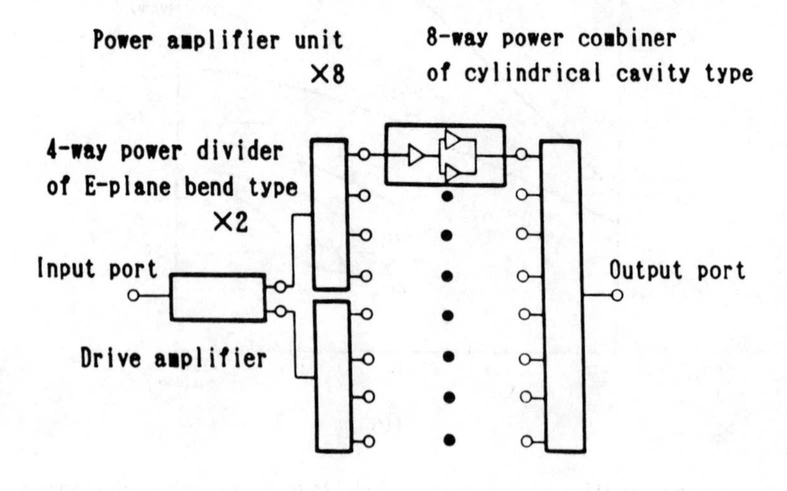


**Figure 7.7** A 50-W 12-GHz combined amplifier.

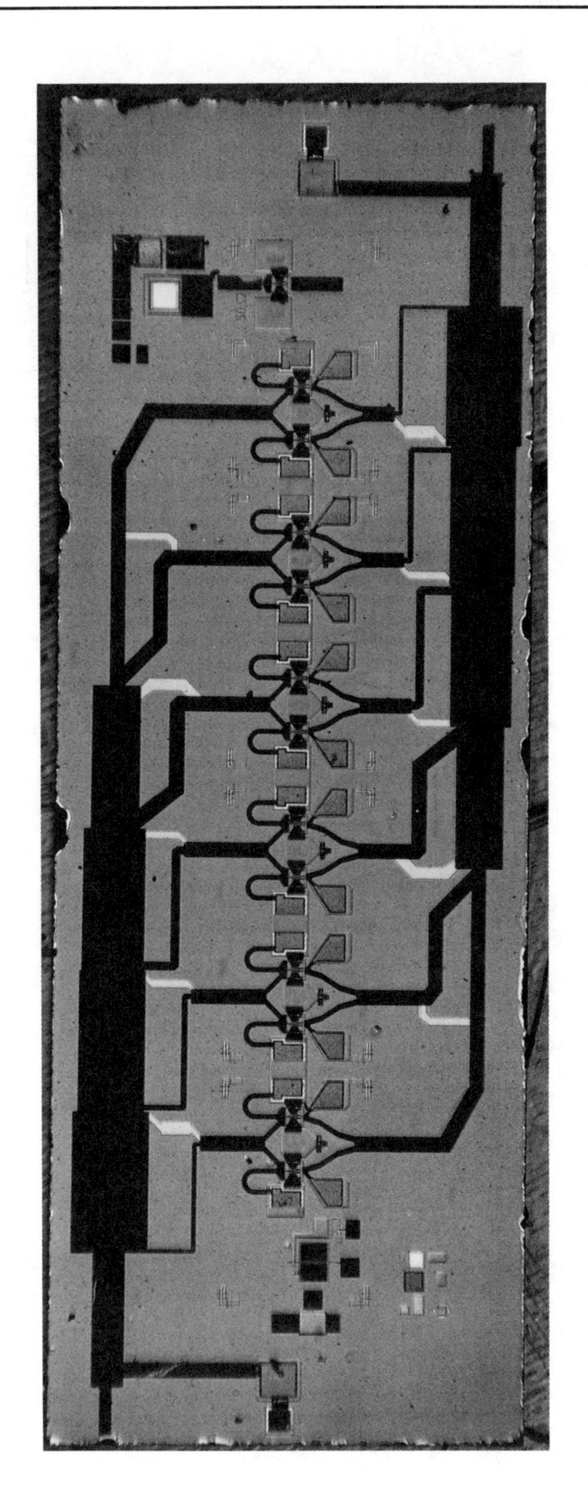

**Figure 7.8** A 0.5-W, 33-GHz MMIC. (Photo courtesy of Texas Instruments [15].)

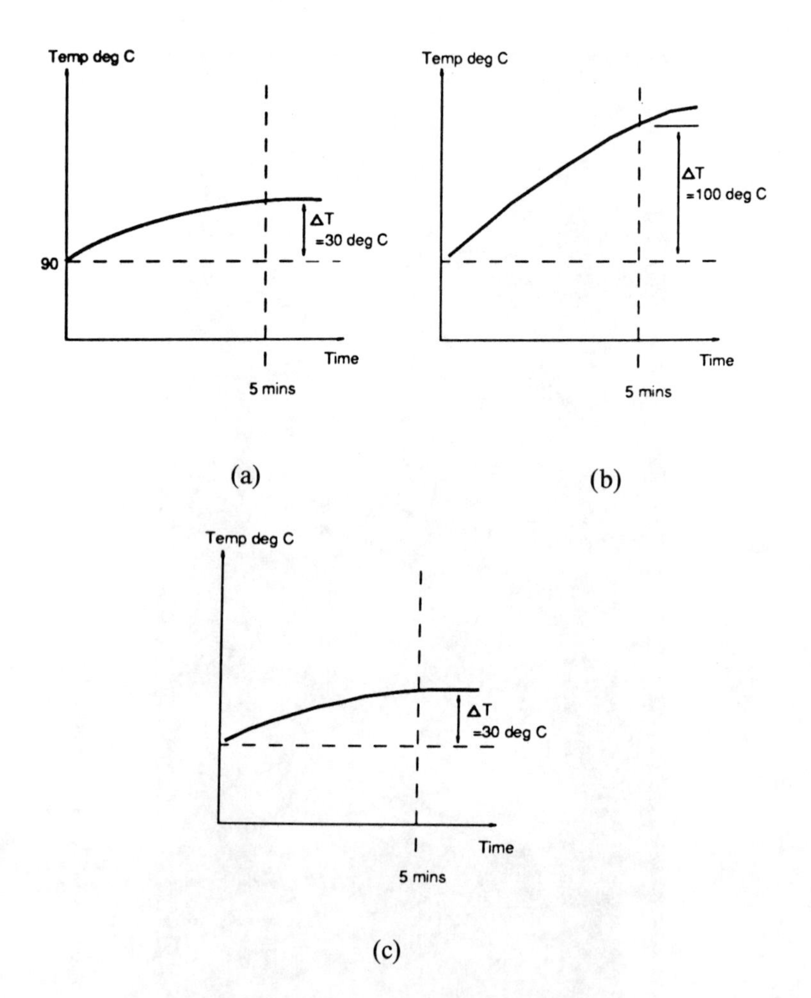


**Figure 7.9** PCM heat sink performance: (a) with PCM and total mass of ~3 lb. (0.8 lb PCM); (b) n PCM, total mass of ~2 lb, and no heat sink; and (c) no PCM, total mass of ~7 lb, an aluminum heat sink (Reproduced with permission from [16]).

200W of heat during a 5-min mission is required. The temperature profile obtaine using 5 lb of aluminum heat sink is replicated with only 0.8 lb of PCM.

Figure 7.10 provides the schematic of a 7.9- to 8.4-GHz, 30-W SSPA designe for military and naval satellite use [17]. Five stages of Class A amplification ar employed to give more than 40 dB of gain. The output stage consists of four nomin 8-W FETs in a planar traveling wave combining circuit (see Sec. 6.4.1) as show in Figure 7.11, which, due to its phase staggered ports, offers a very compact an

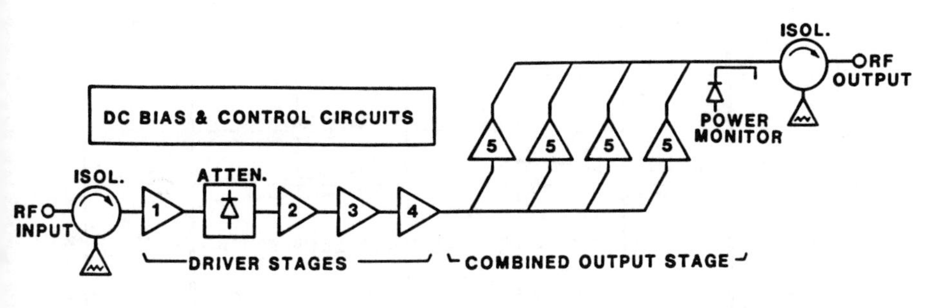


**Figure 7.10** A 30-W, 8-GHz SSPA schematic.

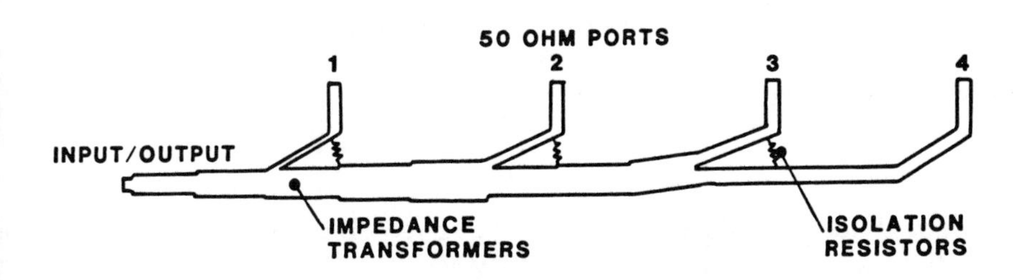


**Figure 7.11** Four-way traveling wave divider/combiner.

efficient layout. A microstrip isolator is used at the input and a low-loss stripline type at the output. A coupler and Schottky diode detector provide output power monitoring and internal *automatic level control* (ALC) facilities in conjunction with a PIN diode attenuator. Output power versus frequency characteristics are shown in Figure 7.12. Saturated power peaks at 35W with a dc to RF efficiency of 24% from a 12-V supply, and the power exceeds 30W at $-1$ dB of gain compression. Amplitude, phase, and two-tone third-order intermodulation data for both the SSPA and a TWTA with similar saturated output power are shown in Figure 7.13. The SSPA exhibits a sharp gain compression knee and some increase in intermodulation near saturation due to simultaneous compression of the output stage and the driver device. To achieve a similar degree of linearity using a TWTA, at

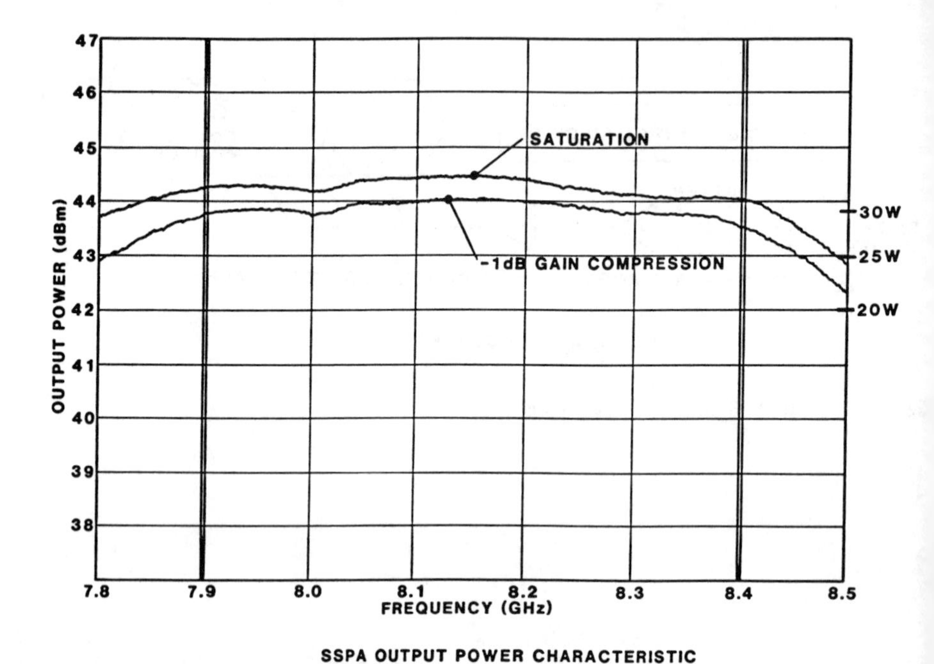


**Figure 7.12** Output power versus frequency.

least a 60-W tube would be required operating at 3-dB output power back-off fror saturation, with a consequent loss of efficiency.

A reliability analysis based on MIL-HDBK 217E yields a predicted MTTF i excess of 200,000 hr. The GaAs FET reliability data were obtained in 10,000-h accelerated life tests carried out by the vendor. This SSPA has demonstrate successful replacement of a 60-W TWTA in a ground mobile DSCS satellite com munications system. Additionally, trials via the SKYNET system demonstrated th feasibility of a highly integrated solid-state transmission unit in association with compact experimental Luneberg lens antenna module.

### 7.2.5 Earth Terminals

In major satellite earth terminals, 50-W C-band SSPA subsystems are currently i use [12]. These rack-mounted units incorporate a comprehensive range of feature including local gain adjustment, RF output power and dc current consumptior and local and remote alarm status displays, power supply and temperature an

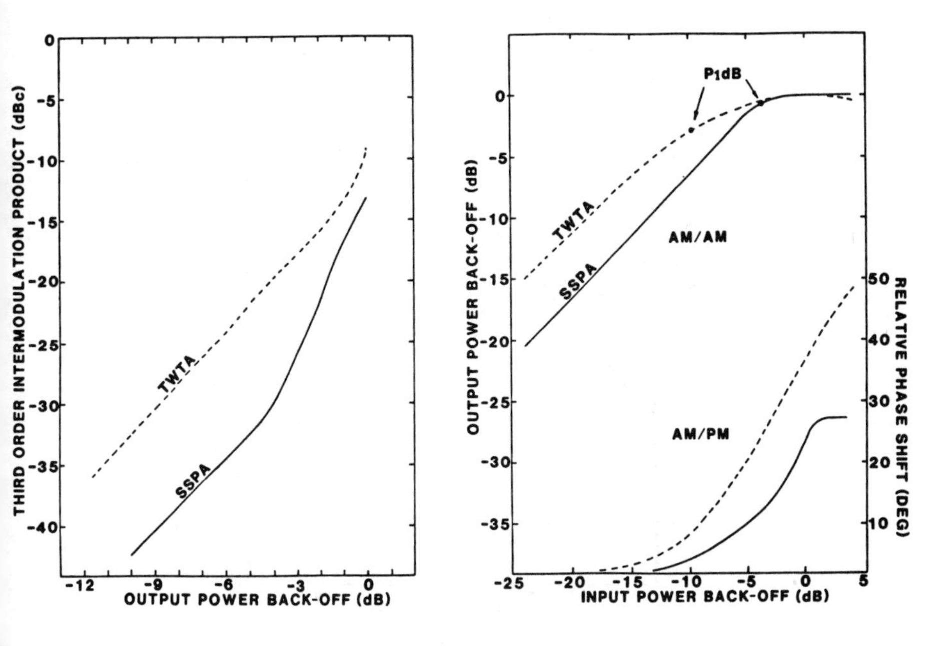


**Figure 7.13** Linearity characteristics.

amplifier bias monitors, reverse RF power, and overtemperature shutdown protection. Parallel combining of four nominally 15-W GaAs FETs is used to obtain this level of power. Amplitude, phase, and intermodulation characteristics are shown in Figure 7.14. An RF-driven self-biasing technique is used on the output stages providing enhanced performance over a 10-dB operating region up to near saturation levels together with a 15% to 20% reduction in thermal dissipation at lower power operation, hence improving reliability. A shift or dip in intermodulation performance is not uncommon with high-power amplifiers at high drive levels. This highlights the difficulty of using the conceptual third-order intercept point method of intermodulation level prediction. Specifying discrete levels within the operating region, or two notional intercept points together with applicable operating regions, provides more accurate system performance information. A 12-way combined power booster amplifier is shown schematically in Figure 7.15, produced by the further combination of three of the 50-W nominal output stages discussed above. For a 10-W input drive level, 130W is generated at $-1$ dB of gain compression, and the saturated output power is in excess of 160W. The unit consumes 75A of dc current from the 10-V supply; hence, effective thermal management is essential. Typically, the operating environment for such an amplifier is a benign 20° to

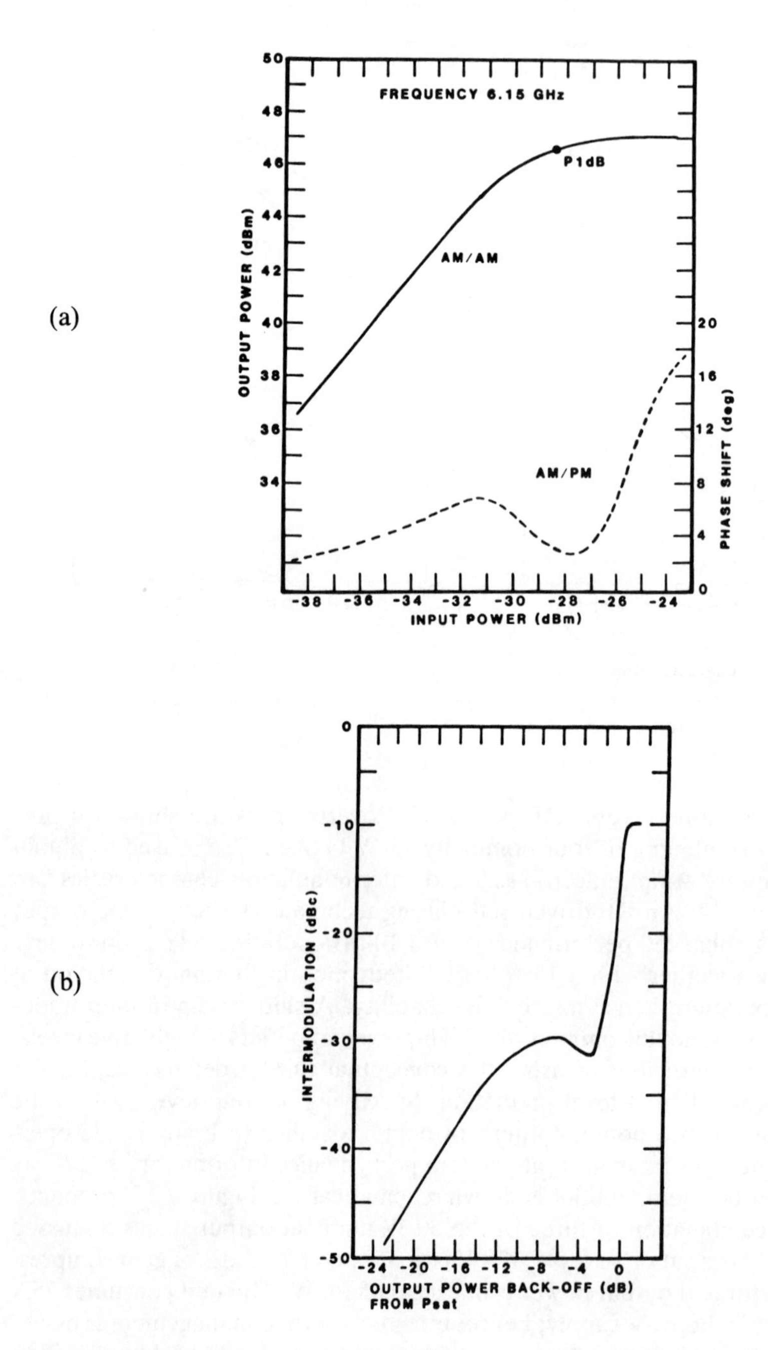


**Figure 7.14** Performance of a 50-W C-band SSPA: (a) amplifier and phase response and (b) intermodulation.

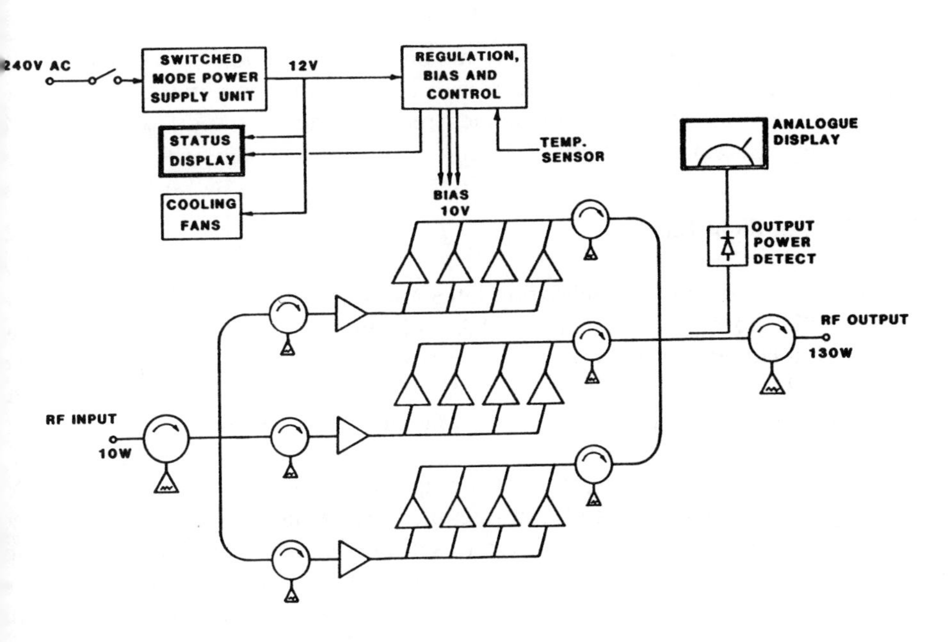


**Figure 7.15** A C-band power booster schematic diagram.

40°C. Multiple dc fan cooling is provided such that fan replacement can occur at prescribed intervals without taking the amplifier out of service.

In the North Sea oil fields, several platforms are provided with solid-state communications terminals. Mounted within the parabolic antenna hub adjacent to the low-noise receiving amplifiers are 5- to 10-W, 14.0- to 14.5-GHz SSPAs providing the primary transmitting power. Hub mounting ensures high efficiency since it eliminates the usual RF power feed losses between a communications cabin and the antenna, which can be up to 10 dB. The use of a low-voltage dc supply between the cabin and the hub is an essential safety feature. A dual-channel "hot standby" system configuration is used where a second amplifier is fully powered and available via automatic channel switching to guarantee communication access. Integral SSPA RF power monitoring and gain adjustment input interfaces allow remote ALC and status monitoring.

Small portable earth terminals used in video conferences and outside broadcast applications take advantage of hub-mounted SSPAs at the C- and Ku-bands where low-voltage supplies, ruggedness, and reliability are of great value.

## 7.3 TERRESTRIAL TELECOMMUNICATIONS

Terrestrial telecommunications are currently dominated by large operators who are responsible for the provision of national networks and intercontinental links.

These networks have relied for many years on TWTA technology, which demands high levels of maintenance. Recent growth in solid-state amplifier capability has allowed an increase in smaller unmanned repeater facilities and low-cost private links in the business community.

### 7.3.1 Line-of-Sight Links

Line-of-sight telephone and television links are having 10- to 25-W SSPA subsystems installed to replace existing TWTA and Impatt diode amplifiers. A typical retrofit kit includes a small switched-mode *power supply unit* (PSU) to convert the nominal −24-V rack unit supply to +12V for the SSPA, a temperature-compensated SSPA with external gain adjustment facility and convection cooling fins, and a mounting plate assembly with SMA/waveguide transitions and interfacing cables. The assembled units including the PSU occupy the position of an existing TWTA, allowing rapid installation and considerable savings in rack real estate. Once installed, the amplifier gain is adjusted to set the channel operating power point. Unlike TWTAs, no complex alignment procedures or 24- to 48-hr stabilization periods are necessary. As well as providing link margin enhancement, a scope for reduced maintenance costs is available, with MTTFs of around 50,000 hr.

In remote unmanned repeater stations, dual-SSPA subsystems in a redundant configuration are used with automatic changeover switching, monitors, and controls. The backup channel SSPA is often arranged in a "warm standby" mode in which the FET gate bias is applied, but the drain supply is switched off under logic control. Hence the backup channel remains cool to enhance system reliability, but is instantly available on changeover command. Where high stability is required in an unstable temperature environment, such as in a remote repeater cabin, proportionally controlled forced air cooling is fitted. The supply voltage and hence speed of a high-reliability (100,000-hr MTTF) brushless dc fan is closed-loop controlled with respect to SSPA case temperature, which, together with the amplifier's integral gain compensation, ensures stability.

### 7.3.2 Linearized Amplifiers

All power amplifiers exhibit nonlinearity in gain and phase transfer characteristics at high RF drive levels. These amplitude (AM/AM) and phase (AM/PM) distortions generate unwanted outputs such as harmonics, intermodulation products, and phase modulation, leading to adjacent channel interference and loss of transmission fidelity. In particular, high-level multicarrier digital systems such as 64 *quadrature amplitude modulation* (QAM), with the difference between adjacent states as low as 1.2 dB in power and 9.5 deg in phase, need very linear operation to reduce spectrum spreading and frequency spacing of adjacent channels [18].

A crude but commonly used method of reducing nonlinear effects is to back-
ff an amplifier's output power operating point relative to saturation (*output back-
ff* or OBO). However, this results in reduced output power and traffic handling
apability and power-added efficiency, which increases demands on power supplies
nd thermal dissipation with a consequent degradation in reliability. For TWTAs,
typical OBO requirement is between 6 and 8 dB, compared to 2 to 4 dB in the
nore linear SSPA. This reduces power-added efficiency from a nominal 30% to
% for a TWTA or from a nominal 20% to 10% with an SSPA.

Several specific linearization techniques are currently in use such as feedfor-
vard, feedback, and predistortion. In addition, adaptive bias control and harmonic
oading are available with SSPA designs. With feedforward the input signal is split
nto two paths as shown in Figure 7.16, with a main channel providing most of the
RF power and a lower power reference channel providing the inverse of all dis-
ortion products present within the main channel. Recombination of the channels
esults in distortion cancellation at the output. The main drawback of this technique
s the need for two fairly well matched amplifiers in parallel, which lowers efficiency
nd increases weight and circuit complexity.

Feedback techniques, shown in Figure 7.17, require special treatment of the
ime delays and bandwidth involved. Simple feedback can be applied via a band-
imiting cavity within the loop. The simplicity of this approach is offset by its narrow
andwidth and the need to avoid destructive oscillations.

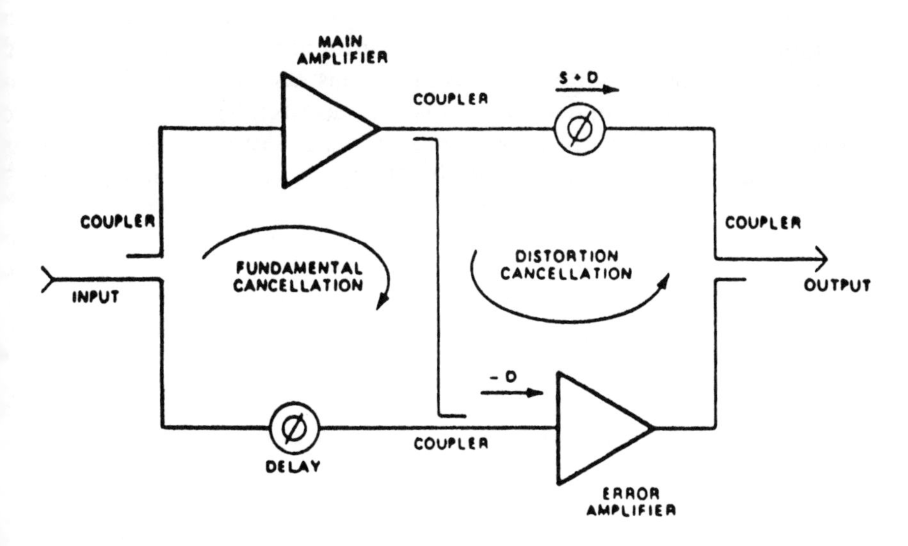


**igure 7.16** Feedforward linearization.

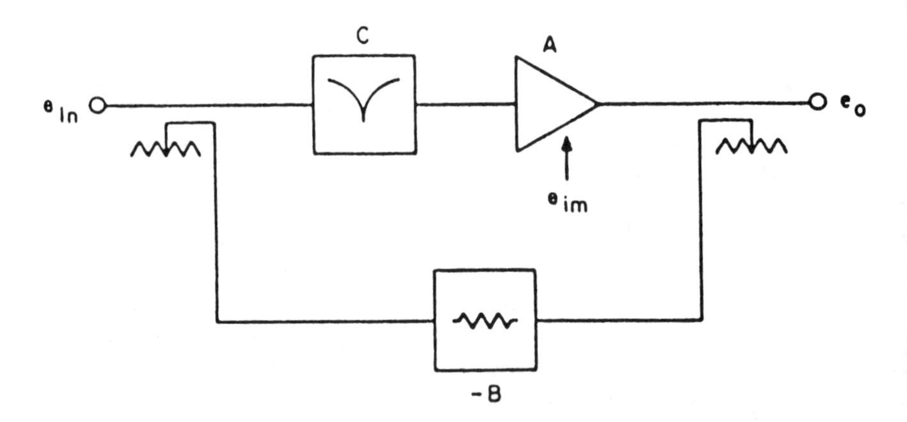


**Figure 7.17** Feedback linearization.

The predistortion approach involves the use of nonlinear elements such a
diodes or low-power FETs to generate amplitude and phase distortion at the amplifier input varying with input drive level. These controlled inverse distortions ar
used to compensate for the amplifier's nonlinear transfer characteristic, henc
linearization occurs. A typical nonlinear function generator arrangement is show
in Figure 7.18. This technique is widely used since efficiency and bandwidth ar
good and it offers the widest flexibility of design and application. Predistortio
circuits are readily integrated with an SSPA, providing interaction with other control functions such as temperature compensation, gain adjustment, and limitin
characteristics as shown in Figure 7.19. Typical performance improvement throug
the use of predistortion for SSPAs and TWTAs is illustrated in Figure 7.20. Recen
moves toward 256 QAM will demand the use of linearized SSPAs.

Adaptive bias control can be achieved through controlled variation of th
FET gate voltage in response to changes in RF input drive, maintaining constan
gain and hence a linearized transfer characteristic. Relatively good performanc
can be achieved with simple schemes; however, care must be taken to avoid introducing excessive time delay in the bias controlling circuit since memory effects ca
introduce hysteresis into the transfer function and degrade linearity in dynami
operation [19]. To maintain dynamic linearity, the bandwidth of the bias networ
needs to be far wider than the maximum modulation frequency.

## 7.4 RADAR AND EW APPLICATIONS FOR HIGH-POWER GaAs FET AMPLIFIERS

High-power GaAs FET amplifiers are finding increasing use in radar and EW
systems, and it is the object of this section to consider their use in such system

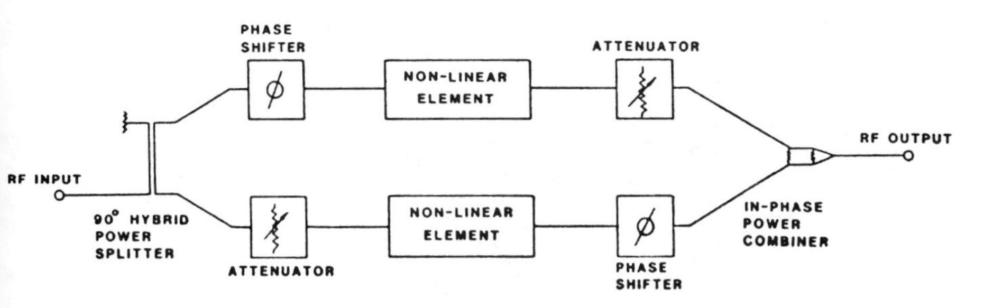


**igure 7.18** General nonlinear function generator.

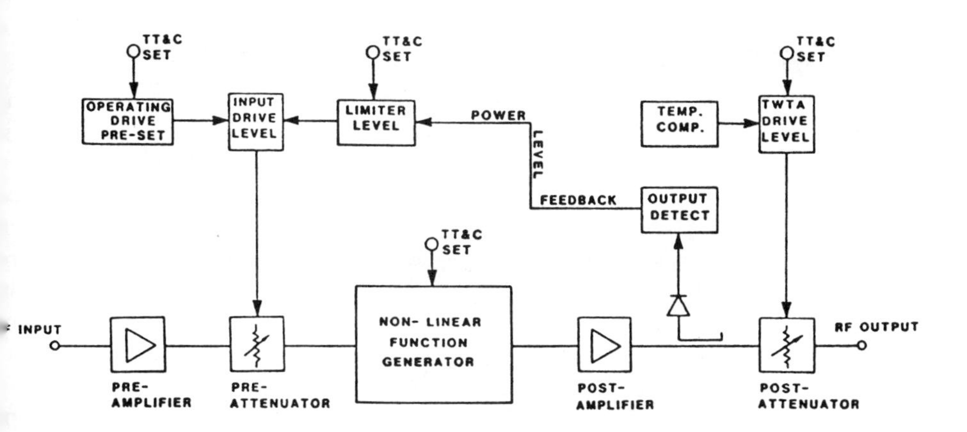


**igure 7.19** Predistortion linearization.

nd to determine the influence of some of the key parameters, such as bandwidth, ain flatness, and phase linearity, on the performance of the system. Modern EW ystems attempt to jam or to deceive the enemy either by retransmitting the enemy's ntercepted radar signal after suitable modification or by generating a false radar ignal. Hence, as far as GaAs FET power amplifiers are concerned, there is no asic difference between radar and EW applications other than that the instanta-eous bandwidth of the amplifier must be much greater for EW (e.g., 6- to 18-Hz) because of the multiplicity of different types of radar operating at different requencies that one might wish to jam. Hence it is sufficient to consider only the adar application in this section.

Radar systems can be divided into two categories, namely CW and pulsed. W radar systems are restricted to low-power levels (<1W) because of the need

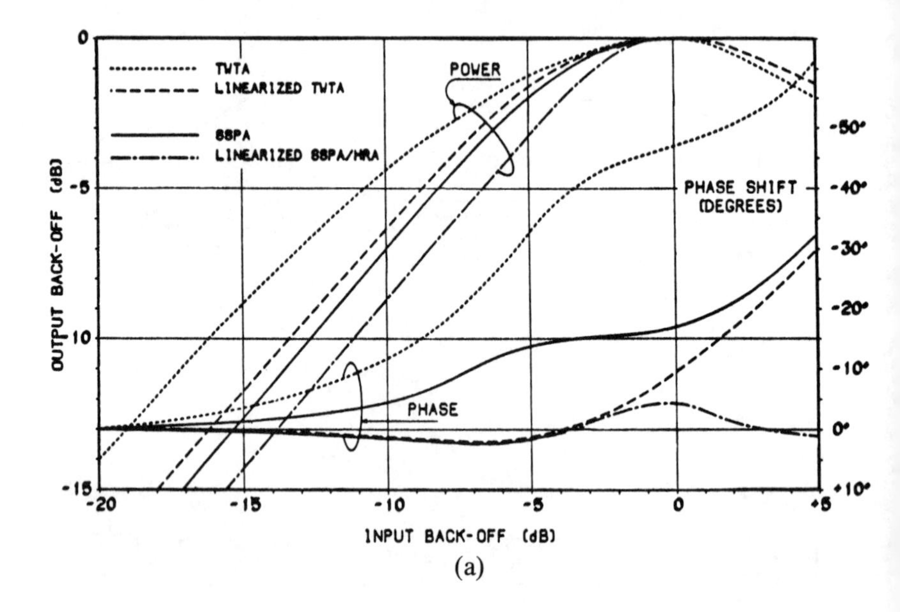


(a)

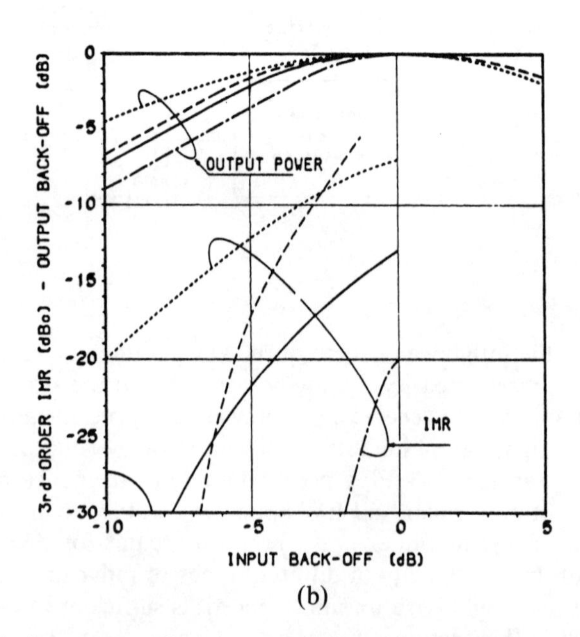


(b)

**Figure 7.20** Linearity improvement through use of predistortion: (a) amplifier AM/AM and AM/PM characteristics and (b) AM/AM and third-order intermodulation ratio (two-carrier) versus input back-off.

o avoid damaging the sensitive low-noise receiving amplifier and because of trans-nitter noise leaking into the receiver and desensitizing it. Hence, high-power GaAs ET amplifiers are not used in CW radars. This problem arises because both the ransmitter and receiver are operating simultaneously in a CW radar and only imited isolation can be achieved between them, assuming a single antenna is used. n a pulsed radar system, the transmitter is turned off when the receiver is turned n, and vice versa, so the receiver can be protected from the transmitter by a high-solation transmitting and receiving switch. Thus only pulsed radar systems need o be considered in this chapter.

The simplest type of pulsed radar is one in which a single-frequency sinusoidal ulse is transmitted. In such a system a narrow pulse must be used in order to btain good range resolution, that is, distinguish between two closely spaced targets. )ne also needs a narrow pulse if the radar is to detect close-in targets since the ystem is blind to any reflected signal received while the transmitter is still operating. he minimum range and resolution are both given by [20]

$$R = c\tau/2 \tag{7.1}$$

vhere $c$ is the velocity of light and $\tau$ is the pulsewidth. As an example, suppose ve want to be able to distinguish between two targets that are 30 m apart, then he maximum pulsewidth that can be transmitted is 0.2 $\mu$s.

To detect distant objects unambiguously, all reflected signals must be received efore the next pulse is transmitted. Thus long-range performance requires a long ime interval between pulses. The unambiguous range is also given by (7.1) if $\tau$ ow represents the time interval between pulses. As an example, if an unambiguous ange of 15 km is required, then a 100-$\mu$s time interval between pulses is needed.

If we define the limit of detectability to be when the reflected signal level rom a distant target equals the thermal noise level, then it can be shown [21] that his implies that the minimum required transmitter power is given by

$$P_T = \frac{4\pi R_{max}^4 \lambda^2 k T_0 F}{A^2 \sigma} \tag{7.2}$$

vhere

$R_{max}$ = maximum required range
$\lambda$ = wavelength
$k$ = Boltzmann's constant
$T_0$ = reference temperature of 290K
$F$ = receiver noise figure
$A$ = antenna aperture area

$\sigma$ = radar cross section of target
$P$ = transmitted peak pulse power
$\tau$ = pulsewidth.

In deriving (7.2), it is assumed that there is zero system loss, that single-puls detection is used, and that the pulsewidth is the reciprocal of the receiver's band width. This condition is a close approximation to the ideal matched filter receive for a simple pulse, when the output signal-to-noise ratio is maximized in the pres ence of white Gaussian noise. As an example, suppose we want to detect an objec having a radar cross section of 1 m$^2$ at a range of 15 km using a 10-GHz rada system having an antenna aperture of 1 m$^2$ and a noise figure of 6 dB, then a pea pulse power of 91W is needed if a 100-ns pulse length is used. In radar desig enormous emphasis is placed on achieving the minimum noise figure so as t minimize the amount of transmitter power needed. This is done not only to eas the cooling and power supply problems but also to reduce the probability of th radar transmissions being intercepted by the enemy.

As these examples show, long-range performance and good resolution take together imply the need for a very high peak power, short pulsewidth, low-duty cycle radar. These requirements are ideally suited to solid-state devices such a Impatt diodes whose output power capability is thermally limited, but they caus considerable problems for GaAs FET and HEMT devices whose output power i electronically limited. As explained in Chapter 1, the output power of a GaAs FE is limited by the drain-gate breakdown voltage and by the maximum fully ope channel current, and neither of these parameters increases by more than a sma amount if the device is operated under pulsed rather than CW conditions.

Some reduction in the required output power can be achieved if the return from a number of pulses are coherently summed [22] prior to detection. For exam ple, if 10 pulses are coherently summed, then the transmitter power can be reduce tenfold for the same probability of detection. This also implies that the radar mu be pointing at the target for 10 times as long before moving to the next bea position. In the case of the example above, the total dwell time would be 1 m Noncoherent integration also reduces the amount of transmitter power require for a given range but by a lesser extent than does coherent integration.

As (7.2) clearly shows, there is a pulse length, peak pulse power trade-of The peak pulse power could be reduced to only 0.9W, which is readily generate at 10 GHz by a GaAs FET amplifier if the pulse length were 10 $\mu$s instead of 0. $\mu$s. However, this would also decrease the resolution capability of the radar on hundredfold. The solution to this problem is to employ the technique of puls compression [23] in which a 10-$\mu$s frequency- or phase-modulated pulse is tran mitted instead of a 0.1-$\mu$s pure single-frequency sinusoid. The reflected pulse fro the target is then applied to a frequency-dependent delay line in the receiver, whic compresses the pulse to 0.1 $\mu$s in this example. As a result pulse compressio nables one to transmit a long pulse with the consequent reduction in peak power, et still achieve the range resolution of a short pulse. Pulse compression is used in early all modern radars, and coherent integration may also be used to further educe the peak power requirement as well as to enable Doppler information to e extracted. Many standard textbooks on radar such as [24–26] include a section escribing pulse compression in detail. Only the elementary principle is described ere so that the reader can understand the implication of its use on the required erformance parameters of a GaAs FET power amplifier.

Pulse compression may be based on linear frequency modulation, nonlinear requency modulation, or phase coding (either binary, such as Barker code, or olyphase, such as Frank code), but it will be assumed that linear FM is used ecause this enables the required amplifier properties to be deduced most readily. n a linear FM system, the frequency of the transmitted pulse is increased (or ecreased) linearly throughout the duration of the pulse as shown in Figure 7.21. 'his signal can be generated, for example, by applying a ramp voltage to a voltage-ontrolled oscillator followed by a power GaAs FET amplifier as indicated in 'igure 7.21.

The compression ratio is, by definition, the ratio of the uncompressed to ompressed pulse lengths and values of 100 are readily achieved. The compression atio is nearly identical to the time-bandwidth product $\tau \Delta f$ where $\tau$ is the transmitted ulse length and $\Delta f$ is the total change of frequency across the pulse (the exact elationship depends on where the start and end of the pulse are defined to be) 23]. The frequency spectrum of a single pulse with linear FM is of course com-licated. For compression ratios of the order of 100 or more, the spectrum has an lmost flat amplitude over a total bandwidth of nearly $\Delta f$ centered on the mean ransmitted frequency [23]. However, some energy also exists outside this region nd so the power amplifier needs a bandwidth of at least $2\Delta f$ if distortion of the pectrum is to be avoided. For the example under consideration of a 10-$\mu$s trans-nitted pulse at X-band having a compression ratio of 100, the amplifier needs a andwidth of at least 20 MHz. In practice, the radar is likely to use a variety of lifferent center frequencies for *electronic counter-countermeasure* (ECCM) pur-oses, so the bandwidth required will be determined by this aspect rather than by ulse compression considerations. It is of interest to note that an unmodulated .1-$\mu$s pulse has a sin $x/x$ spectrum with the width between the first nulls being 0 MHz, and hence the bandwidth required for a given range resolution remains he same regardless of whether or not pulse compression is used. In fact, a more letailed analysis of radar system performance shows that range resolution is fun-lamentally determined by the transmitted bandwidth rather than by the pulse ength.

As shown in Figure 7.21 for a system employing linear FM pulse compression, he reflected pulse from the target is first down-converted in the receiver to IF and hen applied to a dispersive delay line. Such a delay line has a delay which either

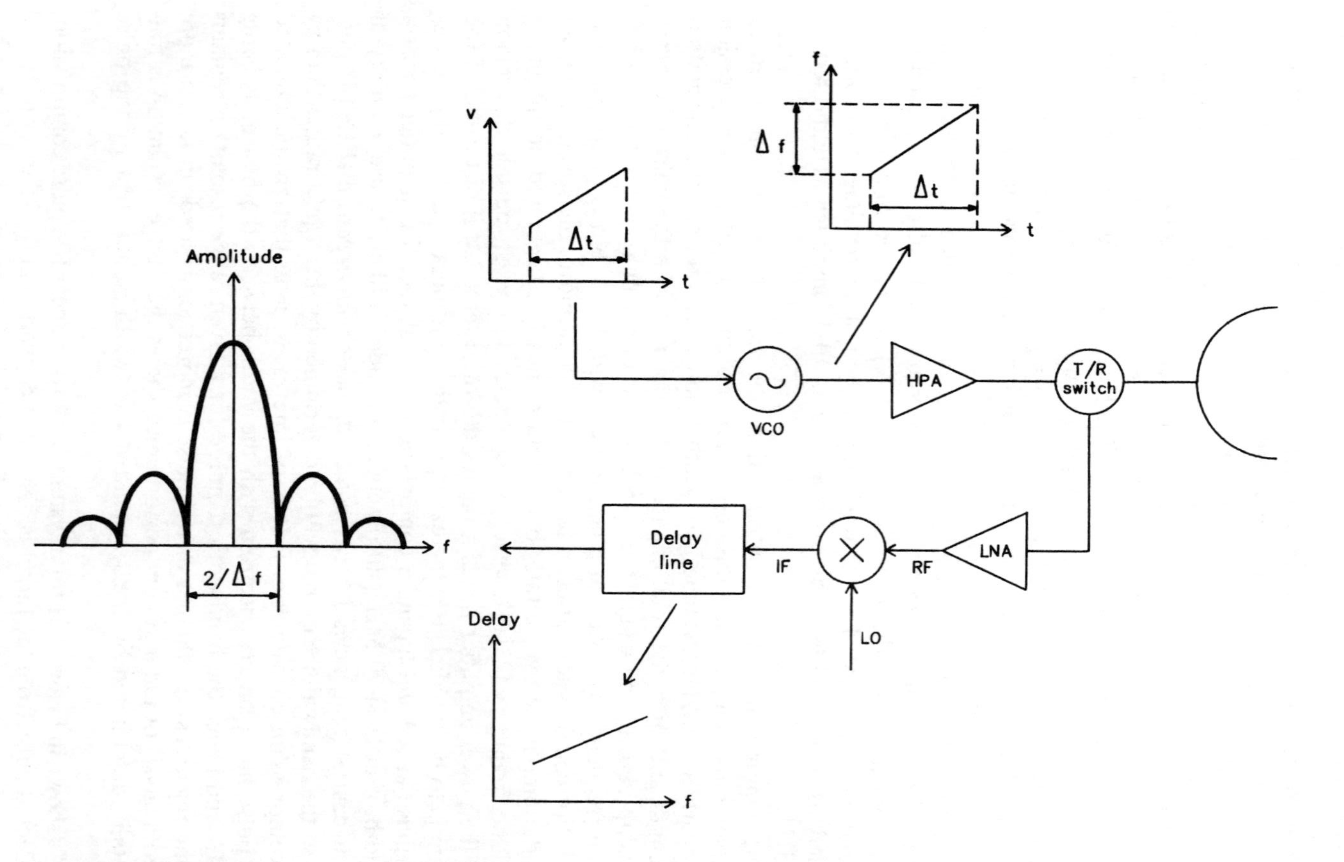


**Figure 7.21** Block diagram of a linear FM pulse compression system.

creases or decreases linearly with frequency depending on whether the trans-
itted frequency increases or decreases with time and on whether the local oscillator
equency is above or below the RF frequency. The output from the delay line is
sin $x/x$ pulse with a width between the first nulls of $2/\Delta f$. Using this criterion for
efining the width of the compressed pulse, the ratio of uncompressed to com-
ressed pulse widths is $\tau\Delta f/2$, which, in this case, has a value of 50. One disadvantage
f this simple form of pulse compression system is that a sin $x/x$ pulse has time
delobes that are only 13 dB below the peak. The detection circuitry in the receiver
as no means of knowing whether these signals, which occur at earlier or later
stants of time than the peak, are the result of echoes from genuine targets at
ifferent ranges or false targets arising from the time sidelobes of the main pulse.
ike antenna sidelobes these time sidelobes, or *range sidelobes* as they are some-
mes called, may obscure genuine targets and present false ones on the display.
ccordingly, the delay line normally has some weighting function or filter associated
ith it to reduce these sidelobes to a respective level, say, 30 to 40 dB down. This
n be readily achieved, albeit with a slight increase in compressed pulsewidth,
hich is equivalent to a slight reduction in range resolution, and a small reduction
peak output from the matched filter.

With this background we can now assess the effect of amplitude and phase
istortion produced by the GaAs FET power amplifier on the performance of the
ulse compression system. The gain of the amplifier is given by

$$S_{21}(\omega) = |S_{21}(\omega)| \exp[j\phi(\omega)] \tag{7.3}$$

or the amplifier to transmit the FM pulse without introducing any distortion,
$_{21}(\omega)|$ must be a constant over the bandwidth of the signal and $\phi(\omega)$ must also
ave a linear phase characteristic [27] or, equivalently, a constant group delay
sponse, since group delay is by definition $-d\phi/d\omega$. Thus, in the ideal case,

$$\begin{aligned} |S_{21}(\omega)| &= G \\ \phi(\omega) &= -T_g\omega \end{aligned} \tag{7.4}$$

o determine the effect of gain and group delay ripple within the amplifier, assume
at they both have a sinusoidal variation given by

$$\begin{aligned} |S_{21}(\omega)| &= G + \alpha \cos\frac{\omega}{\omega_r} \\ \phi(\omega) &= -T_g\omega - \beta \sin\frac{\omega}{\omega_r} \end{aligned} \tag{7.5}$$

where $\omega_r$ is a constant such that $\omega/\omega_r$ determines the number of ripples in th
passband. Then,

$$S_{21}(\omega) = \left(G + \alpha \cos \frac{\omega}{\omega_r}\right) \exp\left[-j\left(T_g\omega + \beta \sin \frac{\omega}{\omega_r}\right)\right] \quad (7.$$

If $\alpha$ and $\beta$ are both small, then $S_{21}(\omega)$ is given approximately by

$$S_{21}(\omega) \approx G\left(1 + \frac{\alpha}{G}\cos\frac{\omega}{\omega_r} - j\beta \sin \frac{\omega}{\omega_r}\right) \exp(-jT_g\omega)$$

$$\approx G\left[1 + \frac{1}{2}\left(\frac{\alpha}{G} - \beta\right) \exp\left(j\frac{\omega}{\omega_r}\right) + \frac{1}{2}\left(\frac{\alpha}{G} + \beta\right) \exp\left(-j\frac{\omega}{\omega_r}\right)\right] \exp(-jT_g\omega) \quad (7.$$

If the incident compressed pulse is represented in the time and frequency domai
by $a_1(t)$ and $a_1(\omega)$, respectively, then the output pulse in the frequency doma
after passing through the amplifier is given by $b_2(\omega) = S_{21}(\omega)a_1(\omega)$ with $S_{21}($
given by (7.7). The output in the time domain is thus given by

$$b_2(t) = G\left[a_1(t - T_g) + \frac{1}{2}\left(\frac{\alpha}{G} - \beta\right) a_1\left(t - T_g + \frac{\omega}{\omega_r}\right) + \frac{1}{2}\left(\frac{\alpha}{G} + \beta\right) a_1\left(t - T_g - \frac{\omega}{\omega_r}\right)\right] \quad (7.$$

which shows that the output consists of three components, namely, the norm
amplified and delayed distortion-free term plus a pair of echoes with one appeari
ahead of the main response and one after it. The radar system will interpret the
echoes as false targets just as it does for the time sidelobes associated with pul
compression. Note that no assumption was made about what form $a_1(t)$ had, a
therefore this analysis applies to any type of pulse compression system or, indee
to a system without pulse compression. If the echo signals are to be 30 dB bel
the main signal and if they are caused only by gain ripples, then (7.5) and (7.
show that the maximum permissible gain ripple in the amplifier is ±0.5 dB, wh
if caused only by group delay variation, then the maximum phase ripple amplitu
about linear is ±3.6 deg. Of course, some of this budget has to be allocated
components other than the power amplifier such as the low-noise amplifier a

own-converter in the receiver. Thus in any practical radar system the power mplifier must have a very flat amplitude and group delay response.

In those situations where the dc supply to the power amplifier must also be witched off when there is no RF input present, it is necessary to have the dc pulse tart before the RF pulse so as to minimize the amplitude and phase changes that vould otherwise occur during the early part of the RF pulse while the GaAs FETs each thermal equilibrium and power supply transients decay. Of course, for true Class B amplifiers the removal of the RF pulse automatically switches off the dc power, which simplifies the control circuitry, but it does lead to a larger amplitude nd phase variation across the pulse.

The power amplifier can usually be operated in the gain compression region vith advantage, provided the required gain and group delay flatness can still be chieved, because this maximizes the efficiency of the amplifier (see Sec. 1.2.2). However, the harmonic output level is inevitably increased if operated in compression and so it is sometimes necessary to insert a filter between the power amplifier nd the antenna to reduce the harmonic output to an acceptable level for EMC easons.

The use of high-power GaAs FET amplifiers is now considered in two specific adar applications, namely, a man-portable radar system and a phased array radar. Figure 7.22(a) shows the MSTAR (*M*an-portable *S*urveillance and *T*arget *A*cquisition *R*adar) developed and produced by Thorn-EMI Electronics. MSTAR is a pulsed Doppler radar employing pulse compression. It weighs 35 kg, operates from batteries, and reportedly [28] has a range of 24 km. Figure 7.22(b) is a photograph of the high-power GaAs FET amplifier used in this radar. A number of general bservations can be made about this radar. First, a small antenna must be used if the radar is to be man portable, which implies the use of a high frequency if an cceptable antenna beamwidth is to be achieved. However, the maximum achievable output power from solid-state amplifiers decreases as the frequency increases nd hence the choice of operating frequency is a compromise between size and power output. MSTAR uses a GaAs FET power amplifier operating in the J-band 12 to 18 GHz).

Second, a surprising aspect to the design of this radar is that the efficiency of the power amplifier is not of primary importance in determining battery life. This arises because state-of-the-art power amplifiers in the J-band are only capable of a few watts of power, but the transmitter is only operating for around 10% of the time. Even if the efficiency is as low as 10% the transmitter can consume only a few watts of power while the antenna motor, controller, and display operate ontinuously and the receiver and signal processing operate for 90% of the time nd thus consume far more power. This situation would no longer be the case if t were possible to substantially increase the output power.

The second application to be considered is the COBRA solid-state phased rray radar being developed by the EUROART consortium (General Electric,

(a)

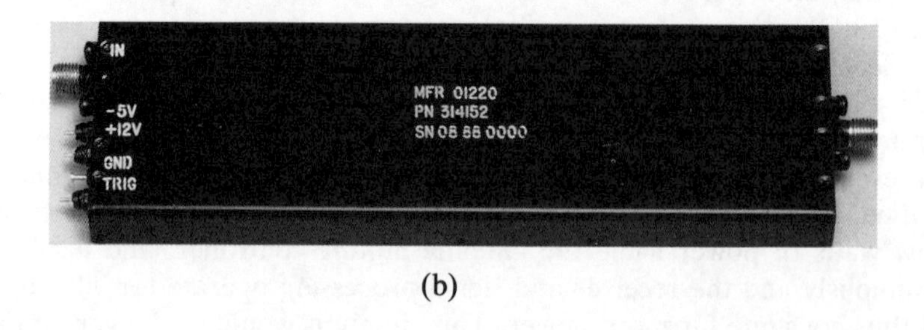


(b)

**Figure 7.22** (a) MSTAR (Photo courtesy of Thorn EMI Electronics) and (b) high-power GaAs FE amplifier used in MSTAR. (Photo courtesy of Systron-Donner Microwave Division.)

iemens, Thomson-CSF, and Thorn-EMI Electronics). COBRA, an acronym for 'ounter *B*attery *Ra*dar, is a C-band weapon-locating radar designed to detect and ack enemy shells and rockets as they approach and then to compute from the rojectile's trajectory the location of the weapon. COBRA is the first fully active olid-state phased array radar to enter full-scale development in Europe. Figure .23 shows an artist's impression of the radar mounted on a lorry. The detection f objects at long range having such small radar cross sections as shells requires ie radar to have a very high power aperture product. However, transportability equires a small antenna, and *low probability of intercept* (LPI) requires a low peak ower. The solution to this dilemma is to transmit a long, low peak power pulse or LPI but at a high duty cycle to achieve the required range performance and ith substantial pulse compression to achieve the required resolution. Low peak ower, high duty cycle, high mean power generation is of course ideally suited to ie use of GaAs FETs in the transmitter and an active phased array radar obtains s total required output power by spatial rather than circuit power combining of large number of such power GaAs FETs. It has been reported [29] that COBRA as around 3000 transmitting and receiving (T/R) modules. High efficiency is very nportant in this application, not only to minimize the amount of dc power required ut also to minimize the infrared (IR) signature presented by the antenna to a eat-seeking missile. The use of a solid-state phased array has several other advan- ages compared to a tube radar such as graceful degradation if one or more modules ail (see Sec. 6.4.1), and increased reliability.

Figure 7.24 shows a block diagram of the T/R module. The power amplifier n Figure 7.24 uses two GaAs MMIC chips, which are corporately combined using ff-chip quadrature couplers (see Sec. 6.4.1). Each MMIC power amplifier chip is two-stage amplifier in which the first stage uses two 1- by 1268-$\mu$m cells, while ie output stage uses four 1- by 1968-$\mu$m cells. Figure 7.25 shows the configuration f the chip. The overall gain of the chip is about 14 dB and the chip size is pproximately 3 by 6 mm. All phased array radar systems require very tight control f the amplitude and phase of the transmitted signal in order to have accurate eam steering and an acceptable sidelobe performance. COBRA, in common with any phased array systems, has a uniform aperture distribution, that is all modules re intended to provide the same output power. To achieve this in the presence f the various factors that can cause variations from module to module in output ower, such as gain variations, temperature variations, input power level variations, ll power amplifiers are operated in saturation and a filter is inserted between the ower amplifier and the antenna to suppress the harmonic output power. Driving he amplifiers into saturation also helps to increase the efficiency (see Sec. 1.2.2).

Future generations of active phased array radar will require a nonuniform perture distribution in order to reduce the sidelobe level in the transmitting mode or better radar performance and reduced probability of intercept. This requires hat the outer elements in the array produce less output power than those in the

Figure 7.33 Artist's impression of the COBRA radar system mounted on a truck. (Photo courtesy of Thorn EMI Electronics.)

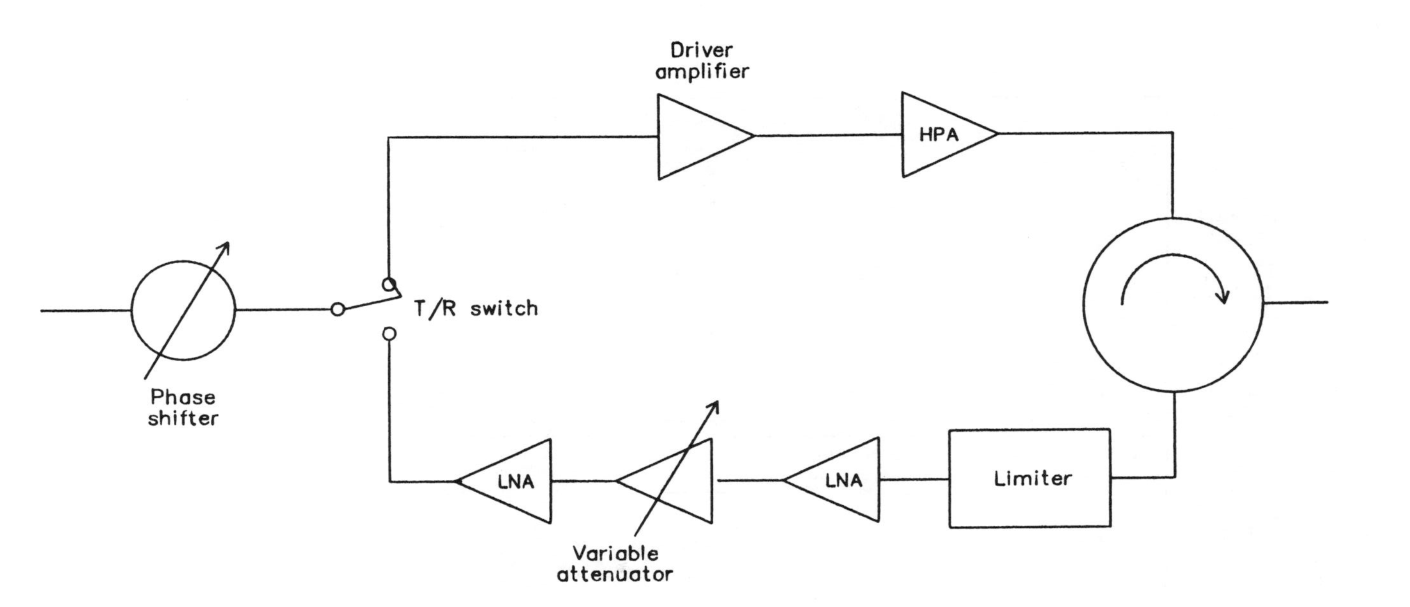


**Figure 7.24** Block diagram of the T/R module used in COBRA.

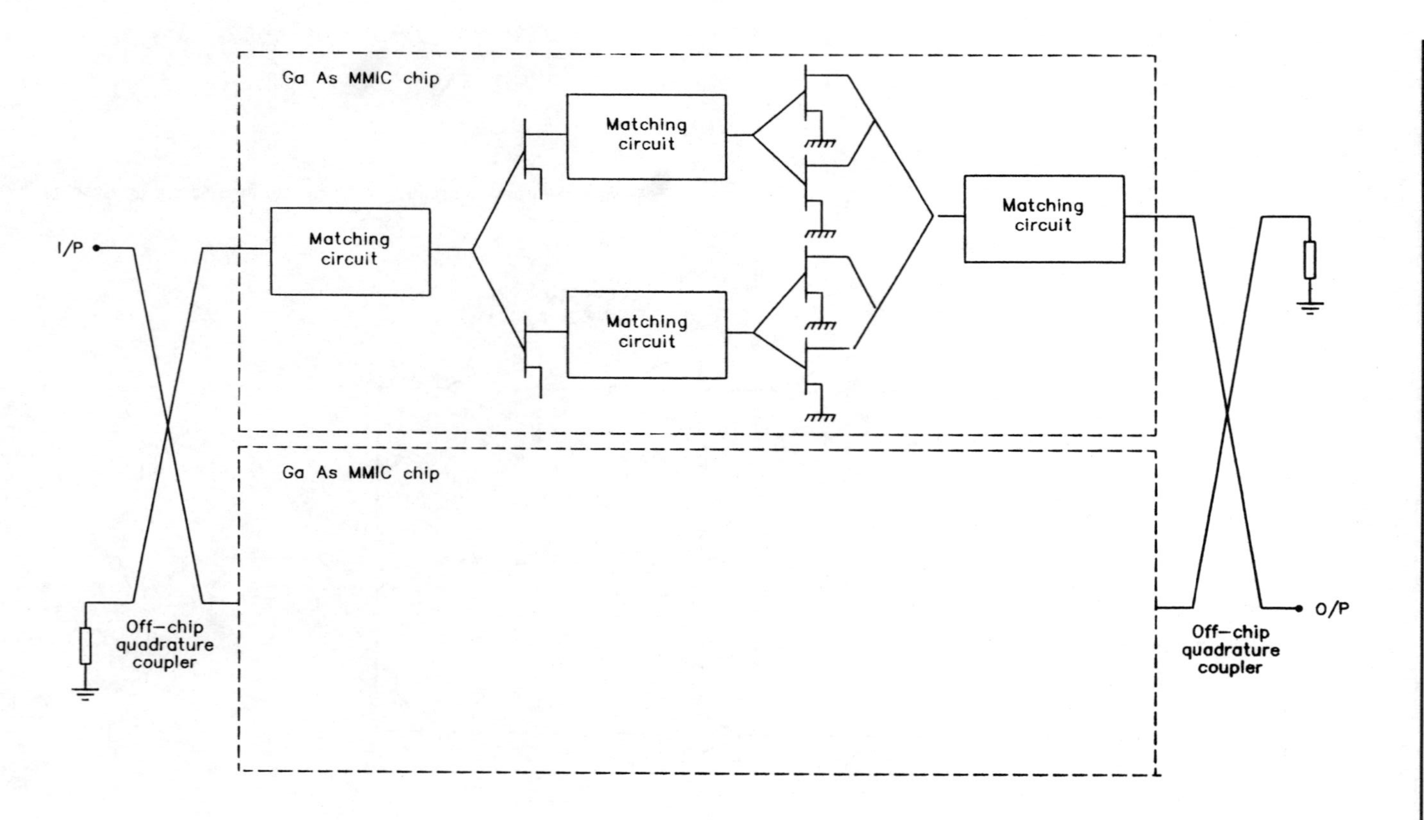


**Figure 7.25** Block diagram of the GaAs FET HPA used in the COBRA T/R module.

enter, which causes a substantial reduction in the peak effective radiated power ompared with the uniform array if the center elements are constrained to produce e same output power in both cases. This large reduction in output power arises ecause the number of elements at a given radial distance from the center of the ray increases in proportion to the distance from the center; hence, a large fraction f the total number of elements has to produce less power to implement the nplitude taper. McQuiddy et al. [30] have shown that the peak effective radiated ower drops by 8.1 dB with the outer elements producing less than one-tenth the ower of the center elements for a 40-dB Taylor aperture distribution. This results the maximum range of the radar being reduced by 37%. To maintain the range erformance, the output power of all elements must be increased sixfold, while ill maintaining the amplitude taper profile.

Thus future phased arrays will require at least some of the T/R modules to roduce substantially higher output power than is currently achievable. One could onstruct the antenna array using only high-power T/R modules and achieve the per by reducing the input signal to the power amplifiers in the off-center positions, else construct the antenna using nonidentical T/R modules with differing output ower capabilities. The former option has several advantages such as flexibility in ansmitted antenna pattern (for example, multiple beams could be generated in e transmitting mode if required), and it may be more economic to manufacture ly identical T/R modules. The disadvantage is that the overall antenna efficiency ould be unacceptably low if Class A amplifiers were used because of the large umber of amplifiers required to deliver less than their maximum output power ee Sec. 1.2.2), and the power dissipated in the antenna would also rise substanally adding to the cooling and IR signature problems. Thus if one wants to take lvantage of using identical T/R modules, then it is imperative that the T/R modules aintain their efficiency under back-off (i.e., Class B operation becomes very tractive [see Sec. 1.2.3] or else adaptive biasing must be used [31]).

The ideal amplifier for solid-state phased array radars is thus one having very gh efficiency and high mean power output under pulsed conditions, which mainins its efficiency under back-off, and whose insertion phase is invariant to drive vel. It is thought that Class B amplifiers using heterojunction bipolar transistors IBTs) come closer to meeting these ideals than do Class A amplifiers using GaAs ETs, so we can expect to see substantial progress over the next few years in HBT chnology for phased arrays. Nevertheless, solid-state phased arrays using GaAs ET power amplifiers are being designed today that generate tens of kilowatts of itput power—a task that would have been unthinkable a few years ago. This emonstrates the enormous progress that has been made in the field of GaAs FET ower amplifiers during the last decade.

## ACKNOWLEDGMENT

The development and production of the MSTAR radar system has been undertak
with the support of the Procurement Executive of the U.K. MOD.

## REFERENCES

1. Shino, T., and K. Kamei, "Microwave Power FETs at Toshiba; Present Status and Future Trenc *Microwave & RF Engineering*, April/May, 1989, pp. S37–S41.
2. Taniguchi, Y., Y. Hasegawa, Y. Aoki, and J. Fukaya, "A C-Band 25 Watt Linear Power FE *IEEE MTT-S Int. Microwave Symp. Digest*, Dallas, TX, May 8–10, 1990, pp. 981–984.
3. Ananasso, F., and M. Fornari, "SSPA and TWTA Complementing in Space Programmes: User Point of View," *ESA/ESTEC Workshop Proc. Conf. on SSPAs for Space Applicati* November 1989.
4. Maggenti, M., T. T. Ha, and T. Pratt, "VSAT Networks—An Overview," *Int. J. of Sate Communications*, Vol. 5, 1987, pp. 219–225.
5. Gatti, G., "ESA's Solid State Power Amplifier Activities," *19th European Microwave Conf. W shop Proc.*, London, UK, September 8, 1989, pp. 153–164.
6. Easton, M., and J. R. Basset, "GaAs Modules Produce Tube-Like Power at Ku-Band," *Microw & RF*, Vol. 29, August 1990, pp. 174–175.
7. Ezzeddine, A., H-L. A. Hung, and H. C. Huang, "High Voltage FET Amplifiers for Satellite Phased Array Applications," *IEEE MTT-S Int. Microwave Symp. Digest*, St. Louis, MO, Jun 6, 1985, pp. 336–339.
8. Bahl, I. J., E. L. Griffin, A. E. Geissberger, C. Andricos, and T. F. Brukiewa, "Class-B Po MMIC Amplifiers with 70% Power Added Efficiency," *IEEE Trans. Microwave Theory and T niques*, Vol. MTT-37, September 1989, pp. 1315–1320.
9. Iso, A., D. Yoshikazu, N. Okubo, S. Kobayakawa, and T. Maniwa, "S-Band High-Efficiency G FET Amplifier," *ESA/ESTEC Workshop Proc. Conf. on SSPAs for Space Applications*, Noven 1989.
10. Ernst, R. L., R. L. Camisa, and A. Presser, "Graceful Degradation Properties of Matched N- Power Amplifier Combiners," *IEEE MTT-S Int. Microwave Symp. Digest*, San Diego, CA, J 21–23, 1977, pp. 174–177.
11. Maas, S. A., *Nonlinear Microwave Circuits*, Norwood, MA: Artech House, 1988, pp. 220–23
12. Railton, G., "Transistor High Power Amplifiers: Capabilities and Applications," *MIOP C Proc.*, Stuttgart, Germany, April 14–16, 1990.
13. Darbandi, A., "C-Band SSPA for TELECOM 2 Satellite," *ESA/ESTEC Workshop Proc. C on SSPAs for Space Applications*, November 1989.
14. Mizuno, H., H. Mitsumoto, and N. Yazawa, "A 12.5 GHz-Band 50W Solid-State Power Ampl for Future Broadcasting Satellites," *IEEE MTT-S Int. Microwave Symp. Digest*, Dallas, TX, 8–10, 1990, pp. 1337–1340.
15. Tserng, H. Q., B. Kim, P. Saunier, H. D. Shih, and M. A. Khatibzadeh, "Millimeter-Wave Po Transistors and Circuits," *Microwave J.*, Vol. 32, April 1989, pp. 125–135.
16. Wormald, R. P., "Latent Heat Exchange for Short Duration Missions," *19th European Micro Conf. Workshop Proc.*, London, UK, September 8, 1989, pp. 80–85.
17. Railton, G., "A 30W Solid State Power Amplifier for Satellite Communications Applicatio *MIOP Conf. Proc.*, Weisbaden, Germany, March 2–4, 1988.
18. Bura, P., D. Gelerman, and P. Ntake, "Power Amplifier for Microwave Digital Radios Inherent Phase Compensation," *IEEE MTT-S Int. Microwave Symp. Digest*, Las Vegas, NV, 9–11, 1987, pp. 479–481.

Bosch, W., and G. Gatti, "Measurement and Simulation of Memory Effects in Predistortion Linearizers," *IEEE Trans. on Microwave Theory and Techniques*, Vol. MTT-37, December 1989, pp. 1885–1890.

Morris, G. V., *Airborne Pulsed Doppler Radar*, Norwood, MA: Artech House, 1988, p. 123.

Stimson, G. W., *Introduction to Airborne Radar*, El Segundo, CA: Hughes Aircraft Co., 1983, p. 198.

Eaves, J. L., and E. K. Reedy, Eds. *Principles of Modern Radar*, New York: Van Nostrand Reinhold, 1987, p. 275.

Klauder, J. R., A. C. Price, S. Darlington, and W. J. Albersheim, "The Theory and Design of Chirp Radars," *Bell System Technical J.*, Vol. 39, July 1960, pp. 745–809.

Nathanson, F. E., *Radar Design Principles*, New York: McGraw-Hill, 1991.

Eaves, J. L., and E. K. Reedy, Eds. *Principles of Modern Radar*, New York: Van Nostrand Reinhold, 1987.

Skolnik, M. I., Ed., *Radar Handbook*, New York: McGraw-Hill, 1990.

Haykin, S., *Communication Systems*, New York: John Wiley & Sons, 1983, pp. 68–70.

*International Defense Rev.*, Vol. 24, 1991, p. 167.

Turbe, G., "COBRA—the Artillery Force Multiplier," *International Defense Rev.*, Vol. 23, 1990, pp. 761–764.

McQuiddy, D. N., R. L. Gassner, P. Hull, J. S. Mason, and J. M. Bedinger, "Transmit/Receive Module Technology for X-Band Active Array Radar," *Proc. IEEE.*, Vol. 79, No. 3, March 1991, pp. 308–340.

Geller, B. D., F. T. Assal, R. K. Gupta, and P. K. Cline, "A Technique for the Maintenance of FET Power Amplifier Efficiency Under Back-Off," *IEEE MTT-S Int. Microwave Symp.*, Long Beach, CA, June 13–15, 1989, pp. 949–952.

# About the Authors

### r. Y. Aoki

oshio Aoki was born in Nagano, Japan, in 1954. He received the B.S. degree in ıysics from Shinsyu University in 1977 and the M.S. and Ph.D. degrees from ohoku University in 1979 and 1983, respectively. He joined Fujitsu Ltd. in 1983. e has been in the Microwave Device Engineering Department of the Compound emiconductor Division and has developed GaAs FETs and HEMTs. He is now sponsible for the development of MMICs.

### . F. Cooke

arry F. Cooke was born in Little Rock, Arkansas, in 1921. After serving with ı American auxiliary to the R.A.F. and the U.S.A.A.F. (1941–45), he received B.S.E.E. in 1948. His senior paper on locked oscillator demodulation won an EEE student prize. In 1957 he joined a Texas Instrument group, which developed e first solid-state TV, microwave transistor, and all solid-state radar. From 1970 was manager of device design and analysis at Avantek, joining Varian in 1979 ıd retiring in 1987. In 1989 he received the Microwave Career Award and currently orks as a consultant specializing in microwave device testing.

### . B. Culbertson

r. Culbertson received his B.S.E.E. degree from the University of Virginia in 76 and the M.S.E.E. degree from Southern Methodist University in 1983. He is microwave design engineer at Texas Instruments in Dallas, Texas. His primary esign responsibility has been associated with power amplifiers for phased-array ıdar transmitting/receiving modules. Mr. Culbertson is a member of IEEE.

### Y. Hirano

Yutaka Hirano was born in Kanazawa, Japan, in 1947. He received his B.S. degre in physics from Kyoto University in 1970. He joined Fujitsu Ltd. in 1970. H developed high-speed and microwave silicon bipolar transistors from 1970 to 197 Since 1976 he has been developing GaAs power FETs, HEMTs, MICs, and MMICs He is currently manager of the Device Engineering Department of the Compoun Semiconductor Division at Fujitsu.

### R. E. Lehmann

Mr. Lehmann received his B.S.E.E. and M.S.E.E. degrees from the University c Illinois in 1974 and 1976, respectively. He is manager of the MMIC Technolog design branch at Texas Instruments in Dallas, Texas. Mr. Lehmann is responsib for developing microwave and millimeterwave GaAs MMICs for phased-arra radar, communication, and wideband EW applications. He has designed numerou hybrid and monolithic components and was involved in the development of th first GaAs transmitting/receiving module on a single chip. Mr. Lehmann holds fou patents in the area of microwave devices and circuits and he is a senior membe of IEEE and a registered professional engineer in the state of Texas.

### O. Pitzalis, Jr.

Octavius Pitzalis has been a senior staff scientist at EEsof since 1987. His majo role in the marketing department is to function as a technical spokesman and als an ombudsman to serve the design engineers' needs. Prior to working at EEso Mr. Pitzalis was a member of the technical staff at Hughes Research Laboratorie in Malibu, California, where he worked on GaAs FET and IMPATT diode powe combiners, the design of MMICs for satellite communications receivers, and th characterization and modeling of low-noise V-band GaAs FETs and HEMT Before 1978, he was with the U.S. Army Research and Development Labs at Fo Monmouth, New Jersey (now called LABCOM). Here he pioneered circuit desig theory in applying bipolar transistors to broad bandwidth power amplifiers an power combining structures. This work began with RF applications at 2 to 30 MH and graduated to microwave work in L- and S-bands to 4 GHz, the upper limit fc transistors at that time. Prior to this, he participated in the earliest developmer of silicon integrated circuits and the first design methods for the micropower ope ation of bipolar transistors in analog and digital circuits.

### G. Railton

Gordon Railton was a founding member and is currently engineering director c Pascall Microwave Ltd., specializing in the design and manufacture of microwav solid-state power amplifiers and subsystems. His background includes 12 years i civil and municipal engineering before changing careers and obtaining an honours degree in electronics at the University of Manchester Institute of Science and Technology. He joined the microwave group of Ferranti International and was responsible for the development of microwave components, including a range of power amplifiers from the L- to Ku-bands and has presented several papers on power amplifier combining, linearization, and control techniques. He is a member of the Institute of Electrical Engineers.

**J. L. B. Walker**

John Walker received his B.Sc. in electrical and electronic engineering in 1970, his M.Sc. in microwaves and communications in 1971, and his Ph.D. in microwave directional filters in 1976, all from the University of Leeds, England. In 1974 he joined GEC Hirst Research Centre where he was engaged initially on the design of silicon bipolar and GaAs FET amplifiers before being appointed group leader of a team engaged in developing millimeterwave silicon and GaAs Impatt diodes and oscillators. Subsequently, he became chief engineer of the Compound Semiconductor Laboratory in the same organization. In 1985 he joined Thorn-EMI Electronics where he was responsible for establishing a microwave hybrid activity for the development and production of microwave amplifiers and subsystems. Dr. Walker is a senior member of IEEE in the United States and a fellow of IEE in the United Kingdom.

# Index


.ctivation energy, for ohmic contact degradation, 249
.ctive array modules, HVFA and, 319-20
.ctive layer
dimensions of, 62
formation of, 97
thickness of, 62
.ctive phased arrays
in radar systems, 347, 351
in satellite applications, 319-21
GC (automatic gain control), 275
.ir-bridge crossover, 47, 88, 123, 125
.LC (automatic level control), 329
.l-gate FET, 123
lloying process, 106
.luminum
for gate electrodes, 123
lattice constant, 139
tetrahedral covalent crystal configuration, 139
luminum gates, 108, 249
.M-AM distortion measurment, 118
.mplifier(s). See also specific types of amplifiers
design technique overview, 30-35
failure during initial biasing, lowering risk of, 194
high-power GaAs FETs. See High-power GaAs FETs
reactively matched, bandwidth limitations of, 35-40
mplitude
combining efficiency of corporate power combining and, 298-99
pulse compression system and, 343-45
mplitude modulation to phase modulation (AM-PM) conversion, 29
AM-PM distortion measurement, 118
Annealing, 100
Applied voltage, 63
Arrhenius plot, 254, 255, 256
Arsenic trichloride, in vapor phase epitaxy, 97
Arsine gas, 99
Attenuator, 117
AuGe/Ni/Au, 106
Automated life testing, 257-60
Automatic gain control (AGC), 275
Automatic level control (ALC), 329

Back IMD, 159
Balanced amplifier, block diagram of, 32
Balun, 21
Bandwidth limitations, of reactively matched amplifiers, 35-40
Barker code, 341
Beryllium, as p/i-type dopant, 99
Bias, 15
Bias points, 194-95
Binary power combining method, 292. See also Corporate power combining
Block diagram
of balanced amplifier, 32
of C-band SSPA, 321, 324
of linear FM pulse compression system, 342
of load-pull measurement system, 122
of single-stage amplifier, 30
of transmitting chain, 189-90
of T/R module, 347, 349
of T/R module GaAsFET HPA, 350
of waveguide phase bridge, 192
Boltzmann's constant, 110, 249
Branch-arm coupler, 285-86

Breakdown current
definition of, 67
test for, 105
Breakdown voltage, 43
Budgeting transmitting chain RF performance, 189-91
Bypass-gate structure, 89

Cable evaluation, in high-power FETs, 114
CAD. See Computer-aided design (CAD)
Capacitance
of depletion region, 25
electrostatic of pads, 87
feedback, 63, 90
gate, 56
gate-drain feedback, 90
gate-source, 28, 63
parallel parasitic, 87
parasitic, protective layers and, 111
spot frequency and, 21
Capacitor, in series, 273
C-band
characteristics, measurement of, 117
FET, 44, 83, 123
SSPA
block diagram of, 321, 324
performance details, 321, 325-26
Chain power combining, 264, 302-5
Channel
cross section, 60, 72
definition of, 48
resistance, 63
temperature, measurement of, 243-48
Chemical vapor deposition (CVD), 138
Chip(s)
backside structure, 90-94, 95
high-power, 128
MESFET. See MESFET chip
mounting techniques, high-power amplifiers and, 227
size
of high-power GaAs FETs, 83-84, 85
of MESFET chip, 123
width of chip vs. gain, 83-84
standard, 123-27
thickness of, 46
CI (confidence interval), 254
CIR (confidence interval ratio), 254
Circuit diagram
for class A or B amplifier, 4
for distributed amplifier, with constant-K gate line and modified m-derived drain line, 309
of distributed amplifier, ideal, 265
for large-signal GaAs FET, 29
for push-pull amplifiers with tuned load, Class B
for push-pull Class B amplifier, 19
for small-signal GaAs FET model, 26
Class A amplifiers
flat doping profile, 3
ideal, analysis of, 3
load impedance, 11
load line, 7, 37
maximum linear power output and, 276
maximum output power, 6-8
nonoptimum load impedances, 37
nonuniform channel doping profile, 3
output power characteristics, 9-10
overdriven, drain current waveform of, 8-9
performance parameters, 22
pinch-off voltage, 3
power-added efficiency, 6-7, 10
single-ended, 279
voltage-current waveforms, 5
vs. Class B amplifiers, 21-23
Class AB bias, 15, 17
Class AB large-signal amplifier simulation, 174, 176-87
input power levels, 183, 184-86
input voltage, 183, 184-85
microstrip layout, 178
multiple-amplitude waveforms for input power, 180, 182-85
output power and power-added efficiency vs. input power, 179-81
output power vs. frequency with input power, 186, 187
power-added efficiency vs. frequency with input power, 186, 187
schematic, 177
simulated output power, 180, 182
Class B amplifiers
flat doping profile, 3
ideal, analysis of, device physics vs. circuit design viewpoint, 3
nonuniform channel doping profile, 3
performance parameters, 22
pinch-off voltage, 3
push-pull, 18-21
with resistive load
disadvantages of, 15
linear output power for, 13-14
load impedance for, 13
power-added efficiency for, 14
voltage-current waveforms for, 11-13
vs. Class A amplifier, 14-15

singled-ended
with resistive load, 11-15
with tuned load, 15-18
vs. Class A amplifiers, 21-23
lass of operation, 194-95
luster matching, 263
OBRA solid-state phased array radar, 345, 347-48
T/R module, 347, 351
block diagram, 347, 349
GaAs FET HPA block diagram for, 350
olumnar heat flow, 228, 229
ombined heat flow, 229
ombining efficiency
in corporate power combining, 298-301
in serial power combining, 304
omb-type gates, 87
in facing rows, 89
ompression ratio, 341
omputer-aided design (CAD), 147-48
of GaAs FET nonlinear models, 148-72
of GaAs FET power amplifiers, 147-87
large-signal amplifier simulation. See Large-signal amplifiers, simulations
onfidence interval (CI), 254
onfidence interval ratio (CIR), 254
ontact resistance, measurement of, 106-7
orporate power combining
balanced amplifier in presence of external mismatch, 294-96
combining efficiency, 298-301
illustration of, 292-94
traveling-wave divider/combiner method, 296-98
two-way power splitters and, 294
vs. serial power combining, 302, 304
oupled-mode analysis, 206-8
oupler
branch-arm, 285-86
Lange, 289, 290
parallel coupled-line directional, 285, 287-88
rat-race, 283, 285
ripps technique, 24, 27
modified, 199-204
urrent continuity, 59
urrent density, 59
distribution of, 62
urtice-Ettenberg model, 167-68, 172
urtice quadratic function MESFET model, 164-65
VD (chemical vapor deposition), 138
-V measurement, 100, 101
-V method, 109, 111
CW radar systems, 337, 339

DBS. See Direct broadcast by satellite (DBS)
DC characteristics, of high-power GaAs FETs, 113-14
Delay time, 63
Device under test (DUT), 119
Die
attachment, using conductive epoxy, 227
mounting, on alumina substrates, 227
thickness, thermal resistance and, 236, 237
Dielectric crossover, 47
Direct broadcast by satellite (DBS), 2
Dissipation, of output network, 209
Distortion features
AM-AM, measurement of, 118
AM-PM, measurement of, 118
harmonic components, measurement of, 119
intermodulation, measurement of, 119-20
measurement techniques, 118
Distributed amplification concept, 264
Distributed amplifier(s)
with constant-K gate line and modified m-derived drain line, 308
circuit diagram, 309
computed performance, 310
equivalent circuit, 266
ideal, circuit diagram of, 265
VSWR problem and, 31, 33
Distributed amplifier power combining
advantages of, 264
concept, origination of, 264
forward gain, 270
large-signal analysis, 275-81
power gain, 267
reverse gain, 269-70
small-signal analysis, 264-70
drain loss in, 273-75
gain loss in, 271-73
gate-line impedance tapering, 273-74
impedance-matching network and, 272
loss within FET and, 271-75
resistive terminations and, 271-75
DMT (doped-channel MIS-like FET), 140-41
Donor density, 62
Donor density half-bandwidth, 62
Donor impurities, introducing, 97
Doped-channel MIS-like FET (DMT), 140-41
Doping profile(s)
power law or hip-up, 50, 51-52
step or hi-lo, 50, 52-56
uniform, 27, 50, 51

wafer evaluation and, 100, 101. See also C-V measurement
Drain
conductance, 63
current, 63
efficiency, 191
resistance, 74
Drain breakdown voltage
for high-power GaAs FETs, 66-69
recessed gate structure and, 67-69
Drain-gate feedback capacitance, 80
Drain-source breakdown voltage, 65
Drain-source conductance, 90
Drain-source potential, 63
Drain-source voltage, 275-76
Drift mobility, 101, 103
Dual-gate FET power amplifiers
interstage network design, 219
large signal insertion phase, 219, 225
lumped element model, 216, 219
output matching network, 217, 222
power-added efficiency, 219
predicted large-signal response, 219, 222
RF configuration, 216
second gate bias-dependent elements, 216, 220
second gate voltage-dependent parameters, 216, 220
second-stage gain flatness, 224
with termination capacitors for second gate, 216, 221
two-stage, 224
output power and, 219, 224
power-added efficiency and, 219, 224
3-W, 216, 218
DUT (device under test), 119

Earth terminals, 330-33
EBIC (electron-beam-induced current), 114, 115
ECCM (electronic counter-countermeasure), 341
Electric field
distribution, 60
temperature and, 96
Electrode layout, 43
Electron-beam-induced current (EBIC), 114, 115
Electron density distribution, 61
Electronic counter-countermeasure (ECCM), 341
Electronic power conditioner (EPC), 322
Electronic warfare (EW), applications for high-power GaAs FETs, 317, 336-51
Electron velocity, 59, 96
Endurance, of passivation film, 111
EPC (electronic power conditioner), 322
Epitaxial wafer growth, 97-100
carrier profile by step-etching, 102
evaluation criteria, 100-03
Epitaxial wafer structure, 50-56
Equivalent circuits
for distributed amplifier, 266
high-power GaAs FETs, channel cross-sectior design, 74-77
S parameters and, 81
EUROART consortium, 345
Evaporation method, for gate metal application, 1
EW (electronic warfare), applications for high-power GaAs FETs, 317, 336-51
Excess insertion loss, 300

Failures
under deep gate bias, 68, 69
mechanisms
in manufacturing, 249
reliability and, 248-60
modes, 69
typical distribution, in satellite applications, 3
Feedback, 31, 33
Feedback capacitance, 63, 90
Feedback linerization, 335-36
Feedforward linerization, 336
Fermi level, 56
FET cell unit, 85
FET chips. See Chip(s)
FET failures. See Failures
FETs. See Field effect transistors
Field effect transistors (FETs). See also GaAs FETs; specific FETs/i
cross sections of, 67
current
internally matched, 128-34
MMIC power amplifiers, 134-36
standard FET chips, 123-27
equivalent, channel cross sections of, 60
high-power
improvements, 45-46
technologies in, 46-47
locations of failures, 67
practical, thermal calculations for, 229-40
technological advances, 142-43
Fishbone gates, 47, 89
Flip-chip structure, 46, 88-89
cross-section, 89
disadvantages of, 89
source inductance and, 92
FLK202XV large-signal model, 173-74, 175-76
Fork divider, 291, 292
Forward diode voltage method, for thermal resistance measurement, 245-47
Fourier series temperature prediction, 232

rank code, 341
ujitsu FLM 3742-10
failed chip, 129, 131
internal view, 128-29
load impedance contours, 132
matching circuit of, 133
power input and output characteristics, 129
ujitsu FLM 4450-25D, 130
internal structure, 132-34
output power and power-added efficiency vs. input power, 135-36
ujitsu 38-GHz MMIC, 134, 136, 138

aAs/AlGaAs heterojunction bipolar transistor, 136
aAs FETs
channel temperature, measurement of, 243-48
Class A amplifiers. See Class A amplifiers
Class B amplifiers. See Class B amplifiers
derivatives. See High-electron mobility transistors (HEMTs)
developmental obstacles, 43
discrete power, 1
first commercially available, 1
first prototype, 43
frequency limit, 227
high-power. See High-power GaAs FETs
ideal, cross section schematic, 24
large-signal, 29
large-signal (nonlinear) model, 26-30
nonlinear models, 148-72
general guidelines for large-signal model extraction, 172
large-signal amplifier simulation, 173-87
MESFET large-signal RF equivalent circuit, 149-61
MESFET static dc model, 162-72
pulsed operation, 240-43, 244
small-signal (linear) model, 24-26
systems applications. See Systems applications,
thermal resistance, measurement of, 243-48
aAs Gunn diodes, 1
aAs Impatt diodes, 1
aAs MESFET buffer layer, 141
aAs monolithic microwave integrated circuits (MMICs). See Monolithic microwave integrated circuits (MMICs)
ain
measurement of, 117
temperature and, 98
ain bandwidth
limitation, of matching networks, 36
for low-pass impedance transformers, 211
Gain compression curves, for large-signal vs. small-signal
tuning, 193
Gallium
lattice constant, 139
tetrahedral covalent crystal configuration, 139
Gallium arsenide field effect transistors. See GaAs FETs
Gate breakdown voltage, 47
gate-drain electrode distance and, 72
for high-power GaAs FETs, 69-72
vs. recess length, 71
Gate capacitance
gate length and, 56-57
in step doping, 56-57
Gate current, in high-power GaAs FET amplifier design, 215-16, 217
Gate-drain feedback capacitance, 90
Gate electrode formation, in manufacturing high-power GaAs
FETs, 108-11
Gate evaluation process, 109-10
Gate fingers
in comb-type gate, 87-88
distributed model using coupled-mode analysis, 206
extending length of, 204, 205
interdigital or comb-type layout, 88-89
layouts, 89
length, 83-84
number per pad, 84-87
Gate formation
gate metal selection and, 108
gate orientation and, 108
gate patterning and, 108-9
methods, 109-11
preprocessing, 109
Gate length, 62
determination in high-power GaAs FETs, channel cross-section design and, 56-58
gate capacitance and, 56-57
thermal resistance and, 236, 237
Gate length/channel thickness, 124
Gate-line impedance, tapering, 273-74
Gate metals
application methods for, 109
selection of, 108
Gate orientation, 108
Gate pads, number determination, 84-87
Gate patterning, 108-9
Gate resistance, 72-73, 212
Gate-source bias voltage, 25
Gate-source capacitance, 63

per unit gatewidth, 80
with input signal level, variation of, 28
Gate-source crossovers, 88, 125
Gate-source potential, 63
Gate spacing, thermal resistance and, 236, 237
Gate-to-drain region, cross section
of planar structure, 70
of recessed structure, 70
Gatewidth, 62
Gauss's theorem, 69
Graceful degradation, 301
of power combined amplifiers, 321, 322
Gunn oscillation effect, 96

Hall bar with Schottky junction metal, 102
Hall coefficient, 103
Hall field, 101, 103
Hall mobility, 100, 101-3
Harmonic balance simulation tools, 147
Harmonic components, measurement of, 119
Harmonic termination effects, in matching network design, 210-12
HBT (heterojunction bipolar transistor), 45, 141-42
Heat. See also Temperature
disposal, for GaAs power FETs, 227-28
dissipation, 43
evaluation, in high-power FETs, 114
flow, 229-30
flow calculations, 228-34
HEMTs. See High-electron mobility transistors (HEMTs)
Heterojunction bipolar transistor (HBT), 45, 141-42
Heterojunction buffer, 47
Heterojunction FET, 139-41
Heterojunction MISFET, 140-41
HET (hot electron transistor), 136-37
High-electron mobility transistors (HEMTs), 1, 45, 141
development, 139
Indium phosphide (InP), 136
multichannel power
development of, 139-40
epistructure, 140
High-frequency harmonics, measuring, 117
High-frequency resistance, 80
High power, 29-30
High-power GaAs FETs
channel cross-section design, 48-77
drain breakdown voltage, 66-69
epitaxial wafer structure and, 50-56
equivalent circuits, 74-77
flow of design process, 48-50
gate breakdown voltage, 69-72
gate length determination, 56-58
objectives, 49-50
parasitic resistance, 72-74
scaling law and, 58-66
chip pattern design, 48
chip backside structure, 90-94, 95
chip size and, 83-84, 85
flow of design process, 77, 78
objectives, 49
output power, 77-79
pad number determination, 84-87
pattern layout, 86, 87-90
total gatewidth, 79-80
unit gatewidth determination, 80-83, 84
current technology for, 45-48
design techniques, 198-204
budgeting transmitting chain RF performar
and, 189-91
for dual-gate FET power amplifier, 216-25
gate current and, 215-16, 217
insertion phase and, 215-16, 217
load-pull, 198-99
matching network, 208-12
modified Cripps method, 199-204
nonlinear CAD, 199
performance characterization and modeling, 192-98
scaling and, 204-8
thermal considerations and, 212-15
development of, 43-45
electronic warfare applications, 336-51
evaluation
of dc characteristics, 113-14
of distortion features, 118-20
of impedance measurement techniques, 12
of load-pull measurement techniques, 120-23, 121-23
of output power measurement, 114-17
manufacturing, 97-113. See Manufacturing
operating characteristics, scaling law and, 63-
output power, annual development and, 44
performance criteria, scaling law and, 64-66
radar applications, 336-51
structural factors, scaling law and, 62
technological advances, 142-43
thermal properties, 94, 96-97
High-voltage FET amplifier (HVFA), 319-20
Hi-lo doping profile. See Step doping
Hip-up doping profile. See Power law doping
Hot electron transistor (HET), 136-37
HVFA (high-voltage FET amplifier), 319-20

Ideality factor, 110

, 4
edance matching
parasitic absorption and, 209
resistance level transformation and, 209
for two-stage power amplifier, 210
edance measurement techniques, of high-power GaAs FETs, evaluation of, 120-23
lantation isolation, 104
urity scattering, 94
lAs/GaAs (indium aluminum arsenide/gallium arsenide) heterojunctions, 45
ium aluminum arsenide/gallium arsenide (InAlAs/GaAs) heterojunctions, 45
ium gallium arsenide/gallium arsenide (InGaAs/GaAs) heterojunctions, 45
ium phosphide (InP), 45
ium phosphide (InP) HEMT, 136
ium phosphide (InP) MISFET, 136-39
uctance, 80
nt failures, 252-53
nite heat sink, 228
ared method, for thermal resistance measurement, 244-45, 246
aAs/GaAs (indium gallium arsenide/gallium arsenide) heterojunctions, 45
hase couplers, in corporate power combining, 294, 296
HEMT, 136
(indium phosphide), 45
MISFET, 136-39
ut matching network
example, 39-40
purpose, 38-39
ut power
Class AB 3.7- to 4.2-GHz, 1-W large-signal amplifier simulation, 179-81
Fujitsu FLM 4450-25D, 135-36
K-band chip, 127
measurement in evaluation of high-power GaAs FETs, 114-17
ertion phase, in high-power GaAs FET amplifier design, 215-16, 217
erdigital FETs
channel cross section, 238
layout, 47
source-drain, 87
dimensions, 86
layout pattern, 86, 87-90
temperature rise and, 231-32
thermal resistance, 238-40
ermodulation distortion
measurement of, 119-20
products, 152
Internally matched FETs
Fujitsu FLM 3742-10, 128-33
Fujitsu FLM 4450-25D, 130, 132-34, 135-36
Interstage network design, for dual-gate FET power amplifier, 219, 223-24
Interstage network load line analysis, 209-10
Intervalley transfer or scattering, 96
Intrinsic channel resistance, 25
Ion implantation, 62, 97, 100
Isolation
implantation, 104
mesa, 104
verification of, 105
in wafer processes, 103-5

K-band FETs, 125-27
K filter sections, 266
Kirchhoff's transformation, 232
Kovar carriers, 227
Ku-band FET, 123

Lange coupler, 289, 290
Laplace equation, 230
Large signal, 29
Large-signal amplifiers
characterization of, 33-34
design process summary, 32
simulation
Class AB, 174, 176-87
by FLK202XV model, 173-74, 175-76
of 3.7- to 4.2-GHz, 1-W Class AB, 174, 176-87
Large-signal analysis, distributed amplifier power combining, 275-81
Large-signal GaAs FETs, 29
Large-signal MESFET models
analytical functions, 148
equivalent circuit
bias nonlinearities, 153-56
bias nonlinearity compensation, 155, 157
charging resistance, 159, 160
drain-to-source channel resistance, 159, 161
gate-source conductance, 151, 155, 158
power gain nonlinearity, 155
in small-signal form, 152
transconductance, 152-55
parts of, 148-49
RF equivalent circuit, 149-61
Large-signal model extraction, general guidelines, 172
Leakage current, test for, 105
Libra™, 147
Life testing, automated, 257-60

Lift-off method, for pattern formation, 109
Linear FM pulse compression system, block diagram, 342
Linear frequency modulation, 341
Linearity, 13
Linearized amplifiers, 334-36
Line-of-sight links, 334
Linerization techniques, 335-36, 337, 338
Liquid crystal method, for thermal resistance measurement, 245
Liquid phase epitaxy, 97
Load line, 7
Load-pull design
  measurement system, block diagram, 122
  measurement techniques
    evaluation of, 121-23
    example of, 122
  process, 34, 198-99
Log-normal graph paper, 253
Log-normal probability density function, 250-52
Lossless output matching network, 32-33
Lossy matching networks, 31, 33
Lossy subnetwork, for dual-gate FET power amplifier, 219, 223-24
Lossy three-port network, 282
Low-loss power combining, 206
Lumped-element model, for dual-gate FET power amplifier, 216, 219
Lumped-element realizations, N-way power combiner/divider networks and, 291-93

MAG (maximum available gain), 30
Manufacturing
  backside processing and, 112-13
  epitaxial wafer growth and, 97-103
  errors, 249
  flow of, 103, 104
  gate electrode formation, 108-11
  isolation and, 103-5
  ohmic electrode formation and, 105-7
  overlay wiring, 112
  protective layers, 111-12
  via-hole connections and, 112-13
Matching impedance, 114, 117
Matching network design, 208
  harmonic termination effects, 210-12
  interstage network load line analysis, 209-10
  output network load line analysis, 209-10
  stability considerations, 212, 213
Matching networks, gain-bandwidth limitation, 36
Matching problem, 35
Material technology, for indium phosphide MISFET, 137-39
Materka MESFET model, 165-66
Maximum available gain (MAG), 30
MBE (molelcular beam epitaxy), 99
Mean time to failure (MTTF), 254-56, 318
  definition of, 252
  for Fujitsu FLM 3742-10, 129-30
  for satellite applications, 318
Mesa isolation, 104
MESFET
  large-signal models, 162, 164-65
  large-signal RF equivalent circuit, 149-61
  models, 148
  soft pinch-off I-V characteristics, 165-66
  square-law behavior near pinchoff, 162
  static dc models, 162-72
    Curtice-Ettenberg model, 167-68, 172
    Curtice quadratic function, 164-65
    Materka model, 165-66
    Statz model, 168-70
    TOM model, 170-72
  static I-V characteristics, 162-63
MESFET chip
  C- and X-band, 123-25
  cross section, 124
  Ku- and K-band, 125-27
    cross section, 126
    output power and power-added efficiency vs. input power, 127
    SEM photograph, 126
  SEM photograph, 124
Mesh-emitter transistor, 87
Metal-insulator-metal (MIM) capacitor, 112
Metal-insulator-semiconductor FET (MISFET), 45, 136
Metal migration, 215-16
Metal-organic chemical vapor deposition (MOCVD), 97, 99
Metal-semiconductor contact, energy-band diagram of, 106
Microstrip-FET analogy, 234, 235
Microwave power dissipation per unit gatewidth, 8
MICs, 148
MIM capacitor, 112
MISFET (metal-insulator-semiconductor FET), 45, 136
MMICs. See Monolithic microwave integrated circuits (MMICs)
Mobile tactical satellite applications, 323-30
  output power vs. frequency, 330
MOCVD (metal-organic chemical vapor deposition), 97, 99
Mode strapping, 212

ɅIodified Cripps design technique, 199-204
Ʌolecular beam epitaxy, 97
Ʌolelcular beam epitaxy (MBE), 99
Ʌonolithic microwave integrated circuits (MMICs), 1-2, 134-36
breadboard prototypes, 148
cost of development, 2
custom design of, 2
development of, 2
power per milimeter of gatewidth concept and, 23-24
power per millimeter of gatewidth and, 23
in satellite applications, 319
0.5-W, 33-GHz, 323, 327
ɅISTAR (Man-portable Surveillance and Target Acquisition Radar), 345, 346
ɅITTF. See Mean time to failure (MTTF)
Ɂultiple-gate FET, 234, 235
Ɂultiple-stage comb-type structure, 89
Ɂultistage amplifiers, 33
ɅwSPICE™, 147

Ɂonalloyed ohmic electrode, 106
onlinear CAD design technique, 199
ɁOnlinear frequency modulation, 341
Ɂonlinear simulation, 34-35
-type dopants, 99
-way power combining, 264, 305-6

BO (output backoff), 335
ffset gate design, 74
hmic contact
degradation, activation energy for, 249
energy diagram of, 106
hmic electrode formation, 105-7
hmic resistance, 106
scillations evaluation, in high-power FETs, 113-14
utput backoff (OBO), 335
utput impedance, 21
utput matching network, 35-38
for dual-gate FET power amplifier, 217, 222
example, 38
utput network load line analysis, 209-10
utput power
Class AB 3.7- to 4.2-GHz, 1-W large-signal amplifier simulation, 179-81
of Fujitsu FLM 3742-10, 129
Fujitsu FLM 4450-25D, 135-36
high-power GaAs FETs, chip pattern design, 77-79
K-band chip, 127
maximum RF, 23
measurement
in evaluation of high-power GaAs FETs, 114-17
in high-power GaAs FETs, 114-17
reduction, unit gatewidth and, 83
temperature and, 98
Output power degradation factor, 300
Over-gate patterning, for pattern formation, 109
Overlay wiring, 112

Pad number determination, in high-power GaAs FETs chip pattern design, 84-87
Pads, electrostatic capacitance of, 87
Parallel coupled-line directional coupler, 285, 287-88
Parallel parasitic capacitance, 87
Parasitic absorption, impedance matching and, 209
Parasitic capacitance
flip-chip structure and, 88
protective layers and, 111
Parasitic elements, 90
Parasitic resistance
for high-power GaAsFETs, 72-74
in series, 25
temperature dependence, 96-97
Passivation, 47
Passivation film, 111-12
Passive power combiner/divider networks, 281-92
N-way power combiner/divider networks, 289, 291-92
simplest type, 282
two-way in-phase power combiner/divider networks, 282-85
two-way quadrature-phase power combiner/divider networks, 285-89
Pattern formation, methods, 109
Pattern layout, in high-power GaAs FETs chip pattern design, 86, 87-90
PBT (permeable base transistor), 136
PCM (phase change material), 324, 328
Peak donor density, 62
Peak donor density location, 62
Performance budget, 189-90
Permeable base transistor (PBT), 136
Phase change material (PCM), 324, 328
Phase characterization, waveguide phase bridge for, 192
Phase coding, 341
Phased-array radars, 2
Phase distortion, pulse compression system and, 343-45
Phase nonuniformities, combining efficiency of corporate power combining and, 298-99
Phase rotation, 82
Phonon scattering, 94
PHS (plated heat sink), 46, 92, 125

Piezoelectric effect, 111
Piezoelectric polarization, gate orientation and, 108
Pinch-off voltage, 3, 63
Plated heat sink (PHS), 46, 92, 125
Poisson's equation, 58-59
Polyphase, 341
Power-added efficiency, 191
 Class AB 3.7- to 4.2-GHz, 1-W large-signal amplifier simulation, 179-81
 Fujitsu FLM 3742-10, 129
 Fujitsu FLM 4450-25D, 135-36
 K-band chip, 127
Power amplifier(s). See also specific power amplifiers
 comparison of, 21, 22
 monolithic, 2. See also Monolithic microwave integrated circuits (MMICs)
 solid-state. See Solid-state power amplifiers (SSPAs)
Power combined amplifiers
 distributed amplifier. See Distributed amplifier
power combining
 satellite applications, 321-23, 326-27
Power combining methods
 corporate, 292-306
 N-way power combining, 305-6
 passive power combiner/divider networks, 281-92
 phase differences and, 263
 serial or chain power combining, 302-5
 technique categories, 264
 Zgo = Zdo = Ao and Og = Od, 307-11
Power divider/combiner network, in corporate power combining, 294-96
Power gain
 of distributed amplifier, 267
 maximum available, 90
Power law doping, 50, 51-52
Power meter, 192
Power pads, number determination, 84-87
Power per milimeter of gatewidth, 23-24
Power supply unit (PSU), 334
Predistortion linearization, 335, 336, 337, 338
Preprocessing, before gate formation, 109
Prescreening, 257
Propagation constant, 82
Protective resistor, 117
Prototype amplifiers, 148
PSU (power supply unit), 334
p-type dopants, 99
Pulse compression, 341
Pulsed amplitude, waveguide phase bridge for, 192
Pulsed operation, 240-43, 244
Pulsed radar systems, 339
Pulsed RF testing, 192-94
Push-pull amplifiers
 Class B, 18-21
  advantages of, 19
  circuit diagram, 19
  compared to other amplifiers, 21, 22
  with tuned load, circuit diagram for, 20
 power splitters and, 282

Quadrature amplitude modulation (QAM), 334
Quadrature couplers, in corporate power combining, 294, 296

Radar systems, categories of, 337, 339
Range sidelobes, 343
Rat-race coupler, 283, 285
Reaction rate (RR), 249
Recessed gate structure
 construction technology, 47
 cross-sectional sketch, 70
 drain breakdown voltage and, 67-69
 gate breakdown voltage calculation, 70-71
Recess gate structure, 109
Reliability
 failure mechanisms and, 248-60
 of FET amplifiers, 47
 in satellite applications, 318-19
 statistics, 250-57
 testing, 257-60
Resistance level transformation, impedance matching and, 209
Reverse gate current, 67
Richardson constant, 110
RR (reaction rate), 249

SARs (synthetic aperture radars), 319, 321
Satellite applications, 318
 active phased arrays and, 319-21
 in earth terminals, 330-33
 mobile tactical, 323-30
 for power combined amplifiers, 321-23, 326-27
 reliability, 318-19
Scalar network analyzer, 117
Scaling, in high-power GaAs FET amplifier design, 204-8
Scaling law, 58-66
 operating characteristics and, 63-64
 performance criteria and, 64-66
 significance of, 66
 structural factors and, 62-62
Schottky barriers, 4, 56
 gate metal selection, 108
 height of, 109

Schottky contack, avalanching at point of
contact, 69
Schottky diode, 4
Schottky junction, 73, 109
gate properties, 109
on Hall bar, 102
sheet resistance measurement and, 100-01
Selenium, as n-type dopant, 99
Semiconductor, failure mechanisms, 248
Semiconductor materials, thermal conductivity
of, 228, 230
Semiconductor potential distribution, 61
Serial power combining, 302-5
combining efficiency, 304
disadvantages, 305
Sheet grounding, vs. via-hole technique, 92-93, 94
Sheet resistance measurement, 100, 107
Side gate pad, 91
Signal attenuation, 82
Silicon, as n-type dopant, 99
Silicon bipolar transistors, 1
Silicon diodes, 1
Singled-ended Class B amplifiers
with resistive load, 11-15
with tuned load, 15-18
disadvantages of, 17
linear power output, 17
power-added efficiency of, 17
voltage-current waveforms for, 15-16
Single-stage power amplifier
block diagram of, 30
gate current and insertion phase, 216, 217
Skin depth, 73
Small-signal amplifiers, 33-34
design process, 30
optimum load impedance for maximum
output power, 34
Small-signal analysis, in distributed amplifier
power combining, 264-70
Small-signal equivalent circuit, 72, 75
Small-signal FETs, materials qualities, 45
Small-signal MESFET model, RF equivalent
circuit, 149-50
Small-signal modeling, 195-98
Smith chart, 202-3
Solid-state phased array radar, ideal amplifier
for, 351
Solid-state power amplifiers (SSPAs), 317
C-band, 321, 324-26
30-W, 8-GHz schematic, 328-29
Source-drain spacing, 62
Source electrode potential, 63
Source-gate spacing, 62
Source grounding inductance, 204
Source inductance, 90
Source induction
reduction methods, 91-92
in sheet grounding vs. via-hole technique, 92-93, 94
Source-island structure
advantages, 93
cross section, 95
SEM photograph, 95
source inductance and, 92
Source-pull measurement, 121
Source resistance, 73-74
S parameters, 80
drain voltage and, 198
equivalent circuit, 81
frequency variation, 75, 76
measurement of, 120
of parallel coupled-line directional coupler, 287
SPICE MESFET large-signal equivalent circuit
model, 150
frequency dispersion of drain-to-source
resistance, 150-51
SPICE model, 172
SPICE time-domain program, 147
Splitter, two-way, 282
Spreading heat flow, 228-29
Sputtering method, for gate metal application, 109
Square-law characteristics
of Curtice quadratic function MESFET
model, 164-65
of MESFET, 162
SSPAs. See Solid-state power amplifiers (SSPAs)
Stability considerations, in matching network
design, 212, 213
Statistics, reliability, 250-57
Statz model, 168-70
Step doping, 50, 52-56
gate capacitance in, 56-57
Step-stress tests, 257
Stored charge under gate, 63
Stress, protective layers and, 111
Student's T distribution table, 254, 257
Surface-state level, protective layers and, 112
Synthetic aperture radars (SARs), 319, 321
Systems applications, 317
in electronic warfare, for high-power GaAs
FET amplifiers, 336-51
radar applications, for high-power GaAs FET
amplifiers, 336-51
on satellites, 318-33
terrestrial telecommunications, 333-36

Technology, trends in, 136-42
Temperature
annealing, 100
boundary conditions and, 236
electron velocity and, 96
failure statistics and, 254-56
gain and, 98
output power and, 98
prediction, Fourier series, 232
rises
repetitive-pulse, 243, 244
transient single-pulse, 242
Terrestrial telecommunications, 333-36
linearized amplifiers, 334-36
line-of-sight links, 334
Test pattern
for measurement of contact resistance, 107
for sheet resistance measurement, 107
Thermal calculations, for practical FETs, 229-40
Thermal properties
conductivity, 228
diffusivity, 230
geometric comparisons, 232, 233
in high-power GaAs FET amplifier design, 212-15
of high-power GaAs FETs, 94, 96-97
Thermal resistance, 228, 234
calculations, 229-40
die thickness and, 236, 237
gate length and, 236, 237
gate spacing and, 236, 237
of interdigital FET, 238-40
measurement, 243-48
apparatus timing chart, 247-48
forward diode voltage method, 245-47
infrared method, 244-45, 246
liquid crystal method, 245
measurement methods, 243-44
substrate thickness and, 239, 241
Thermal time constant, pulsed operation and, 241-43
Thermionic emission, 105
Thermionic field emission, 105-6
Thermodynamics, fundamentals of, 228-29
Titanium, 108
Titanium-tungsten, 108
TOM model, 170-72
Total gatewidth, high-power GaAs FETs, chip pattern design, 79-80
Tournament gate feed pattern, 90, 91
Tournament-type gate signal, 125
Transconductance, 63
Transducer gain, 117
Transmitting chain block diagram, 189-90
Traveling-wave amplifier. See Distributed amplifier
Traveling-wave divider/combiner, in corporate power combining, 296-98
Traveling wave divider/combiner, 329
Traveling-wave tube amplifier (TWTA), 317
Tree method of power combining, 292-301
Tree power combining, 264
Trimethyl gallium, in vapor phase epitaxy, 97
TriQuint MESFET, 151-52
TriQuint's own model (TOM model), 170-72
Tungsten-silicide, 108
Tunneling field emission, 105
Two-dimensional Poisson equation, 59
Two-way in-phase power combiner/divider networks, 282-85
Two-way quadrature-phase power combiner/divider networks, 285-89
Two-way splitters, 282
for corporate power combining, 294
TWTA (traveling-wave tube amplifier), 317

Uniform doping profile, 50, 51
Unit gatewidth determination, high-power GaAs FETs, chip pattern design, 80-83, 84

Vapor phase epitaxy, 97-99
Verification, of complete isolation, 105
Very small aperture terminal (VSAT), 319
Via-hole connections, 47, 125
cross section, 93
source inductance and, 92
source induction, vs. sheet grounding, 92-93, 94
source-island. See Source-island structure
Voltage-controlled current generator, 24
Voltage-current waveforms, for class B amplifiers with resistive load, 11-13
VSAT (very small aperture terminal), 319
VSWR, 31, 283

Wafer processes, 97-113
definition of, 97
epitaxial wafer growth, 97-103
flow of, 103, 104
isolation and, 103-5
ohmic electrode formation and, 105-7
overlay wiring, 112
protective layers, 111-12
Wafers
epitaxial structure, 50-56
growth. See Epitaxial wafer growth
Waveguide phase bridge, block diagram, 192
Width of unit gate, 80

Wilkinson splitter, 282-83
ideal, performance of, 284
N-way, 289, 291
two-way, 282-83, 291

X-band FET, 123

## *The Artech House Microwave Library*

*Acoustic Charge Transport: Device Technology and Applications*, R. Miller, C. Nothnick, and D. Bailey

*Advanced Automated Smith Chart Software and User's Manual, Version 2.0,* Leonard M. Schwab

*Algorithms for Computer-Aided Design of Linear Microwave Circuits,* Stanislaw Rosloniec

*Analysis, Design, and Applications of Fin Lines*, Bharathi Bhat and Shiban K. Koul

*Analysis Methods for Electromagnetic Wave Problems*, Eikichi Yamashita, editor

*Automated Smith Chart Software and User's Manual*, Leonard M. Schwab

*C/NL2 for Windows: Linear and Nonlinear Microwave Circuit Analysis and Optimization, Software and User's Manual*, Stephen A. Maas and Arthur Nichols

*Capacitance, Inductance, and Crosstalk Analysis*, Charles S. Walker

*Design of Impedance-Matching Networks for RF and Microwave Amplifiers*, Pieter L. D. Abrie

*Dielectric Materials and Applications*, Arthur von Hippel, editor

*Dielectrics and Waves*, Arthur von Hippel

*Digital Microwave Receivers*, James B. Tsui

*Electric Filters*, Martin Hasler and Jacques Neirynck

*Electrical and Thermal Characterization of MESFETs, HEMTs, and HBTs*, Robert Anholt

*E-Plane Integrated Circuits*, P. Bhartia and P. Pramanick, editors

*Feedback Maximization*, Boris J. Lurie

*Filters with Helical and Folded Helical Resonators*, Peter Vizmuller

*Frequency Synthesizer Design Toolkit Software and User's Manual, Version 1.0*, James A. Crawford

*Fundamentals of Distributed Amplification*, Thomas T. Y. Wong

*GaAs MESFET Circuit Design*, Robert A. Soares, editor

*GASMAP: Gallium Arsenide Model Analysis Program*, J. Michael Golio *et al.*

*Handbook of Microwave Integrated Circuits*, Reinmut K. Hoffmann

*Handbook for the Mechanical Tolerancing of Waveguide Components*, W. B. W. Alison

*HELENA: HEMT Electrical Properties and Noise Analysis Software and User's Manual*, Henri Happy and Alain Cappy

*HEMTs and HBTs: Devices, Fabrication, and Circuits*, Fazal Ali, Aditya Gupta, and Inder Bahl, editors

*High-Power Microwave Sources*, Victor Granatstein and Igor Alexeff, edtiors

*High-Power GaAs FET Amplifiers*, John Walker, editor

*High-Power Microwaves*, James Benford and John Swegle

*Introduction to Microwaves*, Fred E. Gardiol

*Introduction to Computer Methods for Microwave Circuit Analysis and Design*, Janusz A. Dobrowolski

*Introduction to the Uniform Geometrical Theory of Diffraction*, D. A. McNamara, C. W. I. Pistorius and J. A. G. Malherbe

*LOSLIN: Lossy Line Calculation Software and User's Manual*, Fred E. Gardiol

*Lossy Transmission Lines*, Fred E. Gardiol

*Low-Angle Microwave Propagation: Physics and Modeling*, Adolf Giger

*Low Phase Noise Microwave Oscillator Design*, Robert G. Rogers

*MATCHNET: Microwave Matching Networks Synthesis*, Stephen V. Sussman-Fort

*Matrix Parameters for Multiconductor Transmission Lines: Software and User's Manual*, A. R. Djordjevic *et al.*

*Microelectronic Reliability, Volume I: Reliability, Test, and Diagnostics*, Edward B. Hakim, editor

*Microelectronic Reliability, Volume II: Integrity Assessment and Assurance*, Emiliano Pollino, editor

*Microwave and RF Circuits: Analysis, Synthesis, and Design*, Max Medley

*Microwave and RF Component and Subsystem Manufacturing Technology*, Heriot-Watt University

*Microwave Circulator Design*, Douglas K. Linkhart

*Microwave Engineers' Handbook*, 2 Volumes, Theodore Saad, editor

*Microwave Materials and Fabrication Techniques,* Second Edition, Thomas S. Laverghetta

*Microwave MESFETs and HEMTs*, J. Michael Golio *et al.*

*Microwave and Millimeter Wave Heterostructure Transistors and Applicatons*, F. Ali, editor

*Microwave and Millimeter Wave Phase Shifters, Volume I: Dielectric and Ferrite Phase Shifters*, S. Koul and B. Bhat

*Microwave and Millimeter Wave Phase Shifters, Volume II: Semiconductor and Delay Line Phase Shifters,* S. Koul and B. Bhat

*Microwave Mixers,* Second Edition, Stephen Maas

*Aicrowave Transmission Design Data*, Theodore Moreno

*Aicrowave Transition Design*, Jamal S. Izadian and Shahin M. Izadian

*Aicrowave Transmission Line Couplers*, J. A. G. Malherbe

*Aicrowave Tubes*, A. S. Gilmour, Jr.

*Aicrowaves: Industrial, Scientific, and Medical Applications,* J. Thuery

*Aicrowaves Made Simple: Principles and Applicatons,* Stephen W. Cheung, Frederick H. Levien *et al.*

*AMIC Design: GaAs FETs and HEMTs*, Peter H. Ladbrooke

*Aodern GaAs Processing Techniques*, Ralph Williams

*Aodern Microwave Measurements and Techniques*, Thomas S. Laverghetta

*Aonolithic Microwave Integrated Circuits: Technology and Design*, Ravender Goyal *et al.*

*Jonuniform Line Microstrip Directional Couplers*, Sener Uysal

*'C Filter: Electronic Filter Design Software and User's Guide,* Michael G. Ellis, Sr.

*'LL: Linear Phase-Locked Loop Control Systems Analysis Software and User's Manual,* Eric L. Unruh

*'F Design Guide: Systems, Circuits, and Equations*, Peter Vizmuller

*cattering Parameters of Microwave Networks with Multiconductor Transmission Lines: Software & User's Manual*, A. R. Djordjevic *et al.*

*olid-State Microwave Power Oscillator Design,* Eric Holzman and Ralston Robertson

*errestrial Digital Microwave Communications*, Ferdo Ivanek *et al.*

*'ime-Domain Response of Multiconductor Transmission Lines: Software and User's Manual*, A. R. Djordjevic *et al.*

*'ransmission Line Design Handbook*, Brian C. Waddell

*'ield and Reliability in Microwave Circuit and System Design*, Michael Meehan and John Purviance

*For further information on these and other Artech House titles, contact:*

rtech House
35 Canton Street
orwood, MA 02062
17-769-9750
ax: 617-769-6334
elex: 951-659
nail: artech@world.std.com

Artech House
Portland House, Stag Place
London SW1E 5XA England
+44 (0) 171-973-8077
Fax: +44 (0) 171-630-0166
Telex: 951-659
bookco@artech.demon.co.uk